Fermentation and Food Safety

Editors

Martin R. Adams, MSc, PhD
Reader in Food Microbiology
School of Biomedical and Life Sciences
University of Surrey
Guildford, United Kingdom

M.J. Robert Nout, PhD
Associate Professor
Laboratory of Food Microbiology
Wageningen University
Wageningen, The Netherlands

AN ASPEN PUBLICATION®
Aspen Publishers, Inc.
Gaithersburg, Maryland
2001

The author has made every effort to ensure the accuracy of the information herein. However, appropriate information sources should be consulted, especially for new or unfamiliar procedures. It is the responsibility of every practitioner to evaluate the appropriateness of a particular opinion in the context of actual clinical situations and with due considerations to new developments. The author, editors, and the publisher cannot be held responsible for any typographical or other errors found in this book.

Aspen Publishers, Inc., is not affiliated with the American Society of Parenteral and Enteral Nutrition.

Library of Congress Cataloging-in-Publication Data

Fermentation and food safety/ Martin R. Adams and M.J. Robert Nout
p. cm.
Includes bibliographical references and index.
ISBN 0-8342-1843-7
1. Fermented foods. 2. Food industry and trade—Safety measures. I. Nout, M. J. Robert. II. Adams, Martin, R.
TP371.44.F47 2001
660'.28448—dc21
00-051087

Orders: (800) 638-8437
Customer Service: (800) 234-1660

Editorial Services: Sharyn L. Vitrano/Timothy Sniffin
Library of Congress Catalog Card Number: 00-051087
ISBN: 0-8342-1843-7

Printed in the United States of America

1 2 3 4 5

Contents

Contributors

Martin R. Adams, MSc, PhD
Reader in Food Microbiology
School of Biomedical and Life Sciences
University of Surrey
Guildford, United Kingdom

A. Asante
Scientist
Food Safety Unit
World Heath Organization
Geneva, Switzerland

R.R. Beumer, PhD
Associate Professor
Laboratory of Food Microbiology
Wageningen University Research Center
Wageningen, The Netherlands

Leon Brimer, PhD, DSc
Associate Professor
Department of Pharmacology and Pathobiology
Laboratory of Toxicology
Royal Veterinary and Agricultural University
Copenhagen, Denmark

Michael J. Carter, BA, PhD
Reader in Virology
School of Biomedical and Life Sciences
University of Surrey
Guildford, United Kingdom

Wilhelm H. Hozapfel, PhD
Director and Professor
Institute of Biotechnology and Molecular Biology
Federal Research Centre for Nutrition
Karlsruhe, Germany

Maurice O. Moss, PhD, ARCS, DIC
Visiting Reader
School of Biomedical and Life Sciences
University of Surrey
Guildford, United Kingdom

Yasmine Motarjemi, PhD
Scientist
Food Safety Programme
World Health Organization
Geneva, Switzerland

M.J. Robert Nout, PhD
Associate Professor
Laboratory of Food Microbiology
Wageningen University
Wageningen, The Netherlands

Arthur C. Ouwehand, PhD
Senior Scientist
Department of Biochemistry and Food Chemistry
University of Turku
Turku, Finland

Seppo J. Salminen, PhD
Professor
Degree Programme on Health Sciences
Department of Biochemistry and Food Chemistry
University of Turku
Turku, Finland

M. Hortensia Silla-Santos, PhD
Senior Scientist
Instituto de Agroquímica y Tecnología de Alimentos
Consejo Superior de Investigaciones Científicas
Valencia, Spain

Mike Taylor, BVMS, PhD, MRCVS, CBiol, MIBiol
Head of Parasitology and Ecotoxicology
Veterinary Laboratories Agency
New Haw, United Kingdom

T. Verrips, PhD
Chief Scientist
Unilever Research Laboratory
Vlaardingen, The Netherlands

Atte J. von Wright, PhD
Professor
Institute of Applied Biotechnology
University of Kuopio
Kuopio, Finland

Foreword

In every part of the world, people wage a constant battle against food contamination and the resulting food-borne diseases and food wastage. Efforts to reduce the devastating consequences of food contamination started long before written records. Besides cooking, smoking, and simple sun drying, fermentation is one of the oldest technologies used for food preservation. Over the centuries, it has evolved and has been refined and diversified. Today, a large variety of foods are derived from this technology, which is used in households, small-scale food industries, and large-scale enterprises. Foods so produced form a major part of the human diet all over the world but only a few people are aware of the multitude of fermented products and their importance in the human diet. In fact, all cultures have in the course of their development learned the technique to preserve some of their foods by fermentation. However, the safety of fermented foods is a concern everywhere.

In the past, traditional fermentation technologies were based on experiences accumulated by consecutive generations of food producers, as a result of trial and error. Only relatively recently have science and technology started to contribute to a better understanding of the underlying principles of the fermentation process and of the essential requirements to ensure the safety as well as nutritional and sensory quality of fermented foods. Since the days of Louis Pasteur, who pointed to the importance of hygiene in relation to fermentation, it is known that this technology is easily influenced by various factors during processing, and if not applied correctly, the safety and/or quality of the final product may be jeopardized. As a matter of fact, causes and outbreaks of food-borne illness have been traced back to fermented food, in spite of the general ineffectiveness of food-borne disease surveillance programs in most countries.

Fermentation is also of economic importance in areas or for populations where preservation technologies such as cold storage (refrigeration) or hot-holding cannot be used for lack of resources or facilities. In such situations, fermentation may be considered an affordable technology, which—if applied correctly—results in the safe preservation of foods, including complementary foods for infants. Particularly in developing countries, as a result of poor hygiene and incorrect application of fermentation, complementary foods are often contaminated with pathogens and subsequently are a major cause of infant diarrhea and associated malnutrition.

Against this background, the World Health Organization (WHO), jointly with the Food and Agriculture Organization (FAO), organized in 1995 a workshop to assess fermentation as a household technology for improving food safety.[1] This workshop was the first of its kind, highlighting the critical points in the fermenta-

[1]Fermentation: Assessment and Research. Report of a Joint FAO/WHO Workshop, Pretoria, South Africa, 11–15 December 1995. WHO Consultations and Workshops: WHO/FNU/FOS/96.1

tion process to ensure the safety of the resulting products, in line with the Hazard Analysis and Critical Control Point (HACCP) system. In a way, this book is a result of this workshop. Both Aspen Publishers and the two editors, M.J. Robert Nout and Martin R. Adams, deserve applause for this initiative and the unique approach they have adopted: the book focuses on food safety in all its aspects and is largely hazard based, which helps to identify those areas where knowledge is lacking but needed for a satisfactory risk assessment to be made. This is in contrast to the existing literature on fermented foods, which is generally confined to descriptions of the product(s) and the microbiology/biochemistry of their production.

In the interest of public health and food security, I wish this book a large and interested readership and for fermentation to result in safe, nutritionally adequate, and superbly tasting foods with long shelf lives.

Fritz Käferstein, DVM, PhD
Former Director, Programme of Food Safety and Food Aid, WHO, Geneva;
Current Distinguished Visiting Scientist, Joint Institute for Food Safety and Applied Nutrition, Washington, DC

Preface

Fermented foods enjoy worldwide popularity as attractive, wholesome and nutritious components of our diet. They are produced on an enormous scale employing a huge variety of ingredients and manufacturing techniques. Whether traditional home-made foods, or high-tech products derived from genetically modified organisms, the safety of the consumer remains of foremost importance.

Fermented foods have always been generally regarded as safe, but this reputation has been seriously threatened in recent years by incidents such as outbreaks of illness caused by pathogens in soft cheeses and fermented meats, chronic cyanide intoxications from poorly processed cassava tubers, and mycotoxins in fermented cereal foods. In addition, modern techniques of genetic engineering and biotechnology, which offer considerable opportunities in the area of fermented food production, have also raised safety concerns among consumers. It is necessary therefore to have concrete guidelines on the conditions which lead to safe products and to have a realistic view about what "guaranteed safety" means. The massive impact of HACCP as a systematic approach to ensuring food safety has been widely apparent in recent years and the technique is as applicable to food fermentation as to any other food processing operation. At present however, the literature on fermented foods has no focus on safety, but is mainly descriptive, concentrating on microorganisms responsible for fermentation and on the biochemical changes occurring in the food.

This book aims to integrate modern concepts of safety assurance with the sometimes very traditional environment of the production and distribution of fermented foods. In particular, we have taken a largely hazard-based approach rather than one centered on the different commodities used. Introductory chapters aim to provide a broad understanding of the nature of fermented foods, their production, distribution, and use by consumers, and also discuss the general features of fermentation processes that contribute to the product's overall safety and HACCP. For the bulk of the book, we have sought chapters which describe the principal individual hazards, both chemical and microbiological, and try to provide some guidance on how these might be controlled in food fermentations. These hazards are discussed from the point of view of their severity and incidence, how they get into the food, which foods are specifically at risk, and what, if any, are the conditions that remove or inactivate these hazards. In many cases there is a dearth of published material specifically on fermented foods, and contributors have used data obtained in slightly different contexts to give some guidance. This exercise has proved useful in highlighting where information is lacking and identifying areas where more research is desperately needed.

It is hoped that the book will serve as a source of reference to support and help improve the production of safe fermented foods at all scales (household preparation up to large-scale industrial plants) and using all major food groups.

CHAPTER 1

Fermented Foods and Their Production

M.J. Robert Nout

FERMENTATION AND FOOD SAFETY

Fermentation

Fermentation is one of the oldest methods of food processing. Bread, beer, wine, and cheese originated long before Christ. Although modern food technology has contributed to the present-day high standard of quality and hygiene of fermented foods, the principles of the age-old processes have hardly changed. In industrialized societies, a variety of fermented foods are very popular with consumers because of their attractive flavor and their nutritional value (Figure 1–1).

In tropical developing regions, fermentation is one of the main options for processing foods. In the absence of facilities for home refrigeration, freezing, or home canning, it serves as an affordable and manageable technique for food preservation. Fermentation can also increase the safety of foods by removing their natural toxic components, or by preventing the growth of disease-causing microbes. It imparts attractive flavor and nutritive value to many products. Fermentation is an attractive technique because it is low cost and low technology and it can be easily carried out at the household level, often in combination with simple methods such as salting, sun drying, or heating (e.g., boiling, steaming, frying).

Contrary to unwanted spoilage or toxin production, fermentation is regarded as a desirable effect of microbial activity in foods. The microbes that may be involved include molds (mycelial fungi), yeasts (unicellular fungi), and bacteria. Examples of food fermentations and the microbes responsible for the desired changes will be presented in this chapter.

In general, the desirable effect of microbial activity may be caused by its biochemical activity. Microbial enzymes breaking down carbohydrates, lipids, proteins, and other food components can improve food digestion in the human gastrointestinal tract and thus increase nutrient uptake. Several bacteria excrete B vitamins into food. As a result of their growth and metabolism, substances of microbial origin are found in the fermented food, including organic acids, alcohols, aldehydes, esters, and many others. These may have a profound effect on the quality of the fermented product. For instance, lactic and acetic acids produced by lactic acid bacteria (LAB) have an inhibitory effect on spoilage bacteria in sourdough bread and yogurt, and the production of ethanol and carbon dioxide determines the acceptability of bread, beer, and wine (ethanol disappears from bread during the baking process).

In addition to enzymes and metabolites, microbial growth causes increased amounts of microbial cell mass. This may be of nutritional and aromatic interest in yeast extract, for instance. The presence of living microbial cells such as in nonpasteurized yogurt may well have advantageous effects on the intestinal microflora and, indirectly, on human health.

Nr.	Name of product	Major ingredient(s)	Functional microorganisms
1	quark	cows' milk	lactic acid bacteria (LAB)
2	yogurt	cows' milk	LAB
3	sauerkraut	white cabbage	LAB
4	cultured milk ("karnemelk")	cows' milk	LAB
5	treated black olives	olives	LAB
6	gouda cheese	cows' milk	LAB
7	raw fermented sausages	pork and/or beef meat	LAB
8	yeast-leavened bread	wheat flour	yeasts (Y)
9	lager beer	barley, hops	Y
10	white wine	grapes	Y
11	sherry	grapes	Y
12	lager beer	barley, hops	Y
13	gruyère cheese	cows' milk	LAB + propionibacteria
14	pumpernickel bread	rye flour	LAB + Y
15	mixed sourdough bread	rye + wheat flour	LAB + Y
16	camembert cheese	cows' milk	LAB + molds
17	raw fermented sausage	pork and/or beef meat	LAB + molds
18	soy sauce	soybeans and wheat	Molds + LAB + Y
19	tempeh	soybeans	LAB + Molds + Y

Figure 1–1 Fermented foods representing different types of fermentations and raw materials

Food Safety

In its widest sense, the safety of food must be achieved through safe production, storage, and handling in order to avoid food-borne illnesses such as food intoxication, infectious diseases, or other detrimental effects. In principle, such illnesses can be caused by agents of biological, chemical, or physical nature, as exemplified in Table 1–1.

In this book, the fermentation of food will be viewed in relation to these safety aspects. The

Table 1–1 Some Examples of Threats to Consumer Safety

Nature	*Type of Causative Agents*	*Specific Examples*	*Illness*
Biological	Pathogenic microorganisms	Bacteria (*Salmonella* spp.)	Enteric infections
		Viruses (Rotavirus)	
		Parasites (*Cryptosporidium* spp.)	
	Toxigenic microorganisms	Mycotoxin-producing fungi (*Aspergillus flavus*)	Liver cancer
		Toxin-forming bacteria (*Clostridium botulinum*)	Respiratory failure
		Biogenic amine-forming bacteria (Enterobacteriaceae)	Hypertension
Chemical	Phytotoxins	Cyanogenic glycosides (linamarin) in cassava tubers	Cyanide intoxication
	Environmental contaminants	Pesticides	Intoxications
		Veterinary drugs	
		Heavy metals	
	Toxic metabolites	Biogenic amines (e.g., in fish)	Hypertension
	Food additives and ingredients	Preservatives	Allergies
		Colorants	
Physical	Foreign matter	Glass	Injuries (cuts, perforations)
		Metal	

book begins with a closer look at fermentation as a food processing technology. Next, the intrinsic safety of fermented foods as well as the principles of the hazard analysis critical control point (HACCP) system are discussed, followed by several examples of hazards that are potential threats to consumer safety.

The use of genetically modified ingredients or microorganisms in food fermentations, as well as the use of microorganisms as probiotics in fermented foods, are aspects that require a systematic and critical assessment of their safety.

In the final synthesizing chapter, the HACCP approach is used to illustrate and compare some of the critical processing steps that affect the safety of fermented foods.

FOOD FERMENTATION COMPONENTS

Food Ingredients

Food ingredients include the raw foods chosen for fermentation which can be of plant or animal origin. These foods contain a variety of nutrients for the consumer but also some of these nutrients will be required for the microbial growth and metabolism during the fermentation process. In order to enable microbial activity, sufficient water must be present. Consequently, water must be added in case the other ingredients are too dry. For several reasons, such as taste or preservation, miscellaneous substances may be added to the ingredient mix. These will be discussed in more detail below.

Food Groups

- **Plant origin:** Fermented foods of plant origin are derived from a variety of raw materials of different chemical composition and biophysical properties. Tuberous roots such as potatoes and cereals and tree crops like breadfruit have a relatively high starch content. Legumes and oil seeds generally have a high protein content. Green vegetables, carrots, beets, tomatoes, olives, cucumbers, okra, and forage crops for animal feed silages have a high moisture content. Fruits contain high concentrations of reducing sugars.

Most fermentative preservations of vegetables and cereals are due to the action of LAB, often in combination with yeasts. But other bacteria, such as *Bacillus* spp., or mycelial fungi, such as *Rhizopus* and *Aspergillus* spp., are equally important in the fermentation of legumes and oil seeds.

- **Animal origin:** Foods of animal origin that are fermented are mainly milk, meat, fish, and seafood. These are all quite perishable, and fermentation has long been an effective method for prolonging the shelf life of these valuable nutrients.

Milk originating from cows, buffaloes, sheep, and, occasionally, other animals has a high moisture content; contains proteins, minerals, and vitamins; and has a neutral pH. A rapid acidification by lactic acid bacterial fermentation to pH values of less than 4.5 strongly inhibits the survival and growth of spoilage-causing or disease-associated bacteria. Milk contains ample amounts of the fermentable carbohydrate lactose, which is an essential ingredient to enable this fermentation. Due to the reduced pH, as well as bacterial glycocalix, the viscosity increases and various liquid fermented milk products are obtained, ranging from fluid cultured milk to stringy viscous Scandinavian milks, mixed sour-alcoholic fizzy fluids, and gel-type products such as yogurt. Milk is also rather voluminous; nomadic tribes developed methods to curdle milk and separate the coagulated casein from the residual liquid (whey). The coagulate represents only approximately 10% of the original milk volume. Fermentation by LAB, and occasional propionibacteria, yeasts, and molds, results in a variety of cheeses.

Meat (cow, pork, goat, etc.) contains very low levels of fermentable carbohydrates, thus posing limitations to lactic acid fermentation. Many of the fermented meats derive their long shelf life from a combination of preservative effects. Acidification by LAB is usually boosted by adding sugar, and added salt, nitrate, or nitrite, as well as some extent of dehydration, are major aspects of the preservation of fermented sausages. It is interesting to note that most fermented sausages have been prepared from raw meat. In order to avoid the risk of raw meat-borne food infections by parasites, it is recommended to freeze the meat prior to processing.

Fish and seafood (e.g., shrimp) pose similar restrictions to bacterial fermentation: They contain only low quantities of fermentable carbohydrates. In practice, this has led to two major types of fish products preserved by fermentation. The first is a mixture of fish and salt that results in liquid protein hydrolysate after several months of fermentation. Halotolerant bacteria and yeasts may play a role in flavor formation, but the high salt content is responsible for the preservation. The second type of fish product is a mixture of fish, little salt, and cooked starchy food (e.g., rice or cassava), the latter providing fermentable carbohydrates; these products are preserved mainly by lactic acid fermentation.

Nutrients Required by Microorganisms

- **Carbon and nitrogen:** Most microorganisms require some form of organic carbon. In natural raw materials, this is found mainly in carbohydrates (e.g., monomers, dimers, etc., and a number of polymeric food-storage and structure-giving polysaccharides such as starch, pectins, hemicellulose, and cellulose). Molds and certain bacteria (e.g., *Bacillus* spp.) are good producers of extracellular carbohydrate-degrading enzymes that can release fermentable mono- and oligomeric substances. Accounts of the response of yeasts to sources of carbon, nitrogen, phosphorus, and sulfur are available.[34,41] Yeasts and LAB are known to be poor converters of polysaccharides and pentoses. Although the presence of chitin and acacia gum was shown to increase the rate of yeast growth and fermentation,[31] these compounds were not metabolized as nutrients.

In addition to fermentable saccharides, other nutrients required for cell growth and metabolism include inorganic (e.g., NH_4^+) or organic sources of nitrogen (e.g., urea, amino acids, and peptides, but rarely extracellular proteins), phosporus (e.g., inorganic phosphate or phosphate esters), and sulfur (e.g., sulfate, sulfite, methionine, glutathione). Fungi usually do not require additional nutrients in food fermentation. Optimum carbon/nitrogen ratios for growth are 10–100. For use in industrial fermentations, it was shown that *R. oryzae* could grow well with only starch and nitrogen salts; the addition of vegetable juice further stimulated growth.[37]

- **Minerals:** Iron, magnesium, potassium, sodium, and calcium are normally required for cell growth.[41] Of additional interest is the effect of Ca^{2+} ions, which was reported to increase the ethanol tolerance of *Saccharomyces cerevisiae*.[24] Manganese plays an important role in LAB; deficiencies may lead to fermentation failures.
- **Vitamins and other growth factors:** The most common growth factors for yeasts are biotin, pantothenic acid, inositol, thiamine, nicotinic acid, and pyridoxine.[41] Riboflavin and folic acid are synthesized by all yeasts, but are required by some bacteria. Fungal requirements for additional vitamins in the food environment seem to be negligible. On the contrary, they may contribute to the nutritive value of fermented foods by vitamin synthesis.

Water

Water is essential for microbial growth and metabolism. The extent to which water is available for biological metabolism is expressed as water activity (A_w) or, occasionally, as water potential.[15] A_w is the more commonly used terminology and is defined by the ratio of equilibrium water vapor pressure of the food and of pure water, at a defined temperature such as 20 °C. A_w ranges from 0 to 1. Most microorganisms require A_w to be greater than 0.70, with optimum greater than 0.99.

Added Food Ingredients

A variety of added ingredients affect the activity of microorganisms and thus can be used to regulate the rate and extent of fermentation. Salt and sugar are well known for their antimicrobial effect at high concentrations. Salt levels greater than 15% w/w and sugar concentrations greater than 25% w/w reduce the A_w considerably; in addition, NaCl has a specific inhibitory influence caused by Na^+ ions. Herbs, spices, and so forth may contain inhibitory proteins, organic acids, essential oils, pigments, resins, phenolic compounds, caffeine, and so on. On the other hand, some spices are excellent sources of manganese, which is essential for LAB.

Microorganisms

Three groups of microorganisms are used in food fermentation, namely bacteria, yeasts, and molds. Table 1–2 illustrates some prominent species of microorganisms, some of the food fermentations in which they are of importance, and their function.

How do microorganisms enter the fermentation? Different scenarios of increasing complexity can be distinguished,[25] as will be explained in the following sections.

Natural Fermentation in Raw Substrate

Most raw foods of animal or plant origin contain a variety of microorganisms that arrived by chance or that have an ecological association with them, based on preharvest growing conditions. Food processing activities can also add certain microorganisms to food. If food is allowed to ferment without prior heating, most of these microorganisms can multiply. However, their opportunities will be restricted by their ability to grow and compete in the food, and by external conditions. This usually results in a succession of predominant microorganisms, finally stabilizing in a fermented product that contains a mixed microbial population dominated by microorganisms that are particularly suited to the physicochemical conditions prevailing in the final product. This type of fermentation is exemplified by sauerkraut, which is shredded and salted cabbage that is fermented by LAB.[11] Drawbacks of natural fermentations are that they take a relatively long time to complete, and the outcome is always a surprise.

Use of Traditional Mixed Starter Cultures in Raw or Preheated Substrate

The drawbacks of natural fermentation can be reduced when a large quantity of microorganisms that occur in the final product are added. These may be expected to bring about a more rapid domination and a more predictable quality of the fermented product. An example of this approach is the traditional fermentation of sourdough, a mixture of cereal flour (wheat or rye) with water of dough consistency.[12] A small quantity of previously fermented sourdough is mixed into the new dough, and this practice can be successfully carried out for many years to achieve dependable fermentations.

Traditional mixed-culture starters are also applied in the fermentation of precooked ingredients. Examples are the Indonesian fermented soybean food *tempeh* (also known as *tempe*), [18] as well as African alcoholic beverages such as *Ghanaian pito* made from sorghum.[9] For tempeh, a mixed culture of molds (*Rhizopus* and *Mucor* spp.) especially grown for this purpose on plant leaves *(Hibiscus tileaceus)* is used.[27] The fermentation starter for pito is an inoculation belt, a woven rope that is suspended in the fermenting beer and on which the predominating yeasts (*Saccharomyces* spp.) are present as a sediment. The fermentation of a next batch of pito beer is started by immersing the inoculation belt into the sugary liquid made from sorghum, which is a common cereal in tropical climates.

Use of Pure Cultures (Single or Mixed) in Preheated Substrate

With increasing scale of production, more sophisticated technical facilities, and higher investment and operational risks, the use of laboratory-selected and precultured starter cultures becomes a necessity. Because equipment for aseptic processing at a large production scale is extremely expensive, common procedures include pure culture maintenance and propagation under aseptic conditions (e.g., sterile laboratory and pilot fermentors). At production scale, the food ingredient to be fermented is preheated in

Table 1–2 Selected Examples of Microorganisms and their Function in the Fermentation of Foods

Microbial Group	*Species*	*Fermented Foods*	*Importance of Fermentation for Product Characteristics*
Bacteria	Lactic acid bacteria: *Lactobacillus delbrückii* ssp. *bulgaricus* and *Streptococcus salivarius* ssp. *thermophilus*	Yogurt	Contribute to flavor, shelf life, structure, and consistency by the production of lactic acid, acetaldehyde, diacetyl, and polysaccharides
Yeasts	*Saccharomyces cerevisiae*	Alcoholic beverages (beers, wines)	Produce ethanol, CO_2, and flavor
Molds	*Aspergillus sojae*	Soy sauce	Form proteolytic and saccharolytic enzymes, enabling solubilization and flavor production

order to minimize the level of microbial contamination; subsequently, the pure culture starter (single or mixed cultures) is added and fermentation takes place at the highest practicable and affordable level of hygienic protection.

Use of Pure Cultures in Sterilized Substrate

For the propagation of starter cultures, or in other situations requiring the absolute absence of microbial contamination, food ingredients or otherwise suitable nutrients are sterilized and kept in sterile confinement. Pure cultures of starter microorganisms are added using aseptic techniques. Not only is this an expensive technique, but it is also unnecessary in production-scale food fermentation. Moreover, severe heat treatments required to achieve sterility can harm heat-sensitive nutrients for both the microorganism and the consumer.

Enzymes

The importance of enzymes in fermentation processes lies in their ability to degrade complex substrates. Some examples are in *koji* fermentation, where *A. oryzae* enzymes degrade starch, protein, and cell wall components, and in tempeh production, where *Rhizopus* spp. enzymes degrade cell wall components, protein, lipids, phytic acid, and oligosaccharides. Proteases produced by LAB are important degrading proteins in fermented milk products. Some enzymes involved in the degradation of complex substrates include amylolytic, proteolytic, lipolytic, and cell wall degrading enzymes.

Amylolytic Enzymes

Starch is composed of two polysaccharides, amylose and amylopectin, both consisting of glucose units only (Figure 1–2). *Amylose* is a linear glycan in which the sugar residues are connected by α-1,4 bonds; *amylopectin* is a branched glycan in which glucose residues in the backbone and the side chains are α-1,4 linked. The side chains are attached by α-1,6 linkages.

Starch is degraded by several amylases working simultaneously. α-amylases hydrolyze α-1,4 glucosidic linkages. Iso-amylase is a debranching enzyme; it hydrolyzes α-1,6 glucosidic linkages. Amyloglucosidase liberates single glucose units from the nonreducing end,

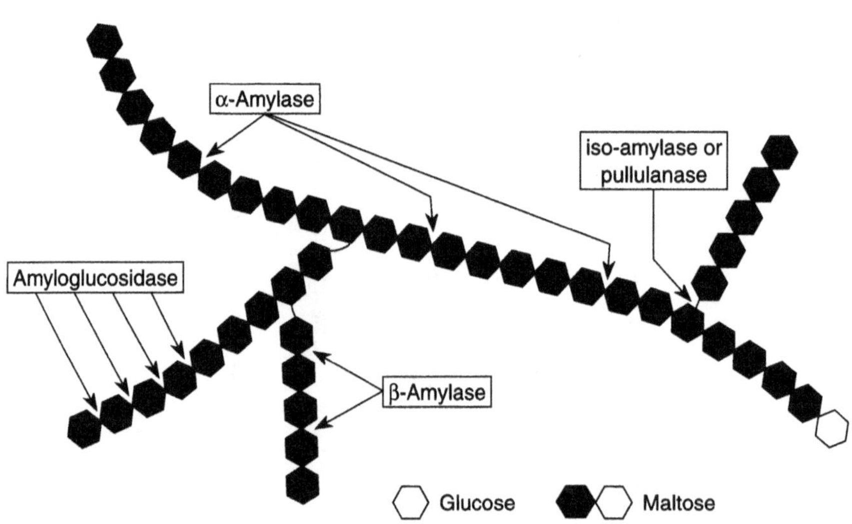

Figure 1–2 Amylopectin molecule and its degradation by several amylases. Adapted with permission from Uhlig, H. (1991). *Enzyme arbeiten für uns. Technische enzyme und ihre Anwendung.* München, Wien: Carl Hanser Verlag.[43]

whereas β-amylase liberates maltose units. These enzymes belong to the class of hydrolases. In industrial-scale production of α-amylases, the enzyme is derived from the pancreatic gland of swine and cattle and from microbial cultures. Microorganisms that produce amylases are *B. subtilis* and *A. oryzae*. *B. licheniformis* produces a heat-stable amylase that can be used at 100–110 °C. Gelatinization of starch by heating enhances enzyme catalysis. Thus, the swollen gelatinized starch substrate is degraded 300 times faster by bacterial amylase and 10^5 times faster by fungal amylase than the native, unswollen, nongelatinized starch.

Proteases

Proteases are classified according to their source (e.g., animal, plant, microbial), their catalytic action (e.g., endopeptidase or exopeptidase), and the nature of the catalytic site. *Endopeptidases* are the proteases that are used most commonly in food processing, but in some cases, they are accompanied by exopeptidases. Endopeptidases cleave polypeptide chains at particularly susceptible peptide bonds distributed along the chain, whereas *exopeptidases* hydrolyze one amino acid from the end of the chain. The four major classes of endopeptidases are carboxyl proteases (i.e., aspartic protease), cysteine proteases, serine proteases, and metallo proteases (Table 1–3). As the names imply, carboxyl, cysteine, and serine proteases have carboxyl, cysteine, and serine side chains, respectively, as essential parts of the catalytic site. The carboxyl proteases generally have maximum catalytic activity at acid pH. The serine proteases have maximum activity at alkaline pH; the closely related cysteine proteases usually show maximum activity at more neutral pH values. The metallo proteases contain an essential metal atom, usually Zn^{2+}, and have an optimum activity at neutral pH. Ca^{2+} stabilizes these enzymes and strong chelating agents (such as ethylenediaminetetraacetic acid, or EDTA) inhibit them.

Enzymes Degrading Cell Wall Components

In order to understand how enzymes affect plant cell walls, the structure of the walls first must be reviewed. Microscopic investigations have revealed that plant cell walls can be divided into three layers: the middle lamella, primary wall, and secondary wall (Figure 1–3). The *middle lamella* acts as an intercellular binding substance and is mainly composed of pectin. *Secondary cell walls* contain less pectin but contain some lignin. *Primary cell walls* consist of cellulose fibers called microfibrils embedded in a matrix of pectins, hemicelluloses, and proteins (Figures 1–3 and 1–4). Pectin is the major binding component of the cell wall, and its degradation by pectolytic enzymes will cause fruit or vegetables to become soft. This is the first step of enzymatic degradation of the cell wall.

- **Pectolytic enzymes:** The basic structure of pectins is a linear chain of α-linked molecules of pyranosyl D-galacturonic acid. Varying proportions of carboxylic groups can be present as methyl esters. The esterification is usually by methanol, in which case the pectin is said to be methylated. When less than 50% of the carboxyl groups are methylated, the pectin is referred to as *low methoxyl pectin;* when more than 50% of the carboxyl groups are methylated, the pectin is referred to as *high methoxyl pectin.* Pectin situated in the middle lamella is removed from the cell wall relatively easily and is most easily degraded by appropriate enzymes. In contrast, enzymes do not easily degrade the pectin within primary and secondary cell walls. Pectinases are defined and classified on the basis of their action toward the galacturonan part of the pectin molecules. Two main groups are distinguished, pectin esterases and pectin depolymerases.
- **Hemicellulases:** Hemicelluloses are polysaccharides that are extracted from plant cell walls by strong alkali. They are composed of four major substances: arabinans, galactans, xyloglucans, and xylans. As an example, arabinans are degraded by the enzymes, arabinanases. Arabinans are branched polysaccharides with a backbone of α-1,5 linked L-arabinose

Table 1–3 Overview of Proteases

Name	Type	Source	pH-Optimum	Optimum Stability pH Range
Proteases of animal origin				
Chymosin	Carboxyl protease	Stomach lining of calves	6.0–7.0	6.5–6.0
Pepsin	Carboxyl protease	Gastric lining of swine or bovine	2.0	
Pancreatic protease*		Pancreas	9.0	3.0–5.0
Proteases of plant origin				
Papain	Cysteine protease	Tropical melon tree (*Carica papaya*)	7.0–8.0	4.5–6.5
Bromelain	Cysteine protease	Pineapple (fruit and stalk)	7.0–8.0	
Ficin	Cysteine protease	Figs (*Ficus carica*)	7.0–8.0	
Bacterial proteases				
Alkaline protease	Serine protease	e.g., *Bacillus subtilis*	7.0–11.0	7.5–9.5
Neutral protease	Metallo protease	e.g., *Bacillus thermoproteolyticus*	6.0–9.0	6.0–8.0
Pronase		*Streptomyces griseus*		
Fungal proteases				
Acid protease	Carboxyl protease	*Aspergillus oryzae*	3.0–4.0	5.0
Neutral protease	Metallo protease	*Aspergillus oryzae*	5.5–7.5	7.0
Alkaline protease	Serine protease	*Aspergillus oryzae*	6.0–9.5	7.0–8.0
Protease	Carboxyl protease	*Mucor pusillus*	3.5–4.5	3.0–6.0
Protease	Carboxyl protease	*Rhizopus chinensis*	5.0	3.8–6.5

*A mixture of trypsin, chymotrypsin, and various peptidases with amylase and lipase as accompanying enzymes.

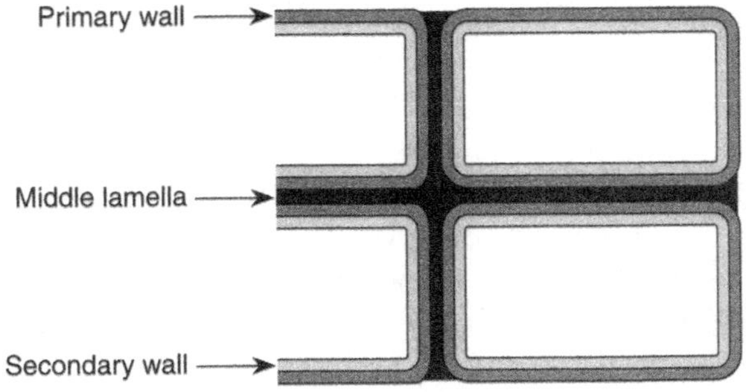

Figure 1–3 Schematic representation of the structure of plant cell walls. Reproduced with permission from Voragen, A. G. J., Van den Broek, L. A. M. (1991). Fruit juices. *Biotechnological Innovations in Food Processing*, pp. 187–210. Edited by Biotol Team. Oxford, UK: Butterworth-Heinemann.[44]

units (Figure 1–5). To approximately every third arabinose molecule, additional arabinose units are attached by α-1,2- or α-1,3 linkages. This can produce complex structures. There are two types of arabinanases: arabinosidase (arabinofuranosidase) and endo-arabinanase. Arabinosidase can be subdivided into two forms, A and B. The B form degrades branched arabinan to a linear chain by splitting off terminal α-1,3 or

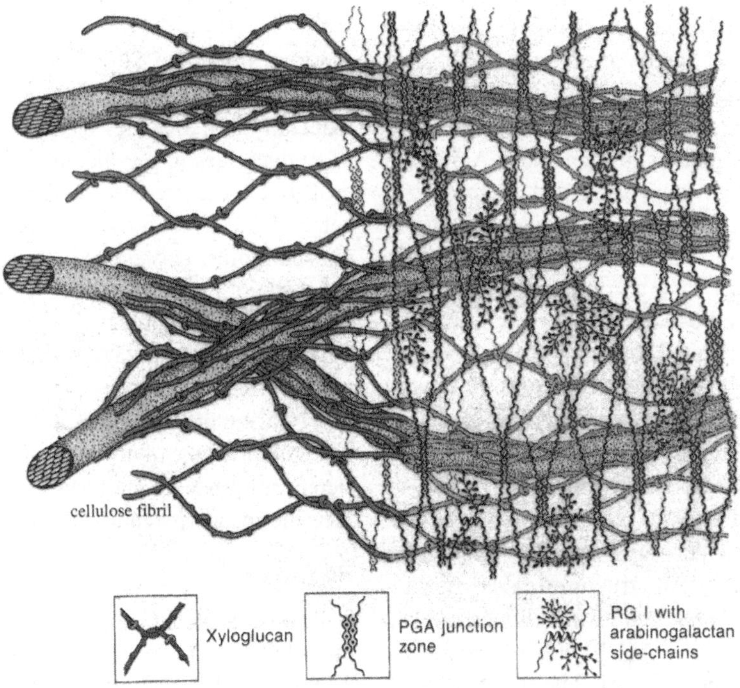

Figure 1–4 Schematic representation of the primary cell wall. PGA and RG are part of pectin. Reproduced with permission from Carpita, N. C., Gibeaut, D. M. (1993). Structural models of primary cell walls in flowering plants: consistency of molecular structure with the physical properties of the walls during growth. *The Plant Journal* **3**, 1–30.[7]

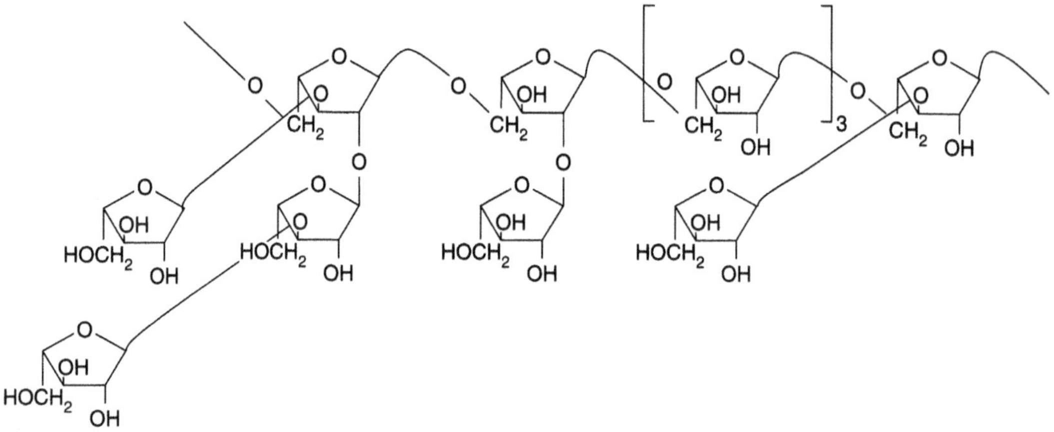

Figure 1–5 The structure of arabinans. Adapted with permission from Beldman, G., Schols, H. A., Pitson, S. M., Searle-van Leeuwen, M. J. F., Voragen, A. G. J. (1997). Arabinans and arabinan degrading enzymes. In *Advances in Macromolecular Carbohydrate Research. Vol. 1,* pp. 1–64. Edited by R.J. Sturgeon. London: JAI Press.[1]

α-1,2 linked side chains. At the same time, this enzyme slowly and sequentially breaks the α-1,5 linkages at the nonreducing end of linear arabinan. Endo-arabinanase hydrolyzes linear arabinan in a random fashion, producing oligomers of shorter lengths. Arabinosidase A degrades the arabinan oligomers to monomers. Arabinanases occur in some plants and microorganisms. Fungal arabinanases have an optimum pH of approximately 4.0; bacterial arabinanases have an optimum pH of approximately 5.0–6.0.

- **Cellulases:** Cellulose is the best known of all plant cell wall polysaccharides. It is particularly abundant in secondary cell walls and accounts for 20–30% of the total dry mass. Cellulose is a linear chain of β-1,4 linked glucose units. In cellulose, these β-1,4 glucan chains aggregate by hydrogen bonds to rigid flat structures called fibrils (Figure 1–4). The degree of polymerization can be very high, up to 14,000 in secondary cell walls. The rigid structure makes cellulose very resistant to degradation by enzymes. Cellulase (often called the cellulase complex) is a multienzyme complex system composed of several enzymes: endo-glucanase, exo-glucanase, cellobiose hydrolase, and cellobiase.

Lipases

Lipases degrade triglycerides. They only act on an aqueous-lipid interface such as a micelle. Generally, the enzymes catalyze the hydrolysis of triglycerides to produce fatty acids and glycerol, but there are also specific enzymes that catalyze the hydrolysis of monoacylglycerides, phospholipids, and esters of sterols. Generally, the mode of action of lipases results in fatty acids being preferentially hydrolyzed from 1- and 3-positions of triglycerides so as to leave 2-substituted monoacylglycerides. Microbial lipases may also catalyze the hydrolysis in all three positions. The composition of the fatty acid (i.e., length, stereoconformation, and degree of saturation) affects the specificity and speed of the lipases.

Phytases

Phytases catalyze the hydrolytic removal of phosphate groups from phytic acid (Figure 1–6), a substance that occurs widely in cereals and legumes and that is known to limit the bioavailability of macro- and micronutrients. The ability to degrade phytate occurs widely among molds *(A. ficuum, R. oligosporus),* yeasts

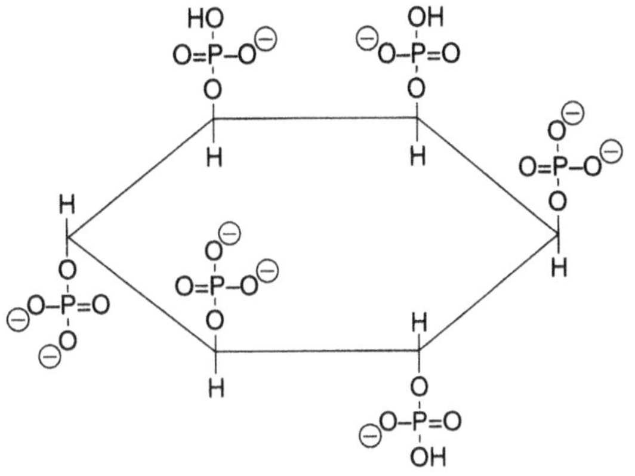

Figure 1–6 Structure of phytic acid

(*Candida krusei, Schwanniomyces castelli*), and LAB (*Pediococcus pentosaceus, Lactobacillus agilis, Lb. confusus*). Because the optimum pH of most phytases is approximately 5.5, the phytate degradation can be less effective if a lactic acid fermentation has resulted in low pH values.

Tannases

Tannases catalyze the hydrolysis of polyphenols (tannins). Fungal enzymes (e.g., from *A. niger*) are used in the food industry to degrade tannins. Tannin removal results in reduced turbidity in tea-based beverages. It also benefits the nutrient bioavailability in tannin-containing fermented cereal products.[2]

DIVERSITY OF FERMENTED FOODS

There is a wide variety of fermented foods worldwide. Various excellent reference books on fermented foods are available,[5,40,45] as are source books on industrial aspects of food biotechnology, including food fermentation.[23,33,39] Mention was made already of the various food ingredients that can be fermented, as well as the microorganisms and enzymes that are used in fermentations.

In the strict sense, fermentation refers to a form of anaerobic energy metabolism. In the context of fermented foods, however, the microbial growth and metabolism can take place under aerobic conditions as well. For example, mold fermentations require oxygen to facilitate their growth and enzyme production.

An aspect of interest is the physical state under which the fermentation takes place. This can be in the form of liquid in which the microorganisms are suspended while a relatively simple mixing device is used to ensure the homogeneity of microorganisms, substrates, and products. In these liquid fermentations, or LFs, the continuous phase is the liquid, and the control of temperature and levels of dissolved oxygen can be achieved with immersion coolers or heaters, and aeration. Quite a different situation occurs in a heap of cooked soybeans in which homogeneous growth of strictly aerobic molds is required. Whereas the particulate matter (i.e., soybeans) contains sufficient water to allow microbial growth, water is not the continuous phase; gas (i.e., air) is.This physical state is referred to as solid-state fermentation, or SSF. Because gas is a poor conductor of heat and mass, SSFs have a tendency to develop gradients of temperature and gas composition. Control of homogeneity requires more complex measurement and mixing systems as compared to LF. In practice, intermediate situations such as shredded vegetables or particles such as olives, which are fermented while immersed in brine, can be encountered. These can be considered as immersion LF because the continuous phase is liquid but contains immersed particles. Obviously, these do not always behave like LF.

Table 1–4 illustrates selected examples of food fermentations representing different food groups, functional microorganisms and enzymes, oxygen requirements, physical states, and present levels of industrialization. As mentioned earlier, food ingredients of plant as well as animal origin are used to prepare fermented foods. Of the functional microorganisms, LAB play the predominant role in the prolongation of shelf life because of the antimicrobial effect of their acids. In addition, yeasts are often found as minority companions of LAB; they sometimes contribute to shelf life by scavenging residual assimilable carbon sources. This is the case, for example, in fermented pickles. The major func-

Table 1–4 Examples of Food Fermentation Processes

Food Group	*Fermented Food*	*Mode of Consumption*	*Microorganisms and Major Products**	*Enzymic Modifications of The Product*	*Oxygen Requirement*	*Physical State*	*Industrialization*†	*References*
Starchy crops: cassava	Agbéli mawè	Cooked paste	Lactic acid bacteria forming lactic and acetic acid	Degradation of starch and cell walls	Aerobic	Solid	1–2	30
Cereals: wheat and rye	Sourdough	Bread	Lactic acid bacteria and yeasts forming lactic and acetic acid, and carbon dioxide	Degradation of starch and maltose	Anaerobic	Solid	1–4	16
Legumes: soybean	Tempeh	Fried snacks or cooked side dish	Molds forming mycelium and enzymes	Degradation of protein, cell walls, and lipids	Aerobic	Solid	1–3	26, 28
Vegetables: cabbage	Sauerkraut	Vegetable dish or relish	Lactic acid bacteria forming lactic acid		Anaerobic	Immersed in liquid	1–4	3
Fruits: grape	Wine	Alcoholic beverage	Yeasts (and occasionally lactic acid bacteria) forming ethanol and flavor		Anaerobic	Liquid	1–4	10
Meats: pork and beef	Raw dried sausages	Hearty food ingredient or snack	Micrococcaceae, lactic acid bacteria (and occasionally molds) forming organic acids and flavor and stabilizing meat color	Degradation of lipids	Anaerobic	Solid	1–4	36
Milk: cow, buffalo	Yogurt	Breakfast or dessert	Lactic acid bacteria forming lactic acid, acetaldehyde, and diacetyl and providing structure	Degradation of proteins	Anaerobic	Liquid	1–4	42
Milk: cow, sheep, goat	Soft cheeses	Sandwiches and snacks	Lactic acid bacteria and molds forming acidity, release enzymes for ripening and flavor development	Degradation of proteins and lipids	Aerobic	Solid	1–4	8

*Mention is made only of functional groups of microorganisms.
†1, home-produced; 2, village style; 3, national market; 4, international market.

tion of yeasts in yeast-fermented products is their copious production of carbon dioxide, such as gas in bread and beer and ethanol in alcoholic beverages. The functionality of molds is mainly in their production of enzymes that degrade polymeric components such as cell wall polysaccharides, proteins, and lipids. This can have considerable consequences for the texture, flavor, and nutritional value of mold-fermented foods. Depending on the functional microorganisms that should be propagated, the incubation environment (i.e., availability of oxygen for strict aerobes, anaerobic conditions when mold growth is undesirable) and physical state can be selected and optimized. It is of interest to note that, in contrast to predominantly LFs that are common in the pharmaceutical industry, a considerable number of food products are fermented by SSF.

PROCESS UNIT OPERATIONS

The manufacture of fermented foods is organized in a sequence of activities called unit operations.[22] Irrespective of the scale at which the process takes place, the following types of unit operations can be distinguished.

Physical Operations

Transport

Transport is one of the most important unit operations. It has the purpose of transferring ingredients (i.e., transport of mass) to the desired localities and/or equipment. Its aim is also to assist in heating and cooling (i.e., transport of heat). There is a wide variety of materials and methods that can be used for transport. The choice depends on the type of product, scale of production, economic considerations, and local conditions.

Grading and/or Sorting

Grading and/or sorting have the purpose of achieving homogeneity of size, color, maturity, hardness, and so forth. At the same time, items that are spoiled, infected, or otherwise deteriorated are removed. From a food safety point of view, grading and sorting are important tools that can be used to optimize the quality of inputs.

Cleaning and/or Washing

Cleaning and/or washing are carried out to remove dirt, dust, insects, agricultural residues, and so forth. Depending on the type of ingredients, dry or wet cleaning can be chosen. In cereal processing, for example, dry cleaning of wheat prior to flour milling is performed by a combination of aspiration and sieving. But, in maize processing, maize kernels can be washed in water prior to wet milling. Water is becoming ever scarcer and expensive, causing increased interest in dry operations.

Physical Separations

Dehulling, peeling, trimming, and other separations are aimed at obtaining the desired anatomical parts of plant or animal tissue while removing the undesired ones. Examples are the removal of poorly digestible seedcoats from soybeans by dehulling, the removal of the cortex of the cassava root because of its bitter taste and possible toxicity, and the trimming of fat from red meat.

Moisture Adjustment

Moisture adjustment is required to achieve, for example, the desired consistency and edibility of dry seeds and grains. The need for a sufficiently high A_w for microbial metabolism was mentioned earlier. Water can be added as an ingredient and mixed; soaking or steeping in water is also a common method to increase the moisture content of ingredients. Especially during longer periods of soaking at favorable temperatures, microbial activity can take place. Several cereal fermentation processes combine soaking with fermentation.

Size Reduction

Size reduction of particulate matter is required to obtain meal or flour from seeds, to obtain pulp from tubers or fuits, or to prepare a homogeneous slurry of meat and other ingredients for sausage making. Size reduction is performed by cutting, grinding, impact hammering, and so forth using a wide variety of equipment that has been developed to suit the processing of specific raw materials.

Mixing

Mixing has the purpose of obtaining homogeneity of mass and heat. Mixing of mass is the most common type of mixing and is exemplified by the mixing of ingredients to obtain a homogeneous product. Mixing at a small scale is relatively easy and can be carried out with simple kitchen utensils. The larger the scale of operation, the more complicated mixing becomes. Mixing of dry components can be achieved using specific mixing equipment such as tumblers or augur-type mixers, but it is also feasible to combine mixing with other operations such as grinding or transport. Mixing of liquids is often achieved in stirred tanks. Mixing of wet and dry ingredients can be carried out in kneading machines for doughs or in stirred tanks for less viscous suspensions.

Bioprocessing Operations

For microbial and enzymatic transformations, a first requirement is the presence of the required ingredients, including the desired microorganism(s) and/or enzyme(s), the required substrates and cofactors, and sufficient water. Several of the unit operations mentioned above will be involved to fulfill these requirements.

In order to allow the transformations to take place, incubation under optimum conditions for the correct period of time will be needed. In order to safeguard the constancy of these incubation conditions, unit operations such as mixing, heating, cooling, and transport are needed to ensure even distribution of mass and heat, compensate for heat losses or generated heat, and compensate for deficiencies or overproduction of mass by additions or removal.

Thermal Processing Operations

Heating and cooling are both characterized as the transport of heat. Heat treatments and cooling are of extreme importance in food processing and thus merit specific attention. The primary objective of heating is to render food palatable. It causes the gelatinization of starch, denaturation of protein, and softening of tough tissues, and transforms a number of flavors.

Heat treatments consist of a warming-up phase, a period of holding time, and a cooling-down phase. Temperatures exceeding 70 °C cause enzyme inactivation and kill vegetative microbial cells. The combination of temperature and time determines the lethal effect of heat treatments. In principle, the term *pasteurization* corresponds to mild heat that kills heat-sensitive vegetative cells of bacteria, yeasts, and molds. *Sterilization* signifies killing all living cells, including more heat-resistant spores of bacteria and molds. In food processing practice, sterilization is not always required to ensure long-term shelf life of foods. In this respect, the term *commercial sterility* indicates a situation where some heat-resistant spores may have survived the heat treatment, but the composition of the food prevents their revival during storage.

In view of bioprocessing, the timing of heat treatments is of crucial importance. Fermentation of ingredients without prior heat treatment (e.g., raw cereals, raw meat, etc.) has by definition a mixed character: Microorganisms and enzymes that were present in the ingredient will take part in the fermentation, sometimes as sole actors and sometimes as an accompaniment to added starter microorganisms. At larger scale production, fermentations must be better defined and controlled, and selected or optimized starter cultures must be used. In order to ensure optimum functioning and to remove potentially hazardous microorganisms, the ingredients can be pasteurized or sterilized prior to inoculation and fermentation. In such case, it should be realized that some heat-sensitive growth factors such as enzymes and vitamins may have to be replenished in order to enable effective microbial metabolism.

Cooling in food fermentation can be used as a processing tool. Prior to inoculation, heat-treated ingredients must be cooled to inoculation temperature. During fermentation, the excessive production of metabolic heat must be removed by cooling in order to prevent fermentation failures. In certain products such as yogurt, the fermentation is halted by cooling in order to prevent excessive acid development. Not all fermented foods have a long shelf life. Refrigerated distribution and storage contribute to shelf life and hygienic safety.

Organization in Flow Diagram

A flow diagram is a pictorial scheme illustrating the sequence in which ingredients and unit operations are combined and providing data regarding processing conditions such as temperature, time, mass, pH, and so forth. Figure 1–7 illustrates this principle.

Appendix 1–A provides flow diagrams for selected food fermentations. For each major food group, two commodities have been selected. These flow diagrams are not intended to provide recipes or exclusive methods of preparing the respective products. Their purpose is to provide an insight into the process conditions and timing, as well as relevant environmental antimicrobial conditions that are of importance in food safety.

PROCESS CONDITIONS

Appendix 1–A aims to provide concrete parameter values (or ranges) affecting microbial growth, metabolism, survival, and death as well as enzymatic activity/stability, that are relevant to food safety. These data can be useful in

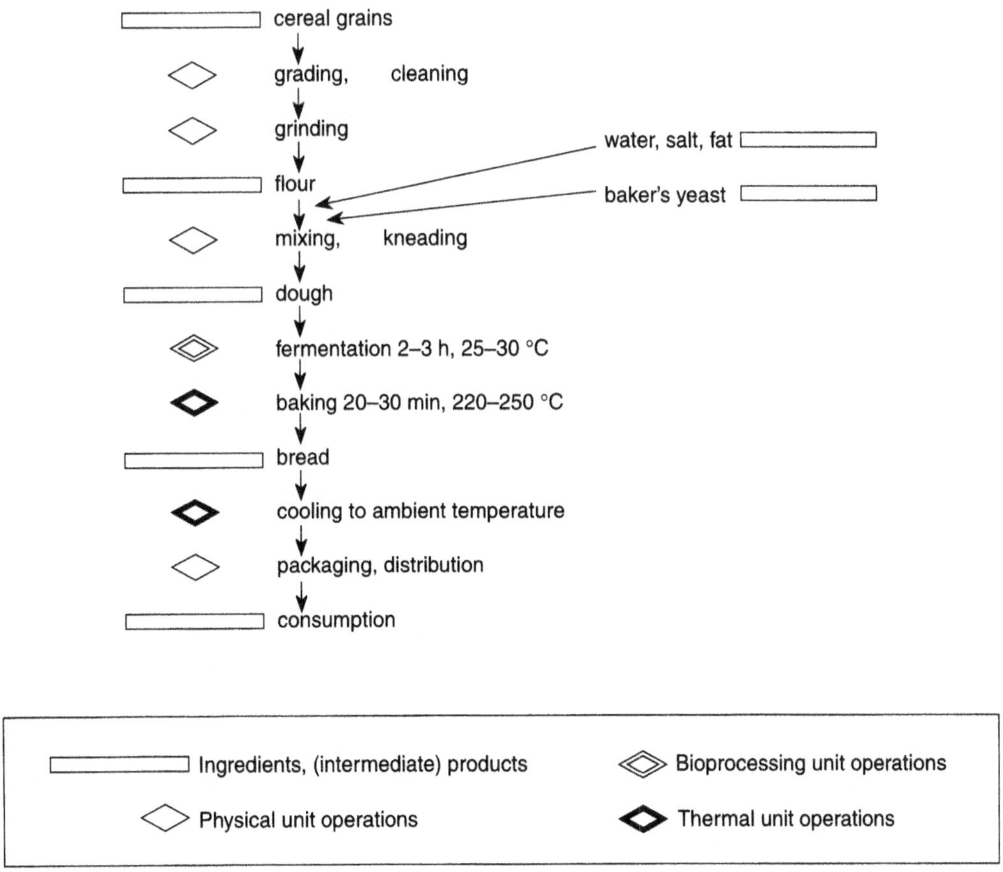

Figure 1–7 Principle of flow diagram (case of bread making)

HACCP approaches. Appendix 1–A provides flow diagrams and process conditions of relevance for the processing of starchy roots and tubers (cassava, Tables 1–A–1 and 1–A–2), cereals (barley, rye, and wheat, Tables 1–A–3 and 1–A–4), legumes (soybeans, Tables 1–A–5 and 1–A–6), vegetables (cabbage, Table 1–A–7 and olives, Table 1–A–8), fruits (grapes, Table 1–A–9 and palm sap, Table 1–A–10), meat (pork, Tables 1–A–11 and 1–A–12), fish (fresh water, Table 1–A–13 and sea water, Table 1–A–14), and milk (yogurt, Table 1–A–15 and cheese, Table 1–A–16). The icons explained in Figure 1–7 are used throughout these flow diagrams to provide a quick overview of the sequence of bio- and thermal unit operations. This is of special importance with regard to safety, as illustrated by the following cases.

- **Case 1:** Is the food heated at all, somewhere during the process? This may be of crucial importance in view of viruses (refer to Chapter 8) and parasites (Chapter 9).
- **Case 2:** If heated, does this take place before or after fermentation? This will have serious implications in view of the activity of endogenous enzymes, which may detoxify endogenous toxic substances (Chapter 4), and in view of naturally occuring pathogens (Chapter 7), as well as the need for added safe starter cultures (Chapters 10 and 11).
- **Case 3:** Is the fermented product usually cooked prior to consumption (e.g., tempeh) or is it eaten uncooked (e.g., most yogurts)? Needless to say, cooking prior to consumption will provide an additional "safety net" to the consumer.

REFERENCES

1. Beldman, G., Schols, H. A., Pitson, S. M., Searle-van Leeuwen, M. J. F. & Voragen, A. G. J. (1997). Arabinans and arabinan degrading enzymes. In *Advances in Macromolecular Carbohydrate Research.* Vol. 1, pp. 1–64. Edited by R. J. Sturgeon. London: JAI Press.
2. Belitz, H. -D. & Grosch, W. (1987). *Food Chemistry.* New York: Springer-Verlag.
3. Bückenhüskes, H. J. (1993). Selection criteria for lactic acid bacteria to be used as starter cultures for various food commodities. *FEMS Microbiol Rev* 12(1–3), 253–272.
4. Bylund, G. (1995). *Dairy Processing Handbook.* Lund, Sweden: Tetra Pak Processing Systems AB.
5. Campbell-Platt, G. (1987). *Fermented Foods of the World: A Dictionary and a Guide.* Guildford, Surrey, UK: Butterworth Scientific.
6. Campbell-Platt, G. & Cook, P. E., eds. (1995). *Fermented Meats.* London: Blackie Academic & Professional.
7. Carpita, N. C. & Gibeaut, D. M. (1993). Structural models of primary cell walls in flowering plants: consistency of molecular structure with the physical properties of the walls during growth. *The Plant Journal* 3, 1–30.
8. Corroler, D., Mangin, I., Desmasures, N. & Gueguen, M. (1998). An ecological study of lactococci isolated from raw milk in the Camembert cheese registered designation of origin area. *Appl Environ Microbiol* 64(12), 4729–4735.
9. Demuyakor, B. & Ohta, Y. (1993). Characteristics of single and mixed culture fermentation of pito beer. *J Sci Food Agric* 62(4), 401–408.
10. Dittrich, H. H. (1995). Wine and brandy. In *Biotechnology.* Vol. 9, *Enzymes, Biomass, Food and Feed,* pp. 463–504. Edited by G. Reed & T. W. Nagodawithana. Weinheim, Germany: VCH Verlagsgesellschaft.
11. Fleming, H. P., Kyung, K. H. & Breidt, F. (1995). Vegetable fermentations. In *Biotechnology.* Vol. 9, *Enzymes, Biomass, Food and Feed,* pp. 629–661. Edited by H. J. Rehm, G. Reed, A. Puhler & P. Stadler. Weinheim, Germany: VCH Verlagsgesellschaft.
12. Foschino, R., Terraneo, R., Mora, D. & Galli, A. (1999). Microbial characterization of sourdoughs for sweet baked products. *Ital J Food Sci* 11(1), 19–28.
13. Fukushima, D. (1989). Industrialization of fermented soy sauce production centering around Japanese shoyu. In *Industrialization of Indigenous Fermented Foods,* pp. 1–88. Edited by K. H. Steinkraus. New York: Marcel Dekker.
14. Garrido Fernandez, A., Garcia, P. G. & Balbuena, M. B. (1995). Olive fermentations. In *Biotechnology.* Vol. 9, *Enzymes, Biomass, Food and Feed,* pp. 593–627. Edited by H. J. Rehm, G. Reed, A. Puhler & P. Stadler. Weinheim, Germany: VCH Verlagsgesellschaft.
15. Gervais, P., Molin, P., Grajek, W. & Bensoussan, M. (1988). Influence of the water activity of a solid substrate on the growth rate and sporogenesis of filamentous fungi. *Biotechnol Bioeng* 31, 457–463.

16. Gobbetti, M. (1998). The sourdough microflora: interactions of lactic acid bacteria and yeasts. *Trends Food Sci Technol* 9(7), 267–274.

17. Gripon, J. C. (1993). Mould-ripened cheeses. In *Cheese: Chemistry, Physics and Microbiology.* Vol. 2, *Major Cheese Groups,* pp. 111–136. Edited by P. F. Fox. London: Chapman & Hall.

18. Ko, S. D. & Hesseltine, C. W. (1979). Tempe and related foods. In *Economic Microbiology.* Vol. 4, *Microbial Biomass,* pp. 115–140. Edited by A. H. Rose. London: Academic Press.

19. Kosikowski, F. V. (1982). *Cheese and Fermented Milk Foods,* 2nd edn. Ann Arbor, MI: Edwards Brothers.

20. Leistner, L. (1995). Stable and safe fermented sausages world-wide. In *Fermented Meats,* pp. 160–175. Edited by G. Campbell-Platt & P. E. Cook. London: Blackie Academic & Professional.

21. Leistner, L. & Dresel, J. (1986). Die Chinesische Rohwurst - eine andere Technologie. *Mitteilungsblatt Bundesanstalt Fleischforschung* 92, 6919–6926.

22. Leniger, H. A. & Beverloo, W. A. (1975). *Food Process Engineering.* Dordrecht, The Netherlands: D. Reidell.

23. Mittal, G. S. (1992). *Food Biotechnology Techniques and Applications.* Lancaster, PA: Technomic.

24. Nabais, R. C., Sa- Correia, I., Viegas, C. A. & Novais, J. M. (1988). Influence of calcium ion on ethanol tolerance of Saccharomyces bayanus and alcoholic fermentation by yeasts. *Appl Environ Microbiol* 54(10), 2439–2446.

25. Nout, M. J. R. (1992). Ecological aspects of mixed-culture food fermentations. In *The Fungal Community: Its Organization and Role in the Ecosystem,* 2nd edn, pp. 817–851. Edited by G. C. Carroll & D. T. Wicklow. New York: Marcel Dekker.

26. Nout, M. J. R. (2000). Tempe manufacture: traditional and innovative aspects. In *Tempe.* Edited by U. Baumann & B. Bisping. Weinheim, Germany: Wiley-VCH.

27. Nout, M. J. R., Martoyuwono, T. D., Bonné, P. C. J. & Odamtten, G. T. (1992). Hibiscus leaves for the manufacture of usar, a traditional inoculum for tempe. *J Sci Food Agric* 58(3), 339–346.

28. Nout, M. J. R. & Rombouts, F. M. (1990). Recent developments in tempe research. *J Appl Bacteriol* 69(5), 609–633.

29. Nout, M. J. R. & Rombouts, F. M. (2000). Fermented and acidified plant foods. In *The Microbiological Safety and Quality of Food,* pp. 685–737. Edited by B. M. Lund, T. C. Baird-Parker & G. W. Gould. Gaithersburg, MD: Aspen Publishers.

30. Nout, M. J. R. & Sarkar, P. K. (1999). Lactic acid food fermentation in tropical climates. *Antonie van Leeuwenhoek* 76, 395–401.

31. Patil, S. G. & Patil, B. G. (1989). Chitin supplement speeds up the ethanol production in cane molasses fermentation. *Enzyme Microb Technol* 11, 38–43.

32. Phithakpol, B. (1993). Fish fermentation technology in Thailand. In *Fish Fermentation Technology,* pp. 155–166. Edited by Ch. -H. Lee, K. H. Steinkraus & P. J. A. Reilly. Tokyo: United Nations University Press.

33. Rehm, H. J., Reed, G., Puhler, A. & Stadler, P. (1995). *Biotechnology.* Vol. 9, *Enzymes, Biomass, Food and Feed.* Weinheim, Germany: VCH Verlagsgesellschaft.

34. Rose, A. H. (1987). Responses to the chemical environment. In *The Yeasts.* Vol. 2, *Yeasts and the Environment,* 2nd edn, pp. 5–40. Edited by A. H. Rose & J. S. Harrison. London: Academic Press.

35. Russell, I. & Stewart, G. G. (1995). Brewing. In *Biotechnology.* Vol. 9, *Enzymes, Biomass, Food and Feed,* pp. 419–462. Edited by G. Reed & T. W. Nagodawithana. Weinheim, Germany: VCH Verlagsgesellschaft.

36. Samelis, J., Metaxopoulos, J., Vlassi, M. & Pappa, A. (1998). Stability and safety of traditional Greek salami—a microbiological ecology study. *Int J Food Microbiol* 44(1–2), 69–82.

37. Seaby, D. A., McCracken, A. R. & Blakeman, J. P. (1988). Experimental determination of requirements for the growth of edible Rhizopus species for use in solid substrate fermentation systems. *J Sci Food Agric* 44(4), 289–299.

38. Spicher, G. & Brümmer, J. -M. (1995). Baked goods. In *Biotechnology.* Vol. 9, *Enzymes, Biomass, Food and Feed,* pp. 241–319. Edited by G. Reed & T. W. Nagodawithana. Weinheim, Germany VCH Verlagsgesellschaft.

39. Steinkraus, K. H. (1989). *Industrialization of Indigenous Fermented Foods.* New York: Marcel Dekker.

40. Steinkraus, K. H. (Ed.). (1995). *Handbook of Indigenous Fermented Foods,* 2nd edn. New York: Marcel Dekker.

41. Suomalainen, H. & Oura, E. (1971). Yeast nutrition and solute uptake. In *The Yeasts.* Vol. 2, *Physiology and Biochemistry of Yeasts,* pp. 3–74. Edited by A. H. Rose & J. S. Harrison. London: Academic Press.

42. Tamime, A. Y. & Robinson, R. K. (1985). *Yoghurt: Science and Technology.* Oxford, England: Pergamon.

43. Uhlig, H. (1991). *Enzyme Arbeiten für Uns. Technische Enzyme und Ihre Anwendung.* München, Wien: Carl Hanser Verlag.

44. Voragen, A. G. J. & Van den Broek, L. A. M. (1991). Fruit juices. In *Biotechnological Innovations in Food Processing. Biotol Series,* pp. 187–210. Edited by Biotol Team. Oxford, England: Butterworth-Heinemann.

45. Wood, B. J. B. (Ed.). (1998). *Microbiology of Fermented Foods,* 2nd edn. London: Blackie Academic and Professional.

Appendix 1–A

Flow Diagrams for Selected Food Fermentations

Table 1–A–1 Food Group: Starchy Roots and Tubers Product Name: Gari (Fermented and Dried Cassava) Reference: 40

Flow Diagram	*Ingredients and Microorganisms*	*Thermal Data (Time at Temp)*	*Other Conditions of Antimicrobial Relevance (Salt, pH, Preservatives, etc.)*
cassava tubers	cassava (*Manihot esculenta*)		
wash			
peel (remove cortex)			
grate to coarse pulp			
put in jute or polypropylene bag			
pressurize by weight to enable de-watering			from 65% moisture content to approx 50% moisture content
ferment	natural fermentation dominated by *Lactobacillus plantarum*, *Coryne-bacterium* spp., *Geotrichum candidum*	12–96 h at 25–35 °C	organic acids (lactic, acetic) 0.6–1.2% w/w initial pH 6.2 final pH 4–4.5
break lumps			
garify (toasting for dehydration and starch gelatinization)		15–20 min at 80–85–100 °C	dehydration to approx 8% moisture content
sieve			
package and store			in polythene bags or hermetic tins
reconstitute with cold water or milk			
consume			

Note: Factors contributing to shelf life: Dehydration to about 8% moisture content, combined with hermetic storage allows shelf life of several months.

Other remarks: During this process, the grating facilitates the enzymatic degradation and detoxification of naturally occurring cyanide (mainly in the form of linamarin). HCN (ppm) levels in cassava roots, peeled roots, grated peeled roots, fermented pulp, and finished gari were 306, 184, 104, 52, and 10, respectively.

Table 1–A–2 Food Group: Starchy Roots and Tubers Product Name: Tape Ketella (Fermented Cassava) Reference: 40

Flow Diagram	*Ingredients and Microorganisms*	*Thermal Data (Time at Temp)*	*Other Conditions of Antimicrobial Relevance (Salt, pH, Preservatives, etc.)*
cassava tubers	cassava (*Manihot esculenta*)		
wash			
peel (remove cortex)			
cut to 5–10 cm pieces			
steam		10–30 min at 80–100 °C	
cool			
inoculate	ragi tapé (inoculum on rice powder) containing *Amylomyces rouxii, Endomyces fibuligera, Endomyces burtonii,* lactic acid bacteria		
ferment		2–3 d at 25–30 °C	ethanol to approx 2% v/v lactic acid 0.1–0.3% final pH approx 5.0
consume			

Note: Factors contributing to shelf life: Tape ketella has no shelf life. It is consumed as a snack or used occasionally as an ingredient in baked goods (cakes).

Table 1–A–3 Food Group: Cereals Product Name: Lager Beer Reference: 35

Flow Diagram	*Ingredients and Microorganisms*	*Thermal Data (Time at Temp)*	*Other Conditions of Antimicrobial Relevance (Salt, pH, Preservatives, etc.)*
1. Malting:	barley (*Hordeum vulgare*)		
soaking		1–2 d at 10–20 °C with intermittent airing, until moisture content 45–50% w/w	
germination		4–6 d at 15–20 °C	generation of saccharolytic, proteolytic, and other brewing enzymes; partial modification of the barley grain
kilning		at 71–92 °C to reduce moisture content to 4–5% w/w	
2. Mashing:			
coarse milling			
mashing		infusion mashing (30 min–several h at 65 °C) or decoction mashing (2–3 h at increasing temperatures from 35–100 °C)	
filtration	product: sweet wort byproduct: spent grain		
3. Wort boiling:			
hops addition	hops (*Humulus lupulus*)		hops have antimicrobial effects
boiling		1–3 h at 100 °C	
filtration			
cooling		to 2–5 °C	
filtration	product: wort		

continues

Table 1–A–3 continued

Flow Diagram	*Ingredients and Microorganisms*	*Thermal Data (Time at Temp)*	*Other Conditions of Antimicrobial Relevance (Salt, pH, Preservatives, etc.)*
4. Fermentation:			
◇ pitching	brewers' yeast (*Saccharomyces cerevisiae* or *S. uvarum*) at approx 10^7 yeast cells per ml wort		
▭			
◈ primary fermentation		1–2 wks at 10–15 °C	
◇ sedimentation	product: young ("green") beer		
▭			
◈ secondary fermentation (lagering)		1–4 wks at 2–6 °C	
◇ filtration	product: mature beer	cooling to 0–2 °C	
▭			
◇ **5. Bottling**			
◆ **6. Pasteurization**		5 min at 60 °C holding temp	
▭ consume			

Note: Factors contributing to shelf life: Lager beer has no shelf life unless it is kept anaerobic and refrigerated. Bottled lager beer derives its shelf life from pasteurization and the exclusion of air.

Table 1–A–4 Food Group: Cereals Product Name: Wheat Mixed Grain Sourdough Bread Reference: 38

Flow Diagram	*Ingredients and Microorganisms*	*Thermal Data (Time at Temp)*	*Other Conditions of Antimicrobial Relevance (Salt, pH, Preservatives, etc.)*
mix ingredients for sourdough	rye flour 15 kg seed sour 0.3 kg water 12.0 l		
fermentation of sourdough	yeasts (*Candida krusei, Pichia saitoi, Saccharomyces cerevisiae, Torulopsis holmii*) and lactic acid bacteria (*Lactobacillus sanfranciscensis, Lb. plantarum, Lb. fermentum*)	15–24 h at 23–31 °C	
seed sour			
mixing and kneading of bread dough ingredients	seed sour 27 kg rye flour 15 kg wheat flour 70 kg bakers' yeast 2 kg salt (NaCl) 2 kg water 51 l		
fermentation of bread dough	see above, but dominated by *Saccharomyces cerevisiae* (added bakers' yeast)	1–2 h at 26 °C	pH 4.6–4.7
divide, make-up			
intermediate proof		5 min	
shape and tin			
tin proof		30–60 min at 30–40 °C	
bake		35–40 min; oven temperature 200–250 °C	0.5–0.7% lactic acid in baked product
cool		to ambient temperature	
consume			

Note: Factors contributing to shelf life: Packaging in paper or other bags permitting moisture migration.
Other Remarks: Storage at 20 °C for several days, limited by staling.

Table 1–A–5 Food Group: Legumes Product Name: Tempe Kedele (Soybean Tempeh) Reference: 27

Flow Diagram	*Ingredients and Microorganisms*	*Thermal Data (Time at Temp)*	*Other Conditions of Antimicrobial Relevance (Salt, pH, Preservatives, etc.)*
▭ ◇ ◈ soak whole soybeans in water	soybeans (*Glycine max*), lactic acid bacteria (*Lactobacillus plantarum*), yeasts (*Saccharomyces dairensis*), Enterobacteriaceae	6–24 h at 20–30 °C	pH soaking water decreased from 6.5 to 4.1
◇ dehull			
◆ cook		20–40 min at 80–100 °C	ph beans 5.0–6.5
◆ cool			
▭ inoculate with mold starter	*Rhizopus microsporus, R. Oligosporus, R. oryzae, Mucor indicus*		
◇ package	plant leaves (banana) or polyethylene sheets or boxes, with perforation holes		limited perforation allows mycelium growth with very little sporulation
◈ ferment		24–48 h at 25–30 °C external temp; internal temperature may become approx 10 °C higher	initial pH 5.0; final pH 6.5–7.5 NH_3 is formed due to proteolysis
▭ harvest fresh tempeh			
◇ slice or dice			
◆ fry or stew		few min in oil of 180 °C, or 5–10 min in water or sauce of 100 °C	
▭ consume			

Note: Factors contributing to shelf life: Fresh tempeh can be stored refrigerated or frozen.

Table 1–A–6 Food Group: Legumes Product Name: Koikuchi-Shoyu (Soy Sauce) Reference: 13

Flow Diagram	*Ingredients and Microorganisms*	*Thermal Data (Time at Temp)*	*Other Conditions of Antimicrobial Relevance (Salt, pH, Preservatives, etc.)*
mixing equal quantities of soybeans and wheat	soybeans (cooked, whole) wheat (roasted, crushed)	cooking: 40–45 min at 130 °C or equivalent; roasting: few min at 170–180 °C	
inoculation	koji starter containing *Aspergillus sojae, Asp. oryzae*		
incubation to obtain molded mass = koji		2–5 d at 25–30 °C	
mixing koji and brine to obtain moromi mash	koji brine		NaCL approximately 18% (w/v)
fermentation	*Tetragenococcus halophila, Zygosaccharomyces rouxii*	6–8 months	ethanol 2–3% (v/v), lactic acid 1–2% (w/w), final pH 4.7–4.8
pressing, residue used as animal feed			
obtain raw soy sauce			
pasteurize		70–80 °C	
bottle			
consume			

Note: Factors contributing to shelf life: Combined effect of high salt concentration, mild acidity, and pasteurization.

Table 1–A–7 Food Group: Vegetables Product Name: Sauerkraut (Sour Fermented Cabbage) References: 11, 29

Flow Diagram	*Ingredients and Microorganisms*	*Thermal Data (Time at Temp)*	*Other Conditions of Antimicrobial Relevance (Salt, pH, Preservatives, etc.)*
white cabbage	*Brassica oleracea*		
harvest, transport			
grading			
coring	byproduct: inner core		
trimming	byproduct: outer leaves		
shredding to 1–2 mm thick			
salting (and homogenous mixing)	add 1.0–3.5 (typically 2.25)% w/w NaCl		
load into fermentation tanks or vats			
cover with sheet			
heading (place weight on cabbage)	using boards and weights or bags filled with water		
fermentation	natural fermentation by microbial succession: *Leuconostoc mesenteroides, Lactobacillus brevis,* dominated by *Lactobacillus plantarum. Lactobacillus bavaricus* occasionally added as a starter culture	3–6 wks at 18–22 °C	at least 1% w/w lactic acid should be formed

Postfermentation Options

1. fresh sauerkraut in bulk	2. fresh packaged sauerkraut	3. fresh packaged long-life sauerkraut	4. canned sauerkraut
	allow secondary fermentation to complete sugar depletion	add preservative (0.1% w/w Na-benzoate)	pasteurize 3 min at 74–82 °C
	fill in plastic pouches	fill in glass or plastic	hotfill in cans

continues

Table 1–A–7 continued

◇ refrigerate and sell shelf life 1–2 wks	◇ refrigerate and sell shelf life >2 months at <7 °C; >3 months under CO_2 modified atmosphere	◇ refrigerate and sell shelf life 8–12 months	◇ refrigerate and sell shelf life 18–30 months
▭ consume	▭ consume	▭ consume	▭ consume

Note: Factors contributing to shelf life: The combination of acidity and depletion of fermentable sugars. In addition, exclusion of air, preservatives, pasteurization, or sterilization, as shown above.

Other remarks: Sauerkraut can be eaten without any heat treatment (e.g., in sandwiches), but it is also popular in cooked dishes with potatoes and meats.

Table 1–A–8 Food Group: Vegetables Product Name: Treated Black Olives in Brine References: 14, 29

Flow Diagram	*Ingredients and Microorganisms*	*Thermal Data (Time at Temp)*	*Other Conditions of Antimicrobial Relevance (Salt, pH, Preservatives, etc.)*
fresh olives	*Olea europaea sativa* of various stages of maturation (yellowish green to purple)		fruits contain 0.5–1.0% organic acids (citric, oxalic, malic), as well as 1–3% phenolic compounds with antimicrobial effect
wash			
brine	5–7% w/v NaCl		
ferment	natural fermentation; no starter added; lactic acid bacteria (dominated by *Lactobacillus plantarum*) and yeasts (*Pichia membranifaciens, Pichia vinii*)		
lye treatment	1–2% w/v NaOH		
air oxidation (blackening)	purging with air; add iron salts to stabilize black color		
wash, neutralize			pH to 5.8–7.9
storage in brine	2.5–5.0% w/v NaCl		
sort, size			
can	add 1.5% w/v NaCl brine		final NaCl concentration approx 1.5% w/w
sterilize			
consume			

Note: Factors contributing to shelf life: The combination of heat treatment and moderate salt and acidity. The antimicrobial effect of phenolic compounds occurring in fresh olives is not relevant for the shelf life of the finished product because the mentioned substances have been degraded during the lye treatment.

Other remarks: Canned olives are often consumed without prior heating.

Table 1–A–9 Food Group: Fruits Product Name: Red Grape Wine References: 5, 10, 25

Flow Diagram	Ingredients and Microorganisms	Thermal Data (Time at Temp)	Other Conditions of Antimicrobial Relevance (Salt, pH, Preservatives, etc.)
grapes	*Vitis vinifera*		fermentation until fermentable
destemming			
crushing, sulfiting	SO_2 100–150 mg/l sometimes additional nitrogen sources		
fermentation "on the skins"	starter cultures are occasionally used; otherwise natural fermentation by succession; wild yeasts (*Kloeckera apiculata, Kloeckera apis, Torulopsis stellata, Candida stellata*), followed by *Saccharomyces cerevisiae* (dominating primary fermentation)	1–3 wks at 20–25 °C	sugars less than 0.1%. Final ethanol 11–17% v/v. Acidity: volatile 0.1–0.15% w/v as acetic acid; total acidity 0.5–0.7% w/v as tartaric acid
racking	separate skins ("lees") from young wine		
secondary fermentation	starter cultures used: *Oenococcus oeni*	few days at pH > 3.5 and temp > 15 °C	conversion of malic acid into lactic acid (typically 1.5–3.5 g/l lactic acid)
clarification			
aging	bulk tanks	3 months–2 years	
blending			
filtration			
bottling			
consume			

Note: Factors contributing to shelf life: The combination of ethanol, moderate acidity, and exclusion of air. Residual sulfite may be present, but this is not an essential preservative. Absence of fermentable sugars ("dry wines") contributes to shelf life.

Table 1–A–10 Food Group: Fruit Product Name: Toddy (Palm Wine) Reference: 40

Flow Diagram	*Ingredients and Microorganisms*	*Thermal Data (Time at Temp)*	*Other Conditions of Antimicrobial Relevance (Salt, pH, Preservatives, etc.)*
collection of palm sap	sap of oilpalm (*Elaeis guineensis*) or coconut palm (*Cocos nucifera*)		
natural fermentation	yeasts (*Saccharomyces cerevisiae, Candida* spp.), lactic acid bacteria (*Lactobacillus* spp., *Leuconostoc* spp.), acetic acid bacteria, *Zymomonas* spp., *Micrococcus* spp.	0–48 h at 25–35 °C	organic acids (lactic, acetic) 0.2–0.4% pH 4.5–3.5 ethanol 4–10%
filtration			
bottle (optional)			
pasteurize (optional)		15 min at 80 °C	
cool (optional)			
consume			

Note: Factors contributing to shelf life: Fresh palm wine has no shelf life except when it is kept refrigerated or bottled and pasteurized.

Table 1–A–11 Food Group: Meat Product Name: Salami (Italian-Style Raw Fermented Sausage) Reference: 6

Flow Diagram	*Ingredients and Microorganisms*	*Thermal Data (Time at Temp)*	*Other Conditions of Antimicrobial Relevance (Salt, pH, Preservatives, etc.)*
medium chopping (cuttering)	pure frozen pork or a pork/beef mix, fat, salt (NaCl) 2.5–4.5% w/w $NaNO_2$ 125 ppm pepper, mace, cardamon, garlic totalling 1% w/w		
filling into casings	natural or artificial casings		
ripening (fermentation)	*Lactobacillus* spp., *Pediococcus acidilactici, Micrococcus* spp., *Debaryomyces hansenii*	15–90 d at 15–25 °C	acidity approx 2.5% w/w as lactic acid, various antimicrobial compounds (e.g., H_2O_2), pH initial 5.8–6.0 to pH final 4.9–6.0
drying		several wks at 12–15 °C to reduce A_w to 0.67–0.92	
consume			

Note: Factors contributing to shelf life: The combination of salt, nitrite, acidity, pH, anaerobic conditions, and reduced water activity. The lower the A_w, the longer the shelf life. This type of sausage is not smoked, in contrast to many other raw fermented sausages.

Other remarks: If no frozen pork was used, it must be heated to 58 °C to kill *Trichinella.* Otherwise, no heat treatment takes place and the product is consumed without prior cooking.

Table 1–A–12 Food Group: Meat Product Name: Lup Cheong (Chinese raw sausage) References: 20, 21

Flow Diagram	*Ingredients and Microorganisms*	*Thermal Data (Time at Temp)*	*Other Conditions of Antimicrobial Relevance (Salt, pH, Preservatives, etc.)*
coarse chopping	pork meat and fat, sugar, salt		
filling into casings	natural casing (intestine)		
predrying		approx 36 h at 48 °C and 65% RH to A_w 0.9	
fermentation	no starter; natural fermentation by lactic acid bacteria, micrococci, etc.	3 d at 20 °C and 75% RH	final pH 5.9 and A_w 0.8 taste preference requires low levels of acidity
cooking			
consume			

Note: Factors contributing to shelf life: The combination of reduced water activity, salt, and moderate acidity.
Other remarks: The product is heated prior to consumption, distinguishing it from the consumption of European-style raw sausage.

Table 1–A–13 Food Group: Fish Product Name: Plaa-Raa (Fish-Rice Paste) Reference: 32

Flow Diagram	*Ingredients and Microorganisms*	*Thermal Data (Time at Temp)*	*Other Conditions of Antimicrobial Relevance (Salt, pH, Preservatives, etc.)*
▭ small fresh water fish (whole)	gourami (*Trichogastea* spp.), snake-head fish (*Ophicephalus striatus*), catfish (*Clarias* spp., *Heterobagrus* sp.), tawes, bards (*Puntius* spp.), butter catfish (*Ompok* sp.), tilapia (*Tilapia* spp.)		
▭ mix with salt	NaCl		salt 11–24% w/w
◇			
◈ ferment	endogenous proteases, *Tetragenococcus halophila, Staphylococcus epidermidis, Micrococcus* spp., *Bacillus* spp.	several days at ambient temperature	
▭ mix with roasted rice	glutinous or normal paddy rice, roast until dark brown, coarse grinding		
◇			
◇ pack in earthen or glass jars			
◈ ferment	*Tetragenococcus halophila* and other lactic acid bacteria, *Staphylococcus epidermidis, Micrococcus* spp., *Bacillus* spp.	several days to a year	acidity 0.7–1.9% as lactic acid; final pH varies from 4.1–6.9
▭ consume			

Note: Factors contributing to shelf life: Combined effect of high salt concentration and hermetic storage.

Table 1–A–14 Food Group: Fish Product Name: Nuoc-Mam (Fish Sauce) Reference: 40

Flow Diagram	*Ingredients and Microorganisms*	*Thermal Data (Time at Temp)*	*Other Conditions of Antimicrobial Relevance (Salt, pH, Preservatives, etc.)*
▭ small sea fish, whole	*Decapterus, Engranlis, Dorosoma, Clupeodes, Stolephorus* spp.		
▭ mix with salt	fish : salt 1:3 to 1:1.5		
◇			
◇ transfer to large vessel			
▭ cover with layer of salt			
◇			
◇ pressurize by weight			
◈ fermentation	endogenous proteases (halotolerant bacteria: *Paracoccus halodenitrificans, Aerococcus haloviridans* are present; their function is not clear)	few months to 1 year at 20–30 °C	lactic acid approx 1% w/w; final pH 5.7–6.0
◇ drain liquid to obtain first-quality Nuoc-Mam			salt content 24–28% w/v
▭			
▭ ◇ wash residue with boiling sea water in several stages to obtain second and subsequent qualities			
◇ bottling			
▭ consume			

Note: Factors contributing to shelf life: The combination of high salt content, organic acids, and hermetic bottling ensure a shelf life of several months.

Table 1–A–15 Food Group: Milk Product Name: Stirred Yogurt Reference: 42

Flow Diagram	*Ingredients and Microorganisms*	*Thermal Data (Time at Temp)*	*Other Conditions of Antimicrobial Relevance (Salt, pH, Preservatives, etc.)*
milk	cow milk		
standardize			
homogenize		55 °C at 20 MPa	
pasteurize (high pasteurization)		5 min at 85 °C	
cool to 30–32 °C			
inoculate	0.025% w/w starter culture containing equal numbers of *Streptococcus salivarius* spp. *thermophilus* and *Lactobacillus delbrueckii* spp. *bulgaricus*		
ferment		16–20 h at 32 °C	organic acids (lactic, acetic) to approx 0.8–1% w/w as lactic acid; final pH 4.1–4.6
stir			
cool to 6 °C			
aseptic packaging			
consume			

Note: Factors contributing to shelf life: The combination of acidity and other antimicrobial compounds produced by lactic acid bacteria, anaerobic conditions in hermetically sealed jar, low storage temperature, and the presence of live (competing) microorganisms.

Other remarks: The product is consumed without prior heating.

Table 1–A–16 Food Group: Milk Product Name: Camembert Surface-Ripened Cheese References: 4, 17, 19

Flow Diagram	*Ingredients and Microorganisms*	*Thermal Data (Time at Temp)*	*Other Conditions of Antimicrobial Relevance (Salt, pH, Preservatives, etc.)*
pasteurization	cow milk 48% fat	20 sec at 70–72 °C	
cooling			
inoculation	mesophilic lactic acid bacteria cheese starter *Lactococcus lactis* spp. *cremoris, Lc. lactis* spp. *lactis, Leuconostoc mesenteroides* spp. *cremoris, Leuc. lactis*		
stir		15–60 min at 34 °C	until 0.2% titratable acidity and pH 4.5–4.6
renetting	rennet 0.01% v/v	2–3 h at 34 °C	
cut, drain whey		24 h at 18 °C	whey pH 5.5
spray-inoculate	spore suspension *Penicillium camemberti*		
superficial drying		5–6 h	
salting	2–3 times by rubbing crystalline salt or in 25% brine	2–3 d at 14 °C 1–1.5 h at 14 °C	
ripening first phase		8 d at 13–16 °C at 95% RH	
ripening second phase		3 wks at 11 °C at 85% RH	pH cheese increases to 6–7
packaging and distribution	aluminium foil, polyethylene film, wooden or cardboard boxes		
consume			

Note: Factors contributing to shelf life: Refrigerated storage at 2–4 °C.

CHAPTER 2

Why Fermented Foods Can Be Safe

Martin R. Adams

INTRODUCTION

Over time, the human diet has evolved to exclude materials that are frankly hazardous and for which there is no simple procedure to render them harmless. As a result, our food supply is generally safe, although it can never be entirely devoid of risk. There are a variety of hazards that can be associated with food materials. These can be broadly classified into three categories: intrinsic, extrinsic, and processing/biogenic (Exhibit 2–1).

Some of these hazards can be severe, as in the case of botulinum toxin or bongkrekic acid poisoning, but they can be controlled, and the risk they pose can be reduced to acceptable levels by effective and hygienic processing and preparation. Safety considerations should always be paramount in modern food processing and are also an implicit aspect of traditional methods of processing and preserving food. This is particularly true of fermented foods, which take potentially hazardous raw materials such as raw meat and milk and transform them into acceptable products with better keeping qualities and reduced risk. The bulk of this book addresses particular groups of hazards, their significance in fermented foods, and ways in which they are controlled. This chapter addresses the mechanisms by which fermentation can improve food safety in more general terms.

PHYSICAL PROCESSING

In Chapter 1, the various physical unit operations associated with fermented food production were described. These operations are common to a number of food production processes and, wherever they occur, they can have some impact on product safety.

Transport

The manner in which raw materials are transported can affect their quality. Physical damage during transport can breach natural antimicrobial barriers, allowing microorganisms access to nutrient-rich underlying tissues, where they can grow rapidly. Excessive delays during transport can also increase time for microbial growth and for natural processes of senescence to occur.

Grading and Sorting

Inspection and sorting to remove damaged or obviously infected raw material is a basic protective barrier that can be useful in reducing risk. It can be used to exclude obviously infected meat, damaged or moldy fruit, or mold-infected grains. For example, mechanical or hand sorting of infected nuts has been shown to reduce overall aflatoxin levels considerably; reduced aflatoxin levels were also evident after hand sorting

Exhibit 2–1 A Classification of Food Hazards

Intrinsic—Properties of the food itself, natural toxins (e.g., cyanogenic glycosides, glycoalkaloids)

Extrinsic—Contaminants posing a direct hazard (e.g., infectious pathogens—*Salmonella, Shigella*, viruses, protozoa; toxic chemical contaminants—pesticide residues, heavy metals)

Processing/Biogenic—Hazards produced as a result of processing and/or generated by microbial contaminants (e.g., nitrosamines, biogenic amines, bacterial toxins, mycotoxins)

of figs.[82] This is, however, a relatively blunt weapon because, in many cases, there is no visual indication to help distinguish foods that pose a hazard from those that do not.

Cleaning and/or Washing

Microbial pathogens can be associated with the surface of fruits and vegetables as a result of contamination in the field. These may be natural soil microorganisms such as the spore-forming pathogens, *Bacillus cereus* and *Clostridum botulinum,* or *Listeria monocytogenes,* or they may be enteric in origin, such as *Salmonella, Escherichia coli,* and *Shigella*, arising from contact with human or animal feces, sewage, or untreated water. Contaminants may also be introduced during harvesting and postharvest handling.

Interior tissues are generally, but not invariably, sterile, and the majority of the surface-associated microflora can be removed by washing. Typically, vigorous washing of plant products in clean water will reduce microbial levels by 10–100-fold, but does not ensure the complete elimination of risk or levels of risk reduction comparable to processes such as pasteurization or canning. The efficacy of washing can be improved slightly by incorporating antimicrobials such as chlorine in the wash water. For example, washing of lettuce leaves with water removed an average of 92.4% of the total microflora. This was increased to 99.5% by including 100 mg l^{-1} of available free chlorine and adjusting the pH to 4.5–5.0.[3] Failure to remove more of the microflora was ascribed to the survival of bacteria in protective hydrophobic pockets or folds in the leaf surface, an effect that will vary between commodities as a result of differences in surface composition and topology. Increased production of ready-to-eat vegetables has heightened interest in the surface disinfection of fruits and vegetables, and the subject has been extensively reviewed.[9]

Washing of animal carcasses with hot solutions of lactic or acetic acid can reduce the total microbial load by 2–3 log cycles, although uniform treatment of all surfaces is difficult and *Salmonella* and *E. coli* show more marked resistance to such treatments.[36]

Physical Separation

Physical separation can include peeling or trimming to remove external layers. Because the external layers are where most contamination will be found, removing them can substantially improve overall microbiological quality provided it is done hygienically, avoiding contamination of the freshly exposed surfaces underneath. The interior tissues of plants are generally considered to be sterile. Although this is not always true, substantial reductions in count can be achieved by using only hygienically removed interior tissues. With leafy vegetables such as cabbage, the bacterial load on inner leaves can be 1000-fold lower than on the outer leaves and the interior tissues of the plant.[55]

Chemical hazards can also be reduced by the physical separation of plant components. The removal of green surface layers from potatoes can reduce their glycoalkaloid content, and detaching the 2–3 mm thick parenchymal layer of cassava that underlies the cortex will reduce the overall concentration of the cyanogenic glycosides, linamarin and lotaustralin.[51]

Moisture Adjustment

Changes in the water activity (A_w) of a material can have a profound effect on microbial hazards. Depending on the circumstances, a decrease in A_w that is produced by drying or the addition of solutes such as salt will reduce the potential for microbial growth but also enhance microbial resistance to adverse conditions, thus potentially enhancing the survival of pathogens that may be present. Increasing the A_w has the potential to allow growth of previously quiescent organisms, perhaps allowing them to grow to numbers that are sufficient to initiate an infection once they are ingested or produce enough toxin in the product to cause illness. Under certain circumstances, an increase in the A_w as a result of soaking could also permit the onset of a lactic fermentation, which will help control or eliminate microbial hazards.

Size Reduction

Comminution of a raw material will disrupt cells, thereby releasing their contents. This, in turn, will increase the supply of nutrients to any microorganisms present, stimulating their growth. This can be beneficial when the growth of those organisms whose dominance is necessary for a successful fermentation is encouraged, but in some circumstances, it could also stimulate growth of any pathogens present. The precise outcome will depend on a number of factors such as relative number of organisms present, their substrate affinities, and their physiological state.

The relative sensitivities of different components of the microflora to antimicrobial factors produced by the plant material may also be important. Many plants have defense systems that are activated by cellular disruption. These systems release compartmentalized enzymes, allowing them to act on their substrates and producing compounds that are active against microorganisms and predators such as insects. Examples of this are the release of isothiocyanate by the myrosinase-glucosinolate system in crucifers[10] and allicin produced during the crushing of garlic.

The reduction of chemical hazards is also sometimes assisted by tissue disruption. The cyanogenic glycosides in cassava are hydrolyzed to produce a cyanohydrin by the endogenous enzyme, linamarase, when cellular damage occurs.[18] Microorganisms appear to play no role in cyanogen reduction in the traditional fermented cassava product, *gari*,[89] although in processes where whole roots or large pieces are fermented, microbial activity does assist the endogenous process by softening the tissues.[91]

Mixing

Mixing will serve to distribute microorganisms throughout a mass of material. Very often, this distribution will accelerate the desirable fermentation and thereby improve safety, although it also has the potential to transfer pathogens to microenvironments where they are protected from antimicrobial factors that prevail elsewhere in the material.

MICROBIAL ACTIVITY

As already noted, the production of fermented foods shares many unit operations with other types of food processing, all of which can contribute to product safety. The unique feature of food fermentations, however, is the central role that microbial activity plays in the overall process, contributing a number of desirable properties such as improved product shelf life, increased safety, and improved flavor or texture. In developed countries today, the availability of modern food preservation techniques, such as an efficient cold chain, have diminished the significance of fermentation as a food preservation technology, although it remains of major importance in developing countries.

Improvements in food safety arising from microbial activity during fermentation are largely due to lactic acid bacteria (LAB), which are a group of organisms that predominate in the majority of fermented foods. Their growth and metabolism inhibit the normal spoilage flora of the food material and any bacterial pathogens that it may contain. This inhibition can act in two ways: It can slow or arrest growth of the organism, or it can inactivate or kill the organism. Both procedures can result in an improvement in safety. With toxigenic pathogens, the inhibition of growth can effectively ensure safety, assuming that initial numbers are below those necessary to produce levels of toxin that can cause illness. With infectious bacterial pathogens, slowing or preventing growth may be insufficient to guarantee safety because the infectious dose of some pathogens can be very low. Inactivation or killing of bacteria does occur, but even quite high levels of inactivation still may not be sufficient to eliminate risk entirely. This will depend on factors such as the pathogen concerned, its initial numbers, and its physiological state.

The LAB

The LAB are a group of gram-positive, nonsporeforming, fermentative anaerobes that are often aerotolerant (Table 2–1). They produce most of their cellular energy as a result of the fermentation of sugars. In the case of hexoses, this can proceed by one of two pathways, providing a useful diagnostic feature as well as playing a critical role in their antimicrobial activity. *Homofermenters* ferment hexoses by the Embden-Meyerhof pathway to produce almost exclusively lactic acid; *heterofermenters* produce less acid overall as a mixture of lactic acid, acetic acid, ethanol, and carbon dioxide using the 6-phosphogluconate/phosphoketolase pathway.[7]

The ability of LAB to inhibit other organisms has been a topic of extensive research over the years, and the field has been reviewed periodically.[4,40,52,72,74] A variety of antimicrobial factors produced by LAB have been identified (Exhibit 2–2); these will be discussed individually. The production of many of these compounds is limited to a few restricted species or strains of LAB. Because these strains/species are not ubiquitous in lactic fermented foods, it is clear that they can only play some contributory role to the overall preservative effect. For so many different LAB to appear in fermented foods, the principal preservative factor must be something that is common to them all. This central and unvarying feature is their use of fermentative pathways to generate cellular energy and the fact that this leads inevitably to the production of organic acids, principally lactic acid, and a decrease in pH. Acidity levels in some fermentations can exceed 100 mM, reducing the pH to less than 4.0 in weakly buffered systems.

Organic Acids and Reduced pH

Low pH and the presence of organic acids are the two principal components of the inhibition of microorganisms under the acid conditions produced by fermentation. The three contributory aspects of acid inhibition were described nearly 50 years ago by Ingram *et al.*[42]

1. the pH
2. the degree of dissociation of the acid
3. the inherent toxicity of the acid anion

Pathogenic bacteria will grow over a pH range of 2–5 units but generally grow best at pH values around neutrality, in the pH range of 6–7. As the

Table 2–1 Principal Genera of Lactic Acid Bacteria Associated with Food

	Rod	*Coccus*	*Homofermenter*	*Heterofermenter*
Lactobacillus	+		+	+
Lactococcus		+	+	–
Enterococcus		+	+	–
Carnobacterium	+		–	+
Leuconostoc		+	–	+
Weissella		+	–	+
Oenococcus		+	–	+
Pediococcus		+	+	–
Streptococcus		+	+	–
Tetragenococcus		+	+	–

Exhibit 2–2 Antimicrobial Factors Associated with Lactic Acid Bacteria

Low pH
Organic acids
Bacteriocins
Carbon dioxide
Hydrogen peroxide
Diacetyl
Ethanol
Reuterin
Nutrient depletion and crowding

pH decreases away from the optimum region, the growth rate declines, eventually reaching zero. At pH values below the minimum that will support growth, the microorganism is progressively inactivated. The rate of inactivation is temperature dependent—the higher the temperature, the faster the rate. This relationship applies for temperatures from chill to lethal, and has been noted in both food and model systems for a number of pathogens such as *Salmonella, E. coli, Listeria,* and *Yersinia enterocolitica*,[33,53,76,88] and also at low A_w.[30]

If microorganisms allowed their internal pH to equilibrate to the same value as their environmental pH, they would only be able to function over a very limited pH range close to neutrality. In fact, for most pathogens, growth only ceases once the pH has dropped to less than pH 4.5. This is because bacteria have the ability to maintain their internal pH higher than that of their acidic environment. Both the cell membrane, which has a low permeability to hydrogen ions, and the intrinsic buffering capacity of the cytoplasm help to maintain the cytoplasmic pH, but the cell also possesses a number of active mechanisms for maintaining pH homeostasis. These include the removal of protons in exchange for the uptake of potassium ions, and the uptake and decarboxylation of amino acids to produce neutralizing amines.[11,12,35]

Although weak organic acids such as those produced by fermentation are less effective at decreasing the extracellular pH, they are more effective at inhibiting bacteria than strong acids such as hydrochloric.[15,20,64,83] For example, recent data have shown that for 19 strains of enterohemorrhagic *E. coli*, the minimum growth pH was 4.25 when hydrochloric acid was used as the acidulant and 5.5 when acetic acid was used.[59] This observation is linked to two important properties of acids such as acetic and lactic acid.

1. They are weak carboxylic acids that only partially dissociate in aqueous solution.
2. In their undissociated form, they carry no net charge and have appreciable lipid solubility, which allows them to diffuse freely through the cell's plasma membrane down a concentration gradient into the cytoplasm.

The fact that the undissociated species is the more active antimicrobial form of these acids is illustrated in Figure 2–1, where the degree of inhibition of *E. coli* is clearly linked with the degree of dissociation of the acid present. In a fermented food, the low pH will increase the proportion of the undissociated form present. When the undissociated acid passes through the cell's cytoplasmic membrane into the higher pH of the cytoplasm, it will dissociate, thereby acidifying the cytoplasm and releasing the acid anion. The increased leakage of protons into the cytoplasm will place a metabolic burden on the cells, which will divert resources away from growth-related functions, thus slowing growth. The cell will also accumulate the acid anion, which can disrupt cellular processes.[79] As the extracellular pH decreases and the total concentration of acid increases, so the burden will increase until growth is no longer possible.

The pK_a of an acid is therefore an important determinant of an acid's antimicrobial activity because it describes the proportion present in the undissociated form at any given pH. Differences in the lipophilicity of the undissociated acid and the intrinsic toxicity of the anion will account for differences in the observed toxicity of acids with similar pK_a values. Acetic acid (pK_a 4.76) is a weaker acid than lactic acid (pK_a 3.86); this could account for the frequent observation that it is a more effective antimicrobial.[12] It may also account for the apparent synergy in the antimicrobial effect of mixtures of lactic and acetic

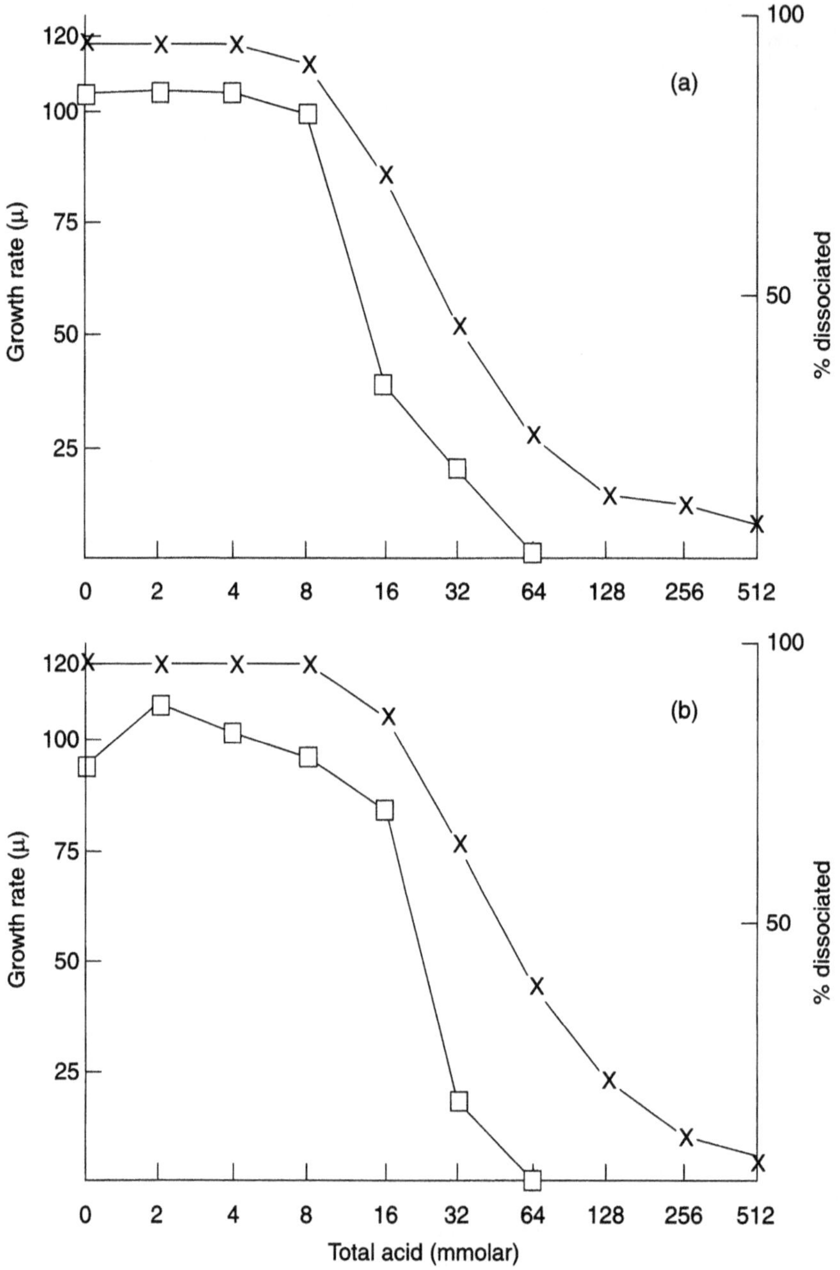

Figure 2–1 Effect of total acid concentration on the percentage acid dissociation (x) and the specific growth rate of *Escherichia coli* (a) with acetic acid and (b) with lactic acid.

acid, where lactic acid, the stronger acid, decreases the pH, thereby increasing the proportion of acetic acid in the undissociated form and thus potentiating its antimicrobial effect.[2,78] Heterofermentative LAB can produce mixtures of lactic and acetic acid when they have alternative electron acceptors such as oxygen or fructose to help regenerate their nicotinamide adenine dinucleotide (NAD). Although heterofermenters produce less overall acidity than homofermenters, their early dominance in several natural vegetable fermentations could be very important in providing the rapid inhibition of other organisms present and setting the fermentation on its subsequent course.

The antimicrobial effect will depend on the amount of acid produced, which will depend on the numbers of LAB present. Numerous studies have shown that bacterial pathogens do not survive well when they are added to a prefermented food where the LAB have had the opportunity to grow to large numbers and the pH is already low.[60,61,70,71,81,87] However, when pathogens are present at the start of fermentation, inhibition will be delayed until the LAB have achieved a numerical dominance and produced sufficient acid to achieve an effect. In one instance, when the LAB (in this case, *Lactococcus lactis*) outnumbered the pathogen (*E. coli*) by more than 5 log cycles, the pathogen was still able to grow for five hours, increasing in numbers by 2 log cycles during that time.[95]

Bacteriocins

Bacteriocins are polypeptide antimicrobials that are produced by bacteria and are bactericidal to other, normally very closely related, organisms. In recent years, a considerable research effort has been devoted to the identification and characterization of bacteriocins produced by LAB. The widespread consumption of LAB in foods without any adverse health effects is taken as an indication that they can generally be regarded as safe[1] and therefore their bacteriocins might have potential as "natural" food preservatives. Tangible practical results from this work have been slight, though it has led to the description of numerous new bacteriocins.[43,50,66]

These new bacteriocins have been classified into four main groups (Table 2–2). Under this scheme, Class I bacteriocins are the lantibiotics, which include nisin, the only bacteriocin that is currently approved for use in foods.[29] Nisin is a small (3.4 kDa) heat-stable polypeptide containing some unusual amino acids introduced by post-translational modification of a ribosomally synthesized precursor. It is distinguished from many other bacteriocins by its relatively broad range of activity, being inhibitory to most gram-positive bacteria. It has been used as a food preservative for 50 years and has, more recently, been given "generally regarded as safe" (GRAS) status in the United States.[69] Nisin's principal use in food processing has been to inhibit the outgrowth of spores, which show the greatest sensitivity to nisin. Gram-positive nonsporeforming pathogens such as *Staphylococcus aureus* and *L. monocytogenes* are inhibited, although *S. aureus* is one of the most nisin-resistant gram positives and *L. monocytogenes* has been shown to acquire nisin resistance quite readily.[26,39,62]

Early work suggested that nisin resistance in several *Bacillus* species was due to the production of a nisinase enzyme that specifically reduced the C-terminal dehydroalanyl lysine to

Table 2–2 Classification of Bacteriocins Produced by Lactic Acid Bacteria

Class	*Subclass*	*Description*
I		Lantibiotics (e.g., nisin); small, heat stable, containing lanthionine
II		Small (<10 kDa), heat stable, non-lanthionine-containing, membrane-active peptides
	IIa	Pediocin-like peptides with Y-G-N-G-V-X-C- near the amino terminus; anti-listerial
	IIb	Two-peptide bacteriocins
	IIc	Thiol-activated peptides
III		Large (>30 kDa) heat labile proteins
IV		Complex bacteriocins: protein with lipid and/or carbohydrate

alanyllysine.[45–47] More recently, attention has focused on the composition of the cytoplasmic membrane where nisin exerts its effect by forming pores. Studies have seen differences in the phospholipid content of membranes in nisin-resistant and nisin-sensitive strains of *L. monocytogenes*[63,90] and changes in the fatty acid composition leading to less fluid membranes in resistant strains.[57] This appears not to be the complete explanation, however, as several studies have implicated other features of the cell envelope in resistance. For example, it has been shown that the hydrophobicity of *L. monocytogenes* cells correlates with their nisin sensitivity,[27,63] that nisin-resistant strains show different sensitivities to cell wall-acting compounds,[22] and that removal of the cell wall of resistant strains abolishes resistance.[27] Some clarification of these findings and the observation that cells possess limited nisin-binding sites[26] has come from the work of Breukink *et al.*[17] They confirmed an observation made 20 years previously[75] that nisin binds with high affinity to Lipid II, a membrane-anchored cell wall precursor, and proposed that differences in nisin sensitivity may reflect different amounts or availability of Lipid II.[17]

Nisin is used commercially as a partially purified concentrate but may be produced *in situ* during a fermentation. Its production during cheese fermentations and the consequent inhibition of other starter organisms led to its first isolation and description. Studies of the effect of nisin-producing cultures on the safety of fermented foods have produced mixed results. *In situ* nisin production has been shown to inhibit the growth of *L. monocytogenes* in Camembert cheese,[56] although levels of 150 IU g^{-1} produced by *Lc. lactis* in fermented rice made no discernible contribution to the inhibition of *S. aureus,* which was ascribed entirely to the lactic acid produced.[95] However, there have been several reports of work employing LAB-producing bacteriocins other than nisin to inhibit *L. monocytogenes* in fermented sausages, cottage cheese, smoked salmon, and bean sprouts.[8,34,58,68]

The value of nisin and other LAB bacteriocins as aids to bacteriological food safety is limited by their range of activity. Gram-negative bacteria, which constitute the majority of food-borne bacterial pathogens, are generally resistant to nisin due to its inability to penetrate the outer membrane. If, however, the outer membrane permeability is increased by treatment with chelators such as ethylenediaminetetraacetic acid (EDTA) or thermal injury, then they do display sensitivity.[14,23,85,86] Prospects for using this approach to inhibit gram negatives in fermented foods are not promising because the pH drop during fermentation does not appear to produce appreciable outer membrane injury and reduces the ability of EDTA to chelate metal ions and permeabilize the outer membrane.[13]

Carbon Dioxide

Heterofermentative LAB produce 1 mole (22.4 liters) of carbon dioxide for every mole of hexose they ferment. This can help establish anaerobic conditions, thereby preventing the growth of obligate aerobes such as molds and slowing the growth of facultative organisms such as the *Enterobacteriaceae.* Anaerobic conditions will also give the LAB a competitive edge and encourage a successful fermentation. An increased partial pressure of carbon dioxide also has its own specific antimicrobial activity. Molds and oxidative gram-negative bacteria are most susceptible, whereas lactobacilli and some yeasts show high tolerance. There is considerable scope for improving the understanding of the inhibitory mechanisms of carbon dioxide and the relative importance of factors such as the ability of carbon dioxide to decrease intracellular pH, its inhibition of enzymatic reactions, and the disruptive effects of its interaction with cell membranes.

Hydrogen Peroxide

In the presence of oxygen, LAB produce hydrogen peroxide through the activity of flavoprotein oxidases. Because LAB are catalase negative, the hydrogen peroxide can accumulate in the medium. Hydrogen peroxide is a strong oxidizing agent and exerts its antimicrobial effect directly or through its degradation products such as superoxide, O_2^-, and hydroxyl radicals.

LAB appear to be more resistant to hydrogen peroxide than many other bacteria. A reported minimum inhibitory concentration for *L. lactis* was 125 mg l^{-1}, whereas that for *S.aureus* was 5–6 mg l^{-1}.[25,92] However, the amounts accumulated by LAB in culture are quite variable. Lactococci growing in milk and the meat starter organism, *Pediococcus cerevisiae,* both produce levels below 1 mg l^{-1} in the bulk environment.[73,77] Increased amounts of hydrogen peroxide can be produced by improving aeration through shaking and at low temperatures when the oxygen solubility is greater.[21,25] It has been suggested that hydrogen peroxide production by LAB in refrigerated foods is a significant antimicrobial factor.[16,37] However, this seems an unlikely scenario in the production of most fermented foods. It is possible that hydrogen peroxide plays a significant role during the early stages of fermentation while residual dissolved oxygen is removed, or that microenvironments with higher levels may persist. In general, though, the contribution of hydrogen peroxide is likely to be relatively minor.

Diacetyl

Diacetyl (2,3-butanedione) is produced by strains of several LAB and plays an important role in the sensory properties of many fermented foods, particularly dairy products. It can accumulate when the organism reaches the stationary phase of growth or when there are alternative sources available from which to generate the key intermediate pyruvate, such as citrate. Antibacterial activity of diacetyl has been demonstrated against a number of bacteria, including *Aeromonas hydrophila, Bacillus* spp., *Enterobacter aerogenes, E. coli, Mycobacterium tuberculosis, Pseudomonas* spp., *Salmonella, S. aureus,* and *Y. enterococolitica*.[31,44,48,65] Gram negatives generally show greater sensitivity and the LAB are most resistant.

However, diacetyl does have an extremely sharp odor, and the acceptable sensory threshold in dairy products is 2–7 mg kg^{-1}.[54] The levels of diacetyl reported to produce appreciable microbial inhition are usually quite high (e.g., for gram-negative bacteria, approximately 200 mg kg^{-1}), well in excess of this level. Some studies have suggested that lower concentrations of diacetyl can be effective, particularly at lower temperatures,[6] but others have failed to see any appreciable contribution of diacetyl at 10 mg kg^{-1} to the inhibition of *E. coli* in refrigerated dairy foods.[38] Overall, the contribution of diacetyl to bacterial inhibition in fermented foods appears to be relatively minor.

Ethanol

Ethanol is a well-established antimicrobial. It is the principal fermentation product of the yeast *Saccharomyces cerevisiae* and plays a central role in the microbiological safety of alcoholic beverages, where it is often the most important single hurdle in a series that can also include heat treatment, low pH, anaerobic conditions, and elevated carbon dioxide levels. Yeast activity also removes the mycotoxin patulin from apple juice (see Chapter 5).

Heterofermentative LAB will also produce ethanol, but in lower concentrations. The levels produced are inversely related to the amount of acetic acid produced because these compounds represent alternative fates of acetyl phosphate in the heterofermentation pathway. In addition to any direct inhibitory effect, ethanol has also been shown to augment the lethal effect of low pH and lactic acid in *E. coli* O157, though at higher concentrations than would be likely to occur in lactic fermentations.[49] Heterofermenters dominate during the initial phase of several natural fermentations and ethanol may, like other products of heterofermentation, contribute in these early stages to setting the fermentation on its desirable course.

Reuterin and Other Low Molecular Weight Compounds

A number of low molecular weight compounds with antimicrobial activity have been isolated and identified from culture filtrates of LAB. The most studied to date has been reuterin, β-hydroxypropanal. This is present in stationary phase cultures of strains of the heterofermenter, *Lactobacillus reuteri,* when grown anaerobically on a mixture of glucose and glycerol or glyceraldehyde. Reuterin exists in three forms—the aldehyde, its hydrate, and a cyclic dimer—and has very broad spectrum antimicrobial activity against bacteria, fungi, protozoa, and viruses. The relative activity of the different forms is not known, although they are thought to act by inhibiting sulfhydryl-containing enzymes.[24]

The addition of reuterin to foods such as minced beef, milk, and cottage cheese has been shown to control coliform growth and to inactivate *L. monocytogenes* and *E. coli* O157.[24,32] However, its role in fermented foods is uncertain. One study where *L. reuteri* and glycerol were added to herring fillets to stimulate reuterin production *in situ* showed some potential to reduce growth of the gram-negative flora present. In other cases where the incorporation of *L. reuteri* in foods has been described, in a fermented milk and an Emmental-type cheese, its presence has been for its potential probiotic effect rather than for reuterin production in the food.[28,84]

Pyroglutamic acid, 2-pyrrolidone-5-carboxylic acid, is a natural constituent of some plant foods and fermented products such as soy sauce and can be produced by the thermal dehydration of glutamic acid.[5,94] It has been shown to be produced by a range of LAB, including strains of *L. casei, L. helveticus, Streptococcus bovis,* and unspecified pediococci.[19,41,93] Pyroglutamic acid is active primarily against gram-negative bacteria such as *Pseudomonas* and *Enterobacter*. Its activity against gram-negative pathogens has not been reported, although it is inactive against the gram-positive pathogens *L. monocytogenes* and *S. aureus*.[41]

A number of other low molecular weight compounds with antimicrobial properties have been identified in culture filtrates from *L. plantarum* cultures. These include mevalonolactone, methylhydantoin, and benzoic acid.[67] They have been shown to act synergistically with lactic acid against *Pantoea (Enterobacter) agglomerans,* but their significance in food fermentations remains to be established.

Nutrient Depletion and Overcrowding

The absence of inhibitory agents is a necessary but insufficient condition for microbial growth. The microbial cell must also have sufficient space and nutrients available to it. Foods will rarely offer microorganisms a uniform medium, and numerous microenvironments may exist where conditions differ substantially from the bulk properties of the material. This is a critical factor in the microbiological stability of butter and similar spreads, where microorganisms are confined in aqueous droplets dispersed in the continuous fat phase. The presence of large numbers of LAB in a fermented food will deplete it of readily available nutrients and occupy space that might have been available to other less desirable organisms. This is supported by the recent observation that filtration of a fermented broth to remove cells of *Lc. lactis* decreased the subsequent inhibition of *E. coli*.[13] It is an area that is only just beginning to be explored.

OVERALL SIGNIFICANCE OF DIFFERENT ANTIMICROBIAL FACTORS

To describe the combined effect of a number of antimicrobial factors, the analogy of a hurdle

Table 2–3 Shelf Life and Composition of Fermented Fish Products from Thailand[80]

Product	*Salt (%)*	*pH*	*Shelf Life*
Pla-som	2.3–5.9	4.0–4.6	3 weeks
Som-fug	2.5–5.8	4.1–5.0	2 weeks
Pla-chom	3.8–4.8	5.0–6.1	2 weeks
Pla-chao	4.4–9.5	4.0–5.3	1–2 years
Pla-paeng-daeng	4.5–9.2	3.9–5.2	6–12 months
Pla-ra	7.8–17.9	4.7–6.2	1–3 years

race has been used. Perhaps nowhere in food preservation is the concept of multiple barriers or hurdles more appropriate than in fermented foods. The LAB themselves can contribute a number of different antimicrobials, and microbial activity is often combined with other inhibitory factors such as salting or partial drying.

It is possible to make some generalizations regarding the relative importance of these various factors. Certainly the predominant microbial factor is the production of organic acids and the reduction in pH, although the others described here could all contribute to the aggregate effect, particularly to ensuring a successful early dominance of LAB. Reduction in A_w generally plays a much more significant role in the inhibition of microbial growth compared to bacterial acid production. This is illustrated by data on the shelf life of fermented fish products in Thailand and their composition, where differences in salt content have greater impact on product shelf life (Table 2–3). Microbial activity is, however, probably more important and effective at reducing numbers of microbial hazards.

REFERENCES

1. Adams, M. R. (1999). Safety of industrial lactic acid bacteria. *J Biotechnol* 63, 171–178.
2. Adams, M. R. & Hall, C. J. (1988). Growth inhibition of food-borne pathogens by lactic and acetic acids and their mixtures. *Int J Food Sci Technol* 23, 287–292.
3. Adams, M. R., Hartley, A. D. & Cox, L. J. (1989). Factors affecting the efficacy of washing procedures used in the production of prepared salads. *Food Microbiol* 6, 69–77.
4. Adams, M. R. & Nicolaides, L. (1997). Review of the sensitivity of different foodborne pathogens to fermentation. *Food Control* 8, 227–239.
5. Airaudo, C. B., Gayte-Sorbier, A. & Armand, P. (1987). Stability of glutamine and pyroglutamic acid under model system conditions: influence of physical and technological factors. *J Food Sci* 52, 1750–1752.
6. Archer, M. H., Dillon, V. M., Campbell-Platt, G. & Owens, J. D. (1996). Effect of diacetyl on growth rate of *Salmonella typhimurium* determined from detection times measured in a micro-well plate photometer. *Food Control* 7, 63–67.
7. Axelsson, L. (1998). Lactic acid bacteria: classification and physiology. In *Lactic Acid Bacteria: Microbiology and Functional Aspects*, pp. 1–72. Edited by S. Salminen & A. von Wright. New York: Marcel Dekker.
8. Bennik, M. H. J., van Overbeek, W., Smid, E. J. & Gorris, L. G. M. (1999). Biopreservation in modified atmosphere stored mungbean sprouts: the use of vegetable associated bacteriocinogenic lactic acid bacteria to control the growth of *Listeria monocytogenes*. *Lett Appl Microbiol* 28, 226–232.
9. Beuchat, L. R. (1998). *Surface decontamination of fruits and vegetables eaten raw: a review*. Food Safety Unit, World Health Organization, WHO/FSF/FOS/98.2 42pp.
10. Bones, A. M. & Rossiter J. T. (1996). The myrosinase-glucosinolate system, its organisation and biochemistry. *Physiol Plantarum* 97, 194–208.
11. Booth, I. R. (1999). The regulation of intracellular pH in bacteria. In *Bacterial Responses to pH*, pp. 19–37. Edited by D. J. Chadwick & G. Cardew. Chichester, England: John Wiley & Sons.
12. Booth, I. R. & Kroll, R. G. (1989). The preservation of foods by low pH. In *Mechanisms of Action of Food Preservation Procedures*, pp. 119–159. Edited by G. W. Gould. London: Elsevier Applied Science.
13. Boziaris, I. S. & Adams, M. R. (1999). Effect of chelators and nisin produced *in situ* on inhibition and inactivation of gram negatives. *Int J Food Microbiol* 53, 105–113.
14. Boziaris, I. S., Humpheson, L. & Adams, M. R. (1998). Effect of nisin on heat injury and inactivation of *Salmonella enteritidis* PT4. *Int J Food Microbiol* 43, 7–14.
15. Brackett, R. E. (1987). Effects of various acids on growth and survival of *Yersinia enteroccolitica*. *J Food Prot* 50, 598–601.
16. Brashears, M. M., Reilly, S. S. & Gilliland, S. E. (1998). Antagonistic action of cells of *Lactobacillus lactis* toward *Escherichia coli* O157:H7 on refrigerated raw chicken meat. *J Food Prot* 61, 166–170.
17. Breukink, E., Wiedemann, I., van Kraaij, C., Kuipers, O. P., Sahl, H. -G. & de Kruijff, B. (1999). Use of the cell wall precursor Lipid II by a pore-forming peptide antibiotic. *Science* 286, 2361–2364.
18. de Bruijn, G. H. (1973). The cyanogenic character of cassava (*Manihot esculenta*). In *Chronic Cassava Toxicity*, pp 43–48. Edited by B. Nestel & R. MacIntyre. Ottawa, Canada: International Development Research Centre.
19. Chen, G. & Russell, J. B. (1989). Transport of glutamine by *Streptococcus bovis* and conversion of glutamine to pyroglutamic acid and ammonia. *J Bacteriol* 171, 2981–2985.
20. Chung, K. C. & Goepfert, J. M. (1970). Growth of *Salmonella* at low pH. *J Food Sci* 35, 326–328.

21. Collins, E. B. & Aramaki, K. (1980). Production of hydrogen peroxide by *Lactobacillus acidophilus*. *J Dairy Sci* 63, 353–357.
22. Crandall, A. D. & Montville, T. J. (1998). Nisin resistance in *Listeria monocytogenes* ATCC 700302 is a complex phenotype. *Appl Environ Microbiol* 64, 231–237.
23. Cutter C. N. & Siragusa G. R. (1995). Population reduction of gram-negative pathogens following treatments with nisin and chelators under various conditions. *J Food Prot* 58, 977–983.
24. Daeschel, M. A. & Penner, M. H. (1992). Hydrogen peroxide, lactoperoxidase systems and reuterin. In *Food Biopreservatives of Microbial Origin*, pp. 155–175. Edited by B. Ray & M. Daeschel. Boca Raton, FL: CRC Press.
25. Dahiya, R. S. & Speck, M. L. (1968). Hydrogen peroxide formation by lactobacilli and its effect on *Staphylococcus aureus*. *J Dairy Sci* 51, 1568–1573.
26. Davies, E. A. & Adams, M. R. (1994). Resistance of *Listeria monocytogenes* to the bacteriocin nisin. *Int J Food Microbiol* 21, 341–347.
27. Davies, E. A., Falahee, M. B. & Adams, M. R. (1996). Involvement of the cell envelope of *Listeria monocytogenes* in the acquisition of nisin resistance. *J Appl Bacteriol* 81, 139–146.
28. Delespaul, G., Leclerc, G., Lepeltier, M. & Chassagne, M. H. (1994). *Cheesemaking process and the cheeses obtained*. French Patent 2,696,620.
29. Delves-Broughton, J. (1990). Nisin and its use as a food preservative. *Food Technol* 142, 100–117.
30. Deng, Y., Ryu, J. H. & Beuchat, L. R. (1998). Influence of temperature and pH on survival of *Escherichia coli* O157:H7 in dry foods and growth in reconstituted infant rice cereal. *Int J Food Microbiol* 45, 173–184.
31. Elliker, P. R. (1945). Effect of various bacteria on diacetyl content and flavour of butter. *J Dairy Sci* 28, 93.
32. El-Ziney, M. G. & Debevere, J. M. (1998). The effect of reuterin on *Listeria monocytogenes* and *Escherichia coli* O157:H7 in milk and cottage cheese. *J Food Prot* 61, 1275–1280.
33. Entani, E., Asai, M., Tsujihata, S., Tsukamoto, Y. & Ohta, M. (1998). Antibacterial action of vinegar against foodborne pathogenic bacteria. *J Food Prot* 61, 953–959.
34. Foegeding, P. M., Thomas, A. B., Pilkington, D. H. & Klaenhammer, T. R. (1992). Enhanced control of *Listeria monocytogenes* by *in situ*-produced pediocin during dry fermented sausage production. *Appl Environ Microbiol* 58, 884–890.
35. Foster, J. W. & Moreno, M. (1999). Inducible acid tolerance mechanisms in enteric bacteria. In *Bacterial Responses to pH*, pp. 55–74. Edited by D. J. Chadwick & G. Cardew. Chichester, England: John Wiley & Sons.
36. Gill, C. O. (1998). Microbiological contamination of meat during slaughter and butchering of cattle, sheep and pigs. In *The Microbiology of Meat and Poultry*, pp. 118–157. Edited by A. Davies & R. Board. London: Blackie Academic and Professional.
37. Gilliland, S. E. & Ewell, H. R. (1983). Influence of combinations of *Lactobacillus lactis* and potassium sorbate on growth of psychrotrophs in raw milk. *J Dairy Sci* 66, 974–980.
38. Guraya, R., Frank, J. F. & Hassan, A. N. (1998). Effectiveness of salt, pH and diacetyl as inhibitors for *Escherichia coli* O157:H7 in dairy foods stored at refrigeration temperatures. *J Food Prot* 61, 1098–1102.
39. Harris, L. J., Fleming, H. P. & Klaenhammer, T. R. (1991). Sensitivity and resistance of *L. monocytogenes* ATCC 19115, Scott A and UAL 500 to nisin. *J Food Prot* 54, 836–840.
40. Holzapfel, W. H., Geisen, R. & Schillinger, U. (1995). Biological preservation of foods with reference to protective cultures, bacteriocins and food-grade enzymes. *Int J Food Microbiol* 24, 343–362.
41. Huttunen, E., Noro, K. & Zang, Y. (1995). Purification and identification of antimicrobial substances produced by two *Lactobacillus casei* strains. *Int Dairy J* 5, 503–513.
42. Ingram, M., Ottaway, F. J. H. & Coppock, J. B. M. (1956). The preservative action of acid substances in food. *Chem and Ind* 42, 1154–1165.
43. Jack, R. W., Tagg, J. R. & Ray, B. (1995). Bacteriocins of gram positive bacteria. *Microbiol Rev* 59, 171–200.
44. Jalander, Y. W. (1936). Diacetyl als Tuberkelbazillen tötender Bestandteil des finnischen Holzteeres. *Naunyn-Schmiedebergs Arch Experimental Pathol Pharmakol* 180, 628–630.
45. Jarvis, B. (1967). Resistance to nisin and production of nisin-inactivating enzymes by several *Bacillus* species. *J Gen Microbiol* 47, 33–48.
46. Jarvis, B. (1970). Enzymic reduction of the C-terminal dehydroalanyl-lysine sequence in nisin. *Proc Biochem Soc* 119, 56P.
47. Jarvis, B. & Farr, J. (1971). Partial purification, specificity and mechanism of action of the nisin-inactivating enzymes from *Bacillus cereus*. *Biochim Biophys Acta* 227, 232–240.
48. Jay, J. M. (1982). Antimicrobial properties of diacetyl. *Appl Environ Microbiol* 44, 525–532.
49. Jordan, S. L., Glover, J., Malcolm, L., Thomson-Carter, F. M., Booth, I. R. & Park, S. F. (1999). Augmentation of killing of *Escherichia coli* O157 by combinations of lactate, ethanol, and low pH conditions. *Appl Environ Microbiol* 65, 1308–1311.
50. Klaenhammer, T. R. (1993). Genetics of bacteriocins produced by lactic acid bacteria. *FEMS Microbiol Rev* 12, 38–85.

51. Kojima, M., Iwatsuki, N., Data, E. S., Villegas, C. D. V. & Uritani, I. (1983). Changes of cyanide content and linamarase activity in wounded cassava roots. *Plant Physiol* 72, 186–189.

52. Lindgren, S. E. & Dobrogosz, W. J. (1990). Antagonistic activities of lactic acid bacteria in food and feed fermentations. *FEMS Microbiol Rev* 87, 149–164.

53. Little, C. L., Adams, M. R., Anderson, W. A. & Cole, M. B. (1994). Application of a log-logistic model to describe the survival of *Yersinia enterocolitica* at sub-optimal pH and temperature. *Int J Food Microbiol* 22, 63–71.

54. Lücke, F. -K. & Earnshaw, R. G. (1991). Starter cultures. In *Food Preservatives*, pp. 215–234. Edited by N. Russell & G. W. Gould. Glasgow, Scotland: Blackie Academic and Professional.

55. Lund, B. M. (1992). Ecosystems in vegetable foods. *J Appl Bacteriol Symp Supp* 73, 115S–126S.

56. Maisnier-Patin, S., Deschamps, N., Tatini, S. R. & Richard, J. (1992). Inhibition of *Listeria monocytogenes* in Camembert cheese made with a nisin-producing starter. *Lait* 72, 249–263.

57. Mazzotta, A. S. & Montville, T. J. (1997). Nisin induces changes in membrane fatty acid composition of *Listeria monocytogenes* nisin-resistant strains at 10 °C and 30 °C. *J Appl Microbiol* 82, 32–38.

58. McAuliffe, O., Hill, C. & Ross, R. P. (1999). Inhibition of *Listeria monocytogenes* in cottage cheese manufactured with a lacticin 3147-producing starter culture. *J Appl Microbiol* 86, 251–256.

59. McKellar, R. C. & Knight, K. P. (1999). Growth and survival of various strains of enterohemorrhagic *Escherichia coli* in hydrochloric and acetic acid. *J Food Prot* 62, 1466–1469.

60. Mensah, P., Tomkins, A. M., Drasar, B. S. & Harrison, T. J. (1990). Fermentation of cereals for reduction of bacterial contamination of weaning foods. *Lancet* 336, 140–143.

61. Mensah, P., Tomkins, A. M., Drasar, B. S. & Harrison, T. J. (1991). Antimicrobial effect of fermented Ghanaian maize dough. *J Appl Bacteriol* 70, 203–210.

62. Ming, X. & Daeschel, M. A. (1993). Nisin resistance of foodborne bacteria and specific resistance responses of *Listeria monocytogenes* Scott A. *J Food Prot* 56, 944–948.

63. Ming, X. & Daeschel, M. A (1995). Correlation of cellular phospholipid content with nisin resistance of *Listeria monocytogenes* Scott A. *J Food Prot* 58, 416–420.

64. Minor, T. E. & Marth, P. H. (1972). Loss of viability of *Staphylococcus aureus* in acidified media: inactivation by several acids, mixtures of acids and salts of acids. *J Milk Food Technol* 35, 191–196.

65. Motlagh, A. M., Johnson, M. C. & Ray, B. (1991). Viability loss of foodborne pathogens by starter culture metabolites. *J Food Prot* 54, 873–878.

66. Nes, I. F., Diep, D.B., Håvarstein, L. S., Brurberg, M. B., Eijsink, V. & Holo, H. (1996). Biosynthesis of bacteriocins in lactic acid bacteria. *Antonie van Leeuwenhoek* 70, 113–128.

67. Niku-Paavola, M. -L., Laitila, A., Mattila-Sandholm, T. & Haikara, A. (1999). New types of antimicrobial compounds produced by *Lactobacillus plantarum*. *J Appl Microbiol* 86, 29–35.

68. Nilsson, L., Gram, L. & Huss, H. H. (1999). Growth control of *Listeria monocytogenes* on cold-smoked salmon using a competitive lactic acid flora. *J Food Prot* 62, 336–342.

69. Nisin preparation: affirmation of GRAS status. (1988). *Federal Register* 54, 11247–11251.

70. Nout, M. J. R., Rombouts, F. M. & Hautfast, G. J. (1989). Accelerated natural lactic fermentation of infant food formulations. *Food Nutr Bull* 11, 65–73.

71. Nout, M. J. R., Rombouts, F. M. & Havelaar, A. (1989). Effect of accelerated natural lactic acid fermentation of infant food ingredients on some pathogenic organisms. *Int J Food Microbiol* 8, 351–361.

72. Ouwehand, A. (1998). Antimicrobial components from lactic acid bacteria. In *Lactic Acid Bacteria: Microbiology and Functional Aspects*, pp. 139–159. Edited by S. Salminen & A. Von Wright. New York: Marcel Dekker.

73. Raccach, M. & Baker, R. C. (1978). Formation of hydrogen peroxide by meat starter cultures. *J Food Prot.* 41, 798–801.

74. Ray, B. & Daeschel, M., eds. (1992). *Food Biopreservatives of Microbial Origin.* Boca Raton, FL: CRC Press.

75. Reisinger, P., Seidel, H., Tschesche, H. & Hammes, W. (1980). The effect of nisin on murein synthesis. *Arch Microbiol* 127, 187–193.

76. Roering, A. M., Luchansky, J. B., Ihnot, A. M., Ansay, S. E., Kaspar, C. W. & Ingham, S. C. (1999). Comparative survival of *Salmonella typhimurium* DT 104, *Listeria monocytogenes* and *Escherichia coli* O157:H7 in preservative-free apple cider and simulated gastric fluid. *Int J Food Microbiol* 46, 263–269.

77. Ross, G. D. (1981). The inhibition of growth of spoilage microorganisms in milk by *Streptococcus lactis* subsp. *diacetylactis, Leuconostoc cremoris* and *L. dextranicum*. *Aust J Dairy Technol* 36, 147–149.

78. Rubin, H. E. (1973). Toxicological model for a two-acid system. *Appl Environ Microbiol* 36, 623–624.

79. Russell, J. B. (1992). Another explanation for the toxicity of fermentation acids at low pH: anion accumulation versus uncoupling. *J Appl Bacteriol* 73, 363–370.

80. Saono, S., Hull, R. R. & Dhamcharee, B., eds. (1986). *A Concise Handbook of Indigenous Fermented Foods in the ASCA Countries.* Canberra, Australia: The Government of Australia.

81. Simango, C. & Rukure, G. (1991). Survival of *Campylobacter jejuni* and *Escherichia coli* in mahewu,

a fermented cereal gruel. *Trans Royal Soc Med Hyg* 85, 399–400.

82. Smith, J. E. & Solomons, G. L. (1994). *Mycotoxins in human nutrition and health*. DG XII. Brussels, Belgium: European Commission.
83. Sorrells, K. M., Enigl, D. C. & Hatfield, J. R. (1989). The effect of pH, acidulant, time and temperature on the growth and survival of *Listeria monocytogenes*. *J Food Prot* 52, 571–573.
84. Speck, M. L., Dobrogosz, W. J. & Casas, I. A. (1993). *Lactobacillus reuteri* in food supplementation. *Food Technol* 47(7), 90–94.
85. Stevens, K. A., Sheldon, B. W., Klapes, N. A. & Klaenhammer T. R. (1991). Nisin treatment for inactivation of *Salmonella* species and other gram-negative bacteria. *Appl Environ Microbiol* 57, 3613–3615.
86. Stevens, K. A., Sheldon, B. W., Klapes, N. A. & Klaenhammer, T. R. (1992). Effect of treatment conditions on nisin inactivation of gram negative bacteria. *J Food Prot* 55, 763–767.
87. Svanberg, U., Sjgren, E., Lorri, W., Svennerholm, A. -M. & Kjaiser, B. (1992). Inhibited growth of common enteropathogenic bacteria in lactic-fermented cereal gruel. *World J Microbiol Biotechnol* 8, 601–606.
88. Tsai, Y. W. & Ingham, S. C. (1997). Survival of *E. coli* O157:H7 and *Salmonella* spp. in acidic condiments. *J Food Prot* 60, 751–755.
89. Vasconcelos, A. T., Twiddy, D. R., Westby, A. & Reilly, P. J. A. (1990). Detoxification of cassava during gari preparation. *Int Journal of Food Sci Technol* 25, 198–203.
90. Verheul, A., Russell, N. J., Vant Hof, R., Rombouts, F. M. & Abee, T. (1997). Modifications of membrane phospholipid composition in nisin-resistant *Listeria monocytogenes* Scott A. *Appl Environ Microbiol* 63, 3451–3457.
91. Westby A. & Choo, B. K. (1994). Cyanogen reduction during lactic fermentation of cassava. *Acta Hort* 375, 209–215.
92. Wheater, D. M., Hisch, A. & Mattic, A. T. R. (1952). Possible identity of "lactobacillin" with hydrogen peroxide produced by lactobacilli. *Nature* 70, 623–626.
93. Yang, Z., Suomalainen, T., Måyrå-Måkinen, A. & Huttunen, E. (1997). Antimicrobial activity of 2-pyrrolidone-5-carboxylic acid produced by lactic acid bacteria. *J Food Prot* 60, 786–790.
94. Yokotsuka, T. & Sasaki, M. (1998). Fermented protein foods in the Orient: shoyu and miso in Japan. In *Microbiology of Fermented Foods,* 2nd edn, pp. 351–415. Edited by B. J. B. Wood. London: Blackie Academic and Professional.
95. Yusof, R. M., Morgan, J. B. & Adams, M. R. (1993). Bacteriological safety of a fermented weaning food containing L-lactate and nisin. *J Food Prot* 56, 414–417.

CHAPTER 3

An Introduction to the Hazard Analysis and Critical Control Point (HACCP) System and Its Application to Fermented Foods

Yasmine Motarjemi

INTRODUCTION

The Hazard Analysis and Critical Control Point system, often known under its acronym HACCP, is a system that was conceived in the 1960s by the Pillsbury Company, National Aeronautics and Space Administration (NASA), and the U.S. Army Laboratories at Natick to ensure the safety of foods for astronauts. Originally developed to ensure the microbiological safety of foodstuffs, it is now recognized that the system can be equally applied to control and prevent chemical and physical hazards.[11]

In the 30 years since its conception, the HACCP system has grown to become the universally recognized and accepted method for food safety assurance. In appreciation of the importance of the HACCP system for enhancing food safety, and in view of the importance of globalization of the food supply and the need for harmonization of food safety requirements, the Codex Alimentarius Commission (CAC) adopted in 1993 the Guidelines for the Application of HACCP, which have received international recognition. In 1997, on the recommendation of the World Health Organization (WHO), the CAC guidelines were revised and improved.[5]

For more than 20 years, WHO has recognized the importance of the HACCP system for the prevention of food-borne diseases and has played an important role in its development, promotion, and harmonization. One of the most important initiatives of the organization has been the promotion of the HACCP system in small industries and cottage industries and the use of the HACCP system for health education activities.

Before explaining the HACCP system and its application to the preparation of fermented foods, it is essential to understand why the system has received such recognition by the food industry and has been promoted by public health authorities, in particular WHO. For this purpose, the historical development of food safety needs to be reviewed.*

HISTORICAL DEVELOPMENT

Food safety has been of concern to humankind since the dawn of history, and many of the problems encountered in our food supply go back to the earliest recorded years. Many rules and recommendations advocated in religious or historical texts are evidence of the concern to protect people against food-borne hazards and food adulteration. Although advances in science and technology in the last few centuries have increased scientists' understanding of chemical and biological hazards, up until the early 1980s, the food hygiene† systems in many countries were based on empirical knowledge acquired through the surveillance of food-borne diseases

*Food safety is the assurance that food will not cause harm to the consumer when it is prepared and/or eaten according to its intended use.

†Food hygiene is defined as all conditions and measures necessary to ensure the safety and suitability of food at all stages of the food chain.

and, in some instances, on the perception of what is considered a hazard. Often, this perception led to the fact that food hygiene was interpreted simply as cleanliness.

Because of the prevalence of diseases transmitted by the fecal–oral route, such as typhoid fever and cholera, the emphasis of food safety programs for a long time was on the improvement of water supply and sanitation infrastructures and the protection of food from fecal contamination. As a result, regulatory authorities focused their inspection activities on the cleaning and sanitation of food businesses and on the personal hygiene of food handlers. A similar approach was adopted in health education activities, and the focus of food hygiene education programs was based mainly on hand washing, boiling of water, and the protection of food from flies. These measures, where implemented, have been effective in reducing the incidence of fecal-orally transmitted food-borne and waterborne diseases, such as typhoid fever and cholera. However, they have been insufficient to prevent other types of food-borne diseases, such as salmonellosis and campylobacteriosis. With the recent drastic changes in food production and lifestyles, some of these diseases have even increased in incidence. In addition, other types of problems have emerged.[6]

THE NEED FOR CHANGE

The end of the twentieth century was marked by significant changes in the food safety assurance system and food control. Traditionally, the method of food safety assurance was based on two types of measure: (1) actions undertaken during the procurement of raw materials, processing, manufacturing, transport, and distribution, including design, layout, and cleaning of premises to prevent contamination; and (2) actions undertaken to ensure that food, which was produced, was indeed safe. The former were usually prescribed in the Codes of Good Manufacturing Practice (GMP) and/or the Codes of Hygienic Practice (GHP). For the purpose of the latter, industries tested the end product for confirmation of safety. Food control and public health authorities also inspected premises for compliance with GMP/GHP and other regulatory requirements. As mentioned above, these inspections often focused on the cleanliness of premises, food handlers, and the immediate environment and failed to identify shortcomings related to the production and preparation procedure itself. Authorities also carried out independent testing of end products. These methods of food safety assurance and food control showed certain weaknesses: The control of premises was based on random inspection and not on what happened during longer periods of time before or after the inspection. End-product testing performed by industry or food inspectors also proved to be costly and time consuming and provided insufficient assurance of food safety (see the following section on the benefits of the HACCP system).

Concomitant with the recognition of the limitations of traditional approaches to food safety control, the concern for food safety grew. There were many reasons for these concerns.[3]

- increasing incidence of food-borne diseases in many parts of the world
- recognition that one of the major health problems of developing countries, infant diarrhea, is to a large extent food-borne
- emergence of new or newly recognized food-borne pathogens such as verocytotoxin-producing *Escherichia coli*, *Campylobacter* spp., food-borne trematodes, and *Salmonella enteritidis*
- increased knowledge and awareness of the serious and chronic health effects of food-borne hazards
- increased number of vulnerable people, such as the elderly, children, pregnant women, immunocompromised individuals, the undernourished, and individuals with other underlying health problems
- increased awareness of the economic consequences of food-borne diseases
- the possibility of detecting minute amounts of contaminants in food, due to advances in scientific and analytical methods
- industrialization and mass production leading to (1) increased risks of large-scale food

contamination and (2) the considerably larger numbers of people affected in food-borne disease outbreaks as a result
- urbanization, leading to a longer and more complex food chain, and thus greater possibilities for food contamination
- new food technologies and processing methods, causing concern either about the safety of the products themselves or about the eventual consequences due to inappropriate handling in households or food service and catering establishments
- changing lifestyles, depicted by an increasing number of meals eaten out of the home, either in food service and catering establishments or at street food stalls
- increased tourism worldwide and international trade in foodstuffs, both leading to a greater exposure to foreign and unfamiliar food-borne hazards
- increased contamination of the environment
- increased consumer awareness of food safety
- lack of or decreasing resources for food safety

As a result of the above, but also recognizing the limitations of the traditional approaches, public health authorities and the food industry have both recognized the need to move to a more preventive, scientific, and cost-effective food safety assurance approach, namely, the HACCP system.

WHAT IS THE HACCP SYSTEM?

The HACCP system is defined by the CAC as a *system which identifies, evaluates, and controls hazards which are significant for food safety.* The value of the HACCP system lies in the fact that it is a scientific, rational, and systematic approach to the identification, assessment, and control of hazards during production, processing, manufacturing, preparation, and use of food to ensure that food is safe when consumed. The HACCP system is based on seven principles, as introduced in Appendix 3–A.[2]

Through the application of these principles, industries or food establishments that apply the system will be led to review their process of food production or preparation critically, to identify hazards and control measures, to establish objective criteria for ensuring safety, and to monitor that control measures have been properly implemented, in particular at those steps during the food production that are critical for food safety. They will also foresee the necessary corrective actions when monitoring procedures indicate failures in the control of hazards. Through verification and documentation, they can also, at all times, verify the adequacy of their measures. Guidance for the application of these principles has also been provided by the CAC text on HACCP Principles and Guidelines for its Application (Appendix 3–A). It is, however, recognized that for the application of the HACCP system to be successful, some additional measures need to be considered.

1. *Management commitment.* The successful implementation of the HACCP system requires a change in attitude of policy and decision makers in all sectors concerned as well as management commitment. Although implementation of the HACCP system brings many benefits, and in the long term may reduce financial costs, its implementation in the initial stages requires additional resources encompassing qualified personnel, technical support facilities, equipment, training, and so forth. Such conditions can be met only where there is a management commitment.
2. *Prerequisites for HACCP.* Prior to applying the HACCP system to any sector of the food chain, that sector should be operating according to the CAC General Principles of Food Hygiene, the appropriate CAC Codes of Practice, and appropriate food safety legislation.
3. *Training.* Adequate training of personnel is a key to effective implementation of the HACCP system. Such training should include an explanation of the HACCP system, the reasons for and the objectives of its application, as well as the responsibilities of each person involved in the imple-

mentation of the HACCP plan. The tasks of operators working at each critical control point (CCP) should also be clearly defined and the operators should be trained in performing them.[2,11]

BENEFITS OF THE HACCP SYSTEM

The benefits of the HACCP system are summarized in the following paragraphs.[11]

- The HACCP system overcomes many of the limitations of the traditional approaches to food safety control (generally based on "snap-shot" inspection and end-product testing), including
  a. limitations of "snap-shot" inspection techniques in predicting potential food safety problems
  b. the difficulty of collecting and examining sufficient samples to obtain meaningful, representative information in a timely manner and without the high cost of end-product analysis
  c. reduction of the potential for product recall
  d. identification of problems without understanding the causes
- The HACCP system allows for the identification of all conceivable, reasonably expected hazards, even where failures have not previously been experienced. It is therefore particularly useful for new operations.
- The HACCP system is sufficiently flexible to accommodate any changes that might be introduced, such as progress in equipment design, improvements in processing procedures, and technological developments related to the product.
- The HACCP system will help target/direct resources to the most critical part of the food operation.
- With the HACCP system, one can expect an improvement in the relationship between (1) food processors and food inspectors and (2) food processors and consumers. The HACCP system provides a scientifically sound basis for demonstrating that all reasonable precautions have been taken to prevent a hazard from reaching the consumer. In this way, it encourages confidence in the safety of food products and thus promotes both confidence in the food industry and stability of food businesses.
- Data collected facilitate the work of food inspectors for auditing purposes.
- The HACCP system is applicable to the whole food chain, from the raw material to the end product (i.e., growing, harvesting, processing or manufacturing, transport and distribution, preparation, and consumption).
- The application of the HACCP system can promote international trade by increasing confidence in food safety.
- The HACCP system can be readily integrated into quality management systems such as total quality management, ISO 9000, and so forth.

AREAS OF APPLICATION

The HACCP principles can be applied in a variety of ways.

- The HACCP system is a system that is used as a method of food safety assurance in food production, processing, manufacturing, and preparation. The CAC guidelines for the application of the HACCP system provide guidance on how the seven principles of HACCP can be implemented in food industries in order to have the greatest chance of success (Appendix 3–A).
- The HACCP system is amenable to effective food control. It allows for more efficient inspection of food operations because the role of food inspectors is centered on the assessment of the HACCP plan and confirmation that it is designed properly and operating effectively.
- The HACCP principles, in particular principles 1 to 5, can be used to study food preparation practices and to identify and assess hazardous behavior, which should be the focus of health education interventions.

- The HACCP concept can be used in the management of overall food safety programs to identify those problems all along the food chain that are of greatest risk to public health, and to prioritize interventions.

THE HACCP SYSTEM IN FOOD HYGIENE

Today, to achieve food safety, it is recognized that there is a need to apply measures of increased specificity (Figure 3–1). At a more general level, the CAC outlines the general principles of food hygiene. These principles lay the foundation for food hygiene. Second, more product-specific hygienic measures may be applied to focus better on issues that are relevant to specific commodities. These measures, also prescribed by the CAC, are described in specific codes of manufacturing or hygienic practices. The CAC has developed codes for a number of products, such as smoked or salted fish, cured ham, and so forth. Finally, application of the HACCP system can further enhance food safety by providing a mechanism for analyzing the hazards for each food or process, developing a tailor-made plan for ensuring food safety with emphasis on CCPs, and ensuring that the critical limits at these points are met. With each layer of the above measures, the degree of assurance of food safety increases.[4]

To harmonize and promote the application of the HACCP system in food industries, the CAC outlined the principles and elaborated the guidelines for the application of the system. The principles of the HACCP system, as defined by the CAC, set the basis for the minimum requirements for mandatory application of the HACCP system. The guidelines are, on the other hand, a general guidance, and adherence to them is voluntary. The CAC text on the HACCP Principles and Guidelines for its Application are presently annexed to the CAC General Principles of Food Hygiene, and their application is consequently recommended. However, due to the status of the CAC in international trade in food, the application of the HACCP system to the production, processing, or manufacturing of food may in some countries become compulsory for food export. The reason is that since the conclusion of the GATT Uruguay Round of Multilateral Trade Negotiations in April 1994 and the coming into force of the World Trade Organization (WTO) Agreement on Sanitary and Phytosanitary Measures, the work of CAC is recognized as the reference or "yardstick" for national food safety requirements. As a result, members of the WTO need to take the work of CAC into consideration

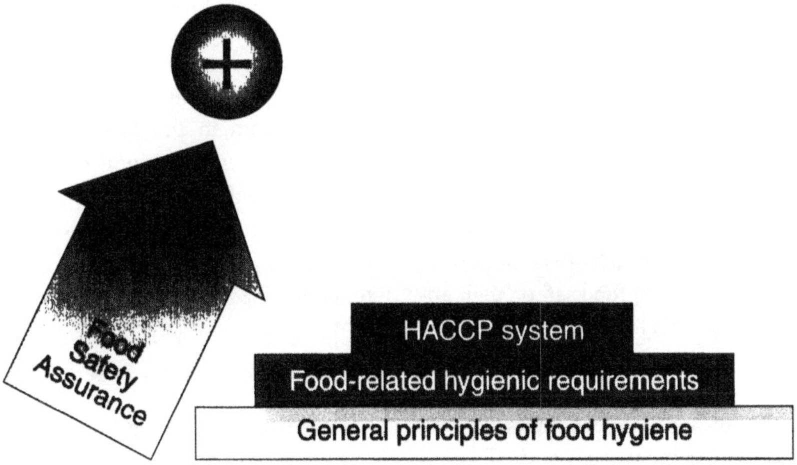

Figure 3–1 HACCP in food hygiene: Each additional measure of food hygiene will increase the degree of food safety assurance.[14]

and adapt their legislation to the provisions provided by CAC. In recent years, some countries such as the United States and the European Union have made application of the HACCP system compulsory in the production and processing of certain foodstuffs such as seafood, including those that are imported.

The implementation of the HACCP system in all food industries is an established goal in many countries. However, most progress made in implementing the HACCP system so far has been achieved in medium- and large-scale food industries, particularly in industrialized countries. Worldwide, analyses of food-borne disease outbreaks show that the greatest majority of food-borne disease outbreaks result from malpractices during food preparation in small businesses, canteens, and homes. In small businesses and homes in both developed and developing countries, the application of the HACCP system meets with greater difficulties.

In recognition of the need for improving the safety of foods that are prepared or processed in homes, food service establishments, street food vendors, and "cottage" industries, WHO proposed the use of HACCP systems to small operations and for health education purposes.[1,7,10]

APPLICATION OF THE HACCP APPROACH TO FOOD PREPARATION

The CAC General Principles of Food Hygiene provide guidance on the basic general requirements that are essential to ensuring food safety as well as food suitability. WHO has adapted these principles to different settings (e.g., food service establishments, street food vendors, and households) and recommended specific hygienic measures for the preparation of food under such conditions.[9,12,13] Although these food hygiene principles are fundamental to achieving food safety, their application for ensuring safety nevertheless has certain limitations.[10]

- They do not provide a mechanism for prioritizing control measures, even though some measures may prove to be more critical than others.
- The guidance is nonspecific to foods, operations, special socioeconomic conditions where the food is prepared, or the cultural factors leading to high-risk practices.

Therefore, it is recommended that HACCP studies be conducted in these settings to identify the measures that are critical for ensuring the safety of foods in the given sociocultural and economic conditions of preparation. The identified measures should be considered CCPs and, depending on the circumstances, they should be the subject either of enforcement and/or of training and education.

In general, small food businesses or street food vendors do not have the expertise necessary for conducting such HACCP studies. The same is also true for domestic food handlers. It is the role of trade associations, food inspectors or, in the case of domestic food handlers, health authorities to conduct such studies and train or educate the food businesses and households in implementing and monitoring control measures at the CCP in an adequate manner. Such an approach provides a mechanism for prioritizing key behavior or practices that should be enforced or promoted in a given professional group or population.

In most cases, a complete HACCP study cannot be conducted for every type of food and preparation method, and priorities must, therefore, be set. Whenever possible, epidemiological data should be used in establishing priorities. High priority should be given to foods that are commonly implicated as vehicles of food-borne diseases and to the type of food operations where outbreaks of food-borne diseases have been reported. Data on food-borne diseases are, however, not always available. In the absence of such data, priorities can be set based on the factors described in the following paragraphs.

Intrinsic Properties of the Food Involved

Some foods may contain toxic chemicals or microbial pathogens or their toxins because of the practices involved in the production of the raw materials. For example, raw meats may be contaminated with microbial pathogens at the slaughtering stage, and raw vegetables may be contaminated with microbial pathogens or toxic chemicals from fertilizers, pesticides, and so forth. Food properties also relate primarily to characteristics of the food that may support the survival and/or growth of microorganisms based on knowledge of microbial ecology and epidemiological history. The characteristics that are most useful are pH, water activity (A_w), and redox potential (Eh). These factors influence the growth of infectious or toxigenic microorganisms. Foods that are possibly hazardous because they readily support rapid and progressive growth of microorganisms should be given high priority. Next, priority should be given to foods that can support the growth of pathogens during prolonged storage periods. Foods that are "relatively stable," in particular with regard to pathogenic organisms, can be assigned lower priority.

Preparation and Handling

Food operations that commonly contribute to the cause of food-borne illnesses are those that (1) prepare hazardous foods in advance of serving, (2) store foods in a manner that might allow microbial growth, and (3) inadequately reheat food to inactivate pathogens or toxins. On the other hand, food that is thoroughly cooked just before consumption is safe from biological hazards although chemicals and certain toxins may not be affected. Food that has been processed, even in a simple form such as fermentation, may be safe when held at ambient temperatures. Similarly, commercially processed foods, especially those that are well packaged, may pose little hazard to the consumer when they are sold by street vendors.

Volume of Food Prepared

The concern about volume of food prepared relates primarily to the amount of food prepared in advance of sale and/or consumption. For instance, in street-vending operations, this amount can be indirectly measured by the average daily sales, the amount of prepared foods on display, and the duration of holding cooked foods on display.

Susceptibility of Consumers

Infants and children, pregnant women, the hospitalized, immunocompromised persons, and the elderly are more susceptible to food-borne diseases than the general population. Foods which are intended for these groups of consumers should receive a higher priority.

Many of the fermented foods, particularly those that are prepared as complementary foods for infants and young children in Africa, fall into one or more of the above categories. In view of this, WHO joined with the Food and Agricultural Organization (FAO) in December 1995 to conduct a workshop on the assessment of fermentation as a household technology to improve food safety, during which the HACCP system was applied to several African fermented foods.[8] CCPs for the preparation of each of these foods were determined and it was recommended that control measures at these points should be included in the health education of food handlers. As an example, the application of the HACCP system to the preparation of *gari* is presented below. HACCP studies have shown that in general, the following measures during the preparation of African fermented foods should be the focus of training and health education.

- rapid and adequate acidification
- sufficient cooking for rendering food safe
- sufficient size reduction of cassava to enable enzymatic degradation of toxic cyanogenic glycosides
- avoidance of moldy raw material in view of possible mycotoxin contamination
- use of safe water

Through these studies, it could also be clearly demonstrated that some preparation practices presented a high food safety risk. For instance, the preparation of *togwa* (see Chapter 12) involves the addition of "power flour" after the food has been cooked. Thus, there is a risk of postcooking contamination, particularly with acid-resistant pathogens. In addition, if the lactic acid fermentation fails, the addition of power flour may lead to the proliferation of bacterial pathogens. To minimize food safety risks, it was recommended to accelerate the fermentation by back-slopping or use of starter cultures, or using power flour of high quality and avoiding postcooking contamination. In addition, the HACCP study showed that the original procedure for power flour preparation needs to be modified. In order to avoid postcooking contamination, it was recommended to boil the water that was used for the purpose of soaking and germination.

APPLICATION OF THE HACCP SYSTEM TO GARI*

1. **Product description**—Gari is a granular starchy food that is made from cassava roots. The processing starts with peeling, washing, and grating the tubers. The grated pulp is then put into bags (often jute or woven polypropylene bags) and left to ferment for several days under weight (pressure), during which time water is also removed. Fermentation is followed by fragmentation, drying, and roasting. During the roasting stage, the core temperature reaches 80–85 °C and the starch is gelatinized. Palm oil is sometimes added during roasting. After the roasting process, the gari, as it is now called, is cooled and stored. The final moisture content will determine its shelf life. When a final moisture content of less than 10% is reached, *gari* may be stored for several months. At higher moisture contents, the shelf life of gari is reduced to a few weeks because of potential mold growth.
2. **Intended use**—Gari is an important part of the staple diet in Nigeria and many other African countries. It is also given to children over one year of age. Gari can be prepared in many different ways. In the following example, soaking in cold water is used.
3. **Flow diagram**—The flow diagram of gari is presented in Figure 3–2.
4. **Hazards of concern**—Hazards considered in this context include biological (e.g., bacteria, viruses, parasites), chemical (e.g., contaminants, mycotoxins), and physical agents.
5. **Identification of hazards, control measures, and CCPs**—Table 3–1 shows the hazards that are associated with each step in the preparation of gari and some possible measures for their control.
   a. *Raw material:* Major hazards of cassava are cyanogenic glucosides (i.e., linamarin and lotaustralin) and contamination by agrochemicals. Cyanogenic glucosides will be hydrolyzed and removed during later stages in the processing and preparation of gari (i.e., grating, fermenting, and roasting). However, in regard to chemical contaminants and agrochemicals, households should obtain assurance from the suppliers regarding the safety of raw products.

      Depending on its source, water may be contaminated. Although the microbiological safety of water is of lesser importance when washing cassava because it will be fermented and heat treated, it is critical that safe water is used in the final preparatory stages before consumption.
   b. *Peeling:* Some foreign matter and pathogens may be introduced at this step. However, because the cassava will be washed and heat treated, the control of hazards, other than peeling in hygienic conditions, is not critical at this stage.
   c. *Washing:* Microbiological hazards may be introduced if the water is not clean. Therefore, as part of a good hy-

*Application of the HACCP system has been simplified and adapted to household conditions. Although the same approach can be used for production on a cottage and industrial scale, the requirements in terms of CCPs, critical limits, and monitoring procedures may be different and more severe. Model HACCP plans are not appropriate for use until they are validated for a specific food and food process.

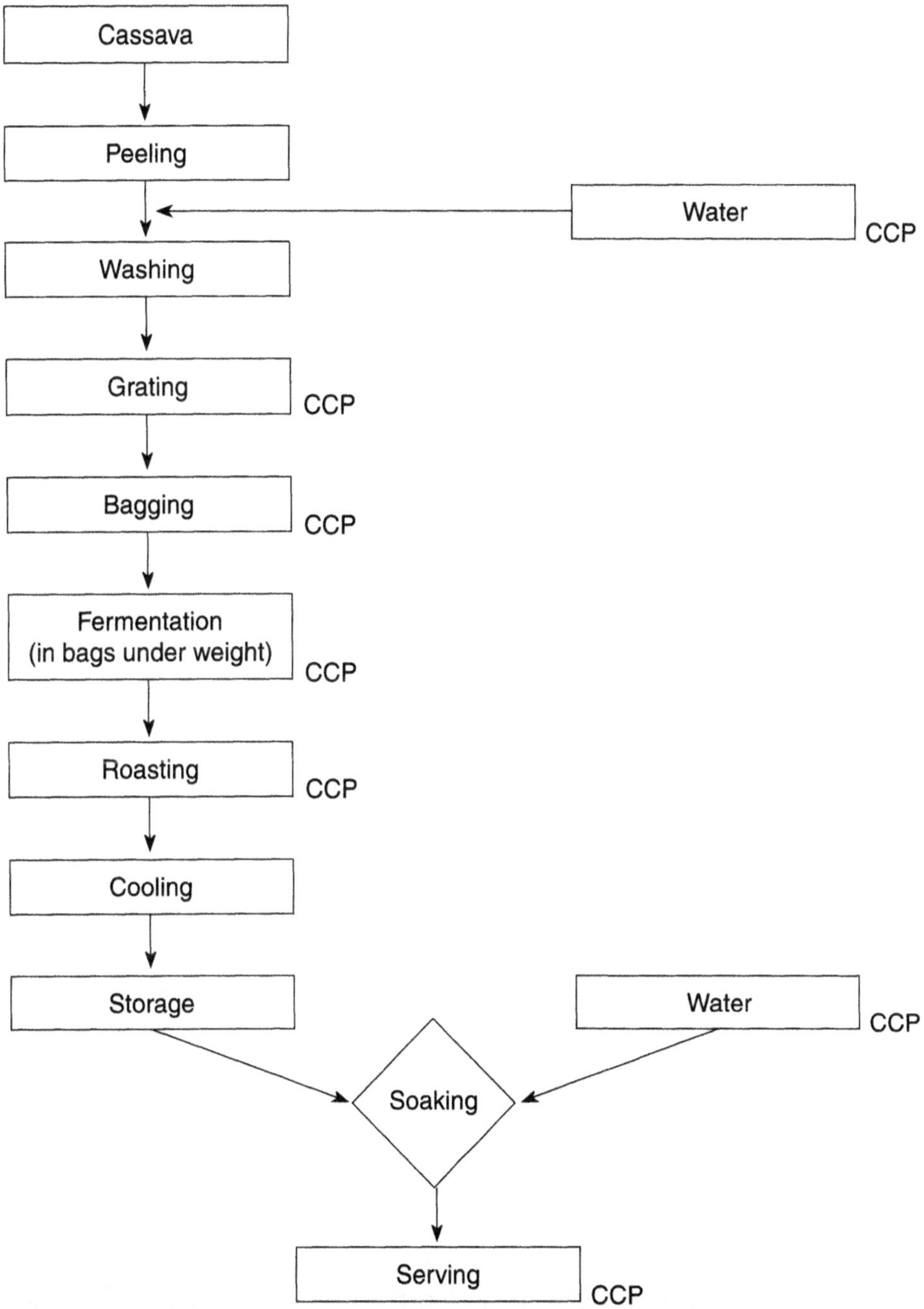

Figure 3–2 Application of the HACCP system to the preparation of *gari* in households[8]

gienic practice, safe water should be used. Hazards introduced at this step can nevertheless be controlled during subsequent steps of gari production. Washing can, though, decrease the amount of foreign matter.

d. *Grating:* This is the most important step with regard to detoxification when the cellular disruption results in the release of linamarase enzymes and greater contact of the enzymes with its substrate linamarin. Therefore, grating

Table 3–1 Application of the HACCP System to the Preparation of *Gari* in Households

Step	*Hazards*	*Control Measures*	*Critical Control Points*	*Critical Limit*	*Monitoring Procedure*	*Corrective Actions*
1. Raw material i) Cassava	a. Agrochemicals	a. Obtain assurance from supplier of adequate preharvest and postharvest handling of roots	a. No			
	b. Cyanogenic glucosides	b. Grating, fermentation, and roasting	b. No			
	c. Pathogens	c. Heat treatment	c. No			
1. Raw material ii) Water	a. Chemical contaminants, depending on the source	a. Obtain assurance of the source of water; use only safe water	a. Yes	a. Clear, free of odor and off-taste	a. Observation, smelling, and tasting	a. Use another source of water
	b. Pathogens (e.g., pathogenic *E.coli*, *Campylobacter*, *V. cholerae*, *Salmonella*, *Cryptosporidium*, *Giardia lamblia*, *Entamoeba histolytica*), Rotavirus	b. i) Use safe water (i.e., filtered and disinfected) or ii) Boil the water	b. Yes for step 10	b. i) Indication of contamination ii) Bubbles	b. i) Inquiry from health authorities ii) Observation	b. i) Boil the water ii) Reboil
2. Peeling	a. Foreign matter	a. Washing	a. No			
	b. Pathogens	b. Heat treatment	b. No			
3. Washing	a. Introduction of pathogens through water	a. Use safe water	a. No			
	b. Residual foreign matter	b. Thorough washing	b. Yes	b. As clean as possible	b. Observation	b. Reclean

continues

Table 3–1 continued

Step	*Hazards*	*Control Measures*	*Critical Control Points*	*Critical Limit*	*Monitoring Procedure*	*Corrective Actions*
4. Grating	a. Residual cyanide	a. Complete grating	a. Yes	a. Absence of coarse particles	a. Observation	a. Regrate
	b. Pathogens	b. Clean the equipment	b. No			
	c. Foreign matter	c. Use well-maintained equipment	c. Yes	c. Free of visible foreign matter	c. Observation	c. Remove foreign material
5. Bagging	Chemical contamination	Use clean bags; obtain assurance from supplier about no previous and hazardous use of the bags	Yes	Absence of chemical contaminants	Monitor the source and other uses of bags	Use other bags
6. Fermentation under weight	Growth of pathogens and production of toxin (e.g., *Staphylococcus aureus*)	Rapid fermentation	Yes	Acid taste and characteristic odors within 24 hours	Observation, smelling, and tasting	Discard
7. Roasting	a. Residual cyanide	a. i) Thorough roasting ii) Breaking up lumps	a. Yes	a. i) Sufficient time for roasting ii) Small size particles	a. i)Time keeping ii) Observation	a. i) Continue roasting ii) Breaking up lumps
	b. Mold growth during storage if high moisture content	b. Same as above	b. Yes	b. Same as above	b. Same as above	b. Same as above

continues

Table 3–1 continued

Step	*Hazards*	*Control Measures*	*Critical Control Points*	*Critical Limit*	*Monitoring Procedure*	*Corrective Actions*
8. Cooling	Contamination through environment	Cool under hygienic conditions (e.g., put in clean container and clean environment)	No			
9. Storing	Mold growth during storage if high moisture content	Thorough roasting (see step 7: roasting); keep in dry conditions	No, see step 7			
10. Soaking and serving	a. Recontamination with water	a. Use safe water	a. Yes see step 1	a. See step 1 (water)	a. See step 1 (water)	a. See step 1(water)
	b. Recontamination by dirty hands, utensils, environment	b. Wash hands and use clean utensils	b. Yes	b. Washing with soap and thorough rinsing with clean water	b. Observation	b. Thorough heating
	c. Growth of pathogens and spores of *Bacillus cereus*, if consumption is delayed for more than four hours	c. Consumption without delay	c. Yes	c. Use within four hours	c. Time keeping	c. Thorough heating

must be thorough to ensure a fast degradation of linamarin.

Foreign matter and pathogens may also be introduced at this stage. Although pathogens can be killed at the roasting step, the prevention and elimination of any foreign matter at this step is essential.

e. *Bagging:* Unclean bags may further contaminate the raw material. Chemical contamination is of particular concern at this step. The bags should not have been used previously for purposes that could jeopardize the safety of gari (e.g., for storage of pesticides).
f. *Fermentation:* Rapid fermentation is important to prevent the growth of undesirable microorganisms and the production of toxins. Fermentation is therefore a CCP for the control of pathogens.

   Fermentation also provides the opportunity (contact time) necessary for the action of linamarase on its substrate. During later stages, however, fermentation may have an antagonistic effect on detoxification because the decrease in pH resulting from fermentation may lead to the stability of cyanohydrins. An optimization of the fermentation process with respect to hydrolysis of linamarin and control of microbial growth is therefore important for ensuring the chemical and microbiological safety of gari.
g. *Roasting:* Further detoxification of cassava occurs during roasting when the hydrogen cyanide is evaporated. Thorough drying at this step is also important for the stability of gari and the prevention of mold growth during storage. It is important to prevent lumps from forming because they may limit the drying and evaporation of hydrogen cyanide.
h. *Cooling:* Cooling should take place under hygienic conditions.
i. *Storing:* To prevent mold growth, gari should be kept under dry conditions and protected from animals and rodents.
j. *Serving:* Water used for soaking gari, as well as hands and utensils, may reintroduce pathogens. It is critical that the water used at this step is safe, and that utensils and hands are washed thoroughly.

REFERENCES

1. Bryan, F. L. (1992). *Hazard Analysis Critical Control Point Evaluations: A Guide to Identifying Hazards and Assessing Risks Associated with Food Preparation and Storage.* Geneva, Switzerland: World Health Organization.
2. Codex Alimentarius Commission. (1997). *Food Hygiene, Basic Texts.* Secretariat of the Joint FAO/WHO Food Standards Program. Rome: Food and Agricultural Organization.
3. Motarjemi, Y. & Käferstein, F. (1999). Food safety, HACCP and the increase in foodborne diseases: a paradox? *Food Control* 10, 325–333.
4. Motarjemi, Y. & Van Schothorst, M. (1999). *HACCP: Principles and Practice.* Geneva, Switzerland: World Health Organization.
5. Motarjemi, Y., Käferstein, F., Moy, G., Miyagawa, S. & Miyagishima, K. (1996). Importance of HACCP for public health and development: the role of the World Health Organization. *Food Control* 7, 77–85.
6. Motarjemi, Y. (1999). The starting point: what is food hygiene? *New Food* 2, 25–30.
7. World Health Organization. (1993). *Application of the Hazard Analysis Critical Control Point (HACCP) System for the Improvement of Food Safety: WHO-Supported Case Studies on Food Prepared in Homes, at Street Vending Operations, and in Cottage Industries.* Unpublished document WHO/FNU/FOS/93.1. Geneva, Switzerland: Author.
8. World Health Organization. (1996). *Assessment of Fermentation as a Household Technology for Improving Food Safety: Report of a WHO Consultation.* Unpublished document WHO/FNU/FOS/96.1. Geneva, Switzerland: Author.

9. World Health Organization. (1996). *Basic Principles for the Preparation of Safe Food for Infants and Young Children.* Unpublished document WHO/FNU/FOS/96.6. Geneva, Switzerland: Author.
10. World Health Organization. (1996). *Essential Safety Requirements for Street-Vended Foods,* rev. edn. Unpublished document WHO/FNU/FOS/96.7. Geneva, Switzerland: World Health Organization.
11. World Health Organization. (1997). *HACCP: Introducing the Hazard Analysis Critical Control Point System.* WHO document WHO/FNU/FOS/97.2. Geneva, Switzerland: World Health Organization.
12. World Health Organization. *The WHO Golden Rules for Safe Food Preparation.* Unpublished document. Geneva, Switzerland: World Health Organization.
13. World Health Organization/BGVV. (1994). *Hygiene in Food-Service and Mass Catering Establishments.* Unpublished document WHO/FNU/FOS/94.5. Geneva, Switzerland: World Health Organization.

Appendix 3–A

HACCP Principles and Guidelines for Its Application

Assemble HACCP team

The food operation should ensure that the appropriate product-specific knowledge and expertise is available for the development of an effective HACCP plan. Optimally, this may be accomplished by assembling a multidisciplinary team. Where such expertise is not available on-site, expert advice should be obtained from other sources. The scope of the HACCP plan should be identified. The scope should describe which segment of the food chain is involved and the general classes of hazards to be addressed (e.g., does it cover all classes of hazards or only selected classes).

Describe product

A full description of the product should be drawn up, including relevant safety information such as composition, physical/chemical structure (including A_w, pH, etc.), microbicidal/static treatments (e.g., heat treatment, freezing, brining, smoking, etc.), packaging, durability, storage conditions, and method of distribution.

Identify intended use

The intended use should be based on the expected uses of the product by the end user or consumer. In specific cases, vulnerable groups of the population such as institutional feeding may have to be considered.

Construct flow diagram

The flow diagram should be constructed by the HACCP team. The flow diagram should cover all steps in the operation. When applying HACCP to a given operation, consideration should be given to steps preceding and following the specified operation.

On-site confirmation of flow diagram

The HACCP team should confirm the processing operation against the flow diagram during all stages and hours of operation and amend the flow diagram where appropriate.

HACCP principles	Guidelines for their application
Principle 1 ***Conduct a hazard analysis***	The HACCP team should list all of the hazards that may be reasonably expected to occur at each step from primary production, processing, manufacture, and distribution until the point of consumption. The HACCP team should next conduct a hazard analysis to identify for the HACCP plan which hazards are of such a nature that their elimination or reduction to acceptable levels is essential to the production of a safe food. In conducting the hazard analysis, wherever possible, the following should be included: • the likely occurrence of hazards and severity of their adverse health effects • the qualitative and/or quantitative evaluation of the presence of hazards • survival or multiplication of microorganisms of concern • production or persistence in foods of toxins, chemicals, or physical agents • conditions leading to the above The team must then consider what control measures, if any, exist that can be applied for each hazard. More than one control measure may be required to control a specific hazard(s) and more than one hazard may be controlled by a specified control measure.
Principle 2 ***Determine the critical control points (CCPs).***	There may be more than one CCP at which control is applied to address the same hazard. The determination of a CCP in the HACCP system can be facilitated by the application of a decision tree (Figure 3–A–1) which indicates a logic reasoning approach. Application of a decision tree should be flexible, given whether the operation is for production, slaughter, processing, storage, distribution, or other. It should be used for guidance when determining CCPs. This example of a decision tree may not be applicable to all situations. Other approaches may be used. Training in the application of the decision tree is recommended. If a hazard has been identified at a step where control is necessary for safety, and no control measure exists at that step, or any other, then the product or process should be modified at that step, or at any earlier or later stage, to include a control measure.
Principle 3 ***Establish critical limit(s).***	Critical limits must be specified and validated if possible for each critical control point. In some cases, more than one critical limit will be elaborated at a particular step. Criteria often used include measurements of temperature, time, moisture level, pH, A_w, and available chlorine, and sensory parameters such as visual appearance and texture.

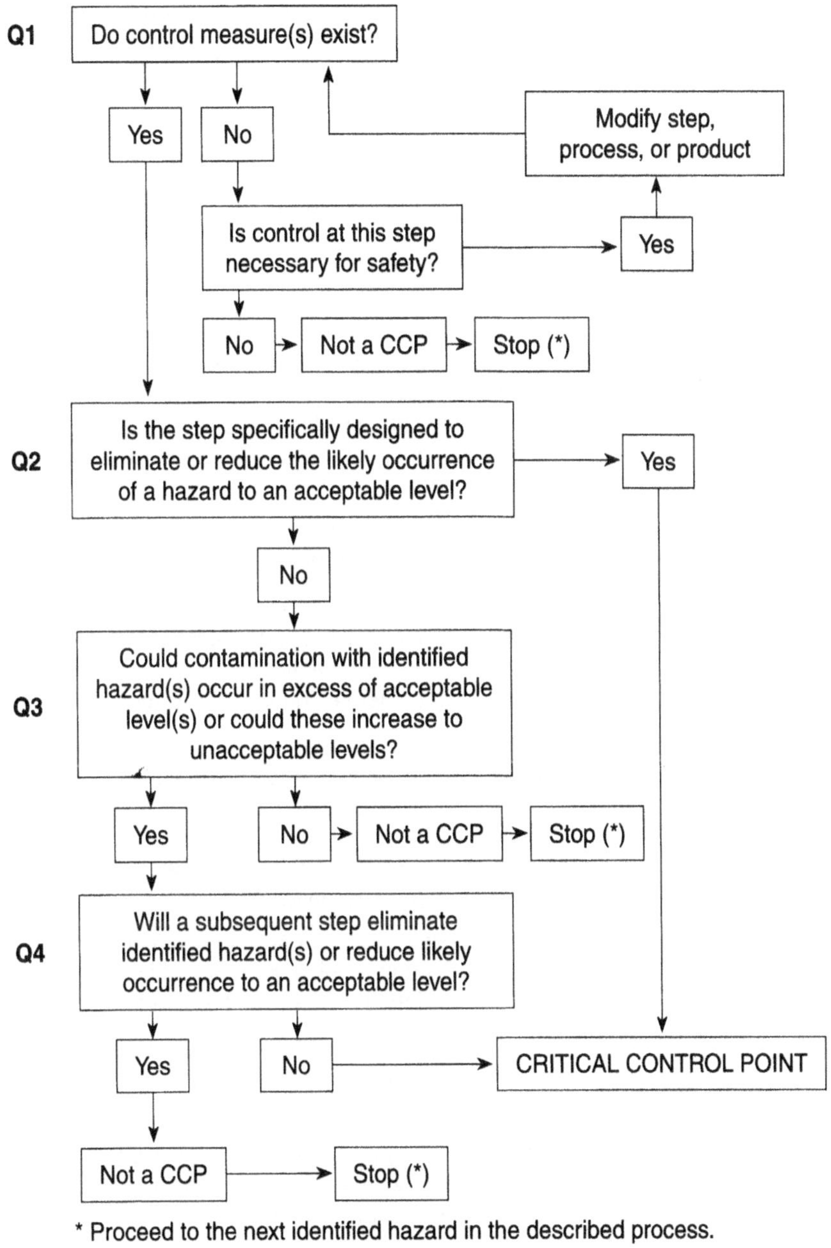

Figure 3–A–1 Example of a decision tree to identify CCPs.[2]

HACCP principles	Guidelines for their application
Principle 4 ***Establish a system to monitor control of the CCPs.***	Monitoring is the scheduled measurement or observation of a CCP relative to its critical limits. The monitoring procedures must be able to detect loss of control at the CCP. Further, monitoring should ideally provide this information in time to make adjustments to ensure control of the process to prevent violating the critical limits. Where possible, process adjustments should be made when

HACCP principles	Guidelines for their application
	monitoring results indicate a trend toward loss of control at a CCP. The adjustments should be taken before a deviation occurs. Data derived from monitoring must be evaluated by a designated person with knowledge and authority to carry out corrective actions when indicated. If monitoring is not continuous, then the amount or frequency of monitoring must be sufficient to guarantee that the CCP is in control. Most monitoring procedures for CCPs will need to be done rapidly because they relate to on-line processes and there will not be time for lengthy analytical testing. Physical and chemical measurements are often preferred to microbiological testing because they may be done rapidly and can often indicate the microbiological control of the product. All records and documents associated with monitoring CCPs must be signed by the person(s) doing the monitoring and by a responsible reviewing official(s) of the company.
Principle 5 ***Establish the corrective action to be taken when monitoring indicates that a particular CCP is not under control.***	Specific corrective actions must be developed for each CCP in the HACCP system in order to deal with deviations when they occur. The actions must ensure that the CCP has been brought under control. Actions taken must also include proper disposition of the affected product. Deviation and product disposition procedures must be documented in the HACCP recordkeeping.
Principle 6 ***Establish procedures for verification to confirm that the HACCP system is working effectively.***	Establish procedures for verification. Verification and auditing methods, procedures, and tests, including random sampling and analysis, can be used to determine if the HACCP system is working correctly. The frequency of verification should be sufficient to confirm that the HACCP system is working effectively. Examples of verification activities include • review of the HACCP system and its records • review of deviations and product dispositions • confirmation that CCPs are kept under control Where possible, validation activities should include actions to confirm the efficacy of all elements of the HACCP plan.
Principle 7 ***Establish documentation concerning all procedures and records appropriate to these principles and their application.***	Efficient and accurate recordkeeping is essential to the application of a HACCP system. HACCP procedures should be documented. Documentation and recordkeeping should be appropriate to the nature and size of the operation. Documentation examples are: • hazard analysis • CCP determination • critical limit determination Record examples are: • CCP monitoring activities • deviations and associated corrective actions • modifications to the HACCP system

CHAPTER 4

Chemical Hazards and Their Control: Endogenous Compounds

Leon Brimer

INTRODUCTION

Raw materials of vegetable origin may contain natural toxic or antinutritional compounds, endogenous constituents that are synthesized by the plant itself. *Antinutritional* means a deleterious effect due to the hindrance of uptake or use of other components in the diet. Examples of antinutritional compounds include tannins, which among others bind to proteins, making them unacceptable as substrates for proteases; proteinase inhibitors, which inhibit proteinases such as trypsin and chymotrypsin; phytate, which binds a number of minerals, making them unavailable for uptake; and thiaminase, which degrades vitamin B_1. An effect may be due to the parent compound or to metabolites that are formed in the gut or in the organism after absorption.

Although a few toxic and antinutritional compounds found in plants are proteins, most are low molecular weight compounds. Examples of proteins are ricin,[53] which is found in the seeds of *Ricinus communis* L. (castor bean); lectins,[213] found especially within the legumes; and the proteinase inhibitors, which are also common in legumes.[62] The smaller molecules with deleterious effects belong to the groups of compounds that are normally classified as secondary metabolites. The number of secondary constituents isolated from plants, fungi, and animals is high. Luckner worked with 26 biosynthetic groups divided into 107 subgroups, many of which contained more than 1,000 structures.[155] The majority of these compounds was found in plants. The alkaloids, for example, have had the attention of phytochemists for more than 150 years. In 1950, approximately 2,000 alkaloids were recognized; by 1970, the number had increased to approximately 4,000; 20 years later, approximately 10,000 were known.[217] It is necessary, then, to focus on the most important endogenous plant toxins as seen from a food and feed point of view.

The most prominent constituents known to restrict the nutritional value of food or fodder include certain nonprotein amino acids, alkaloids and glycosides, together with the tannins. However, knowledge concerning the influence of fermentation on these agents is very limited except for certain of the glycosides. Because a number of very important commodities of food and fodder worldwide do contain toxic glycosides,[46] the occurrence and effects of toxic glycosides, and their fate during food fermentations, will be presented in this chapter.

TOXIC AND ANTINUTRITIONAL GLYCOSIDES IN FOOD AND FEED

Glycosides consist of one or more genins (aglycones) to which one or more mono- or oligosaccharides are linked. The glycosidic linkage(s) may differ (i.e., one differentiates between O-, S-, and C-glycosides) (Figure 4–1). If the sugar part is a glucose moiety, it is called a *glucoside*. A number of different glycosides and oligosaccharides causing physiological effects (toxins) or reduced uptake or use of nutrients after ingestion are known in the plant kingdom. So

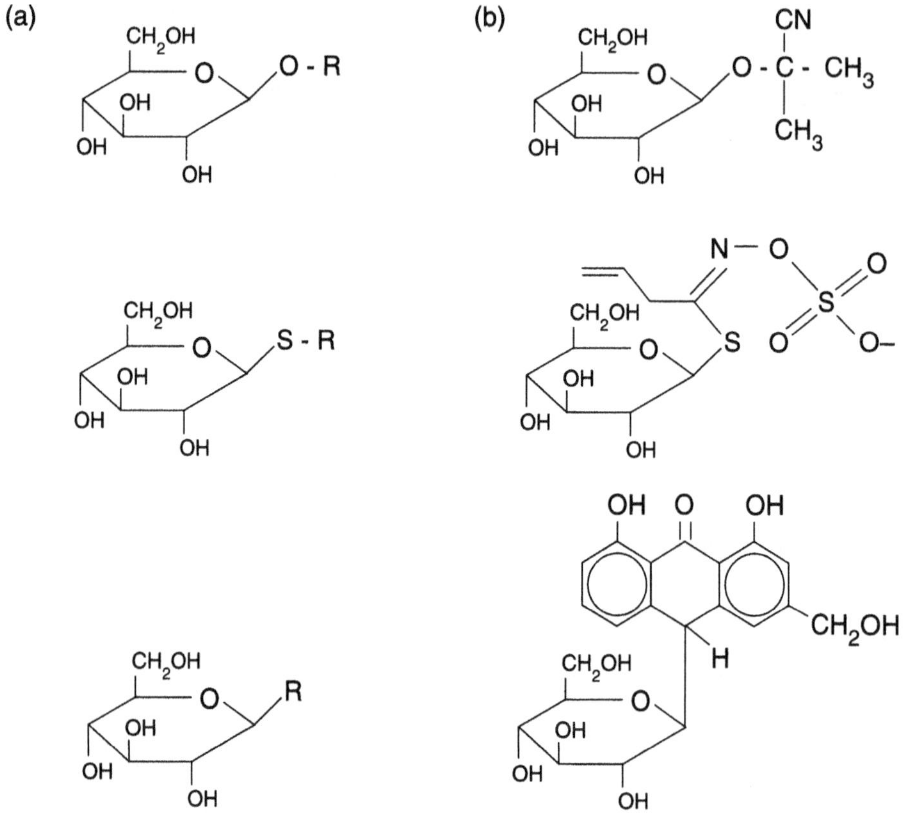

Figure 4–1 (a) The general structure of O-, S-, and C-glucosides as representatives of the broader groups of O-, S-, and C-glycosides, respectively. At top, an O-glucoside; middle, an S-glucoside; bottom, a C-glucoside. (b) Examples of naturally occurring O-, S-, and C-glucosides. Top (Linamarin—a cyanogenic glucoside found in cassava), middle (Sinigrin—a glucosinolate), and bottom (Barbaloin—an anthrone C-glucoside from *Aloe* spp.; laxative). *Note:* R = aglycone (= genin).

are a number of bitter-tasting glycosides that reduce the palatability of the plant (Table 4–1). A few of these glycosides have been shown to be protective to the plant;[129,205] however, most have only been recognized as toxic to domestic animals or humans. The broad range of compounds listed in Table 4–1 illustrates the diversity of chemical structures found even within the restricted field of toxic and antinutritional glycosides and oligosaccharides. Because this diversity also means a broad range of different mechanisms of action, the most important compounds are described in the following paragraphs.

Cyanogenic Glycosides

Cyanogenesis (cyano—Greek [kyanos = blue] and genesis—Greek [creation]) means the formation of cyanide/hydrogen cyanide, or HCN. Organisms that possess the ability to release cyanide may be termed cyanogenic or cyanophoric (phoros—Greek [bearing]). If the cyanide is formed from the breakdown of another compound, this compound is called a cyanogenic (cyanogenetic) compound, or simply, a *cyanogen*. Cyanogens include cyanogenic glycosides, cyanogenic lipids, cyanohydrins, and

Table 4–1 Toxic, Antinutritional, and Bitter-Tasting Glycosides and Oligosaccharides

Compound group or compound	*Examples*	*Toxicity, taste, etc.*
	O-Glycosides, Sugar Esters, and Oligosaccharides In sources of food and feed	
Cyanogenic glycosides	Linamarin in *Manihot esculenta,* Euphorbiaceae, in general widespread in the plant kingdom (Tracheophyta and Spermatophyta)[52,129,182]	Acute and chronic toxicity due to release of HCN; neurotoxicity of intact glycosides discussed; bitter taste
Glycoalkaloids	Chachonin and solanin in *Solanum tuberosum,* Solanaceae (Angiospermae)[140,206]	Corrosive to the gastrointestinal tract; upon absorption, acutely toxic due to several mechanisms; bitter taste
Glycosides of organic nitriles	Simmondsin in *Simmondsia californica* (Jojoba), Buxaceae (Angiospermae)[2,277]	Causes chronic toxicity of unknown mechanism
Glycosides and sugar esters of aliphatic nitrocompounds	Miserotoxin in *Astragalus* spp., Fabaceae (= Leguminosae; Angiospermae)[160,204]	Acutely toxic to ruminants; inhibit the TCA-cycle of the cells
Methylazoxymethanol (MAM) glycosides	Cycasin in *Cycas* spp., Cycadaceae (Gymnospermae)[22,154]	Carcinogenic
Naringin	In *Citrus* spp., espec. *C. paradisi* (grapefruit), Rutaceae (Angiospermae)[227]	Bitter taste
Oligosaccharides	In seeds of several legume spp., Fabaceae (= Leguminosae; Angiospermae)[94,215]	Flatulence-producing
Platyphylloside	In *Betula pendula*, Betulaceae (Angiospermae)[262]	Antinutritional (deterrrent) to several animal species
Polyphenols	2-hydroxyarctiin in *Carthamus tinctorius* (Safflower), Asteraceae (= Compositae; Angiospermae)[98,203]	Cathartic (laxative); bitter taste
Ptaquiloside	In *Pteridium aquilinum*, Polypodiaceae (Tracheophyta)[249,250]	Acutely toxic and carcinogenic
Saponins	Triterpene or steroid saponins in *Quinoa* spp., *Borassus flabellifer, Glycyrrhizae glabra* and *Balanites* spp. (Angiospermae)[74,95,125]	Some atoxic, other mildly to strongly toxic; several are bitter tasting
Vicine and convicine	In *Vicia faba* (faba bean), Fabaceae (= Leguminosae; Angiospermae)[276]	Acutely toxic to glucose-6-phosphate dehydrogenase-deficient individuals
	In medicinal and toxic plants	
Carboxyatractyloside (CAT) and related compounds	CAT in *Atractylis gummifera*, Asteraceae (= Compositae; Angiospermae)[49,187]	Acutely toxic; inhibit mitochondrial oxidative phosphorylation

continues

Table 4–1 continued

Compound group or compound	*Examples*	*Toxicity, taste, etc.*
Cardeno- and bufodienolides	"Digitalis" glycosides in *Digitalis spp.* (cardiac glycosides), Scrophulariaceae (Angiospermae)[161,225]	Acutely toxic to the heart
Cucurbitacins	Cucurbitacin L in *Citrullus colocynthis,* Cucurbitaceae (Angiospermae), some cucurbitacins also present in food plants (ref. text below)[103,126]	Intensely bitter substances, some of which are acutely toxic
Glycosides of Vitamin D_3	Glycosides of 1α,25-$(OH)_2D_3$ in *Solanum glaucophyllum,* Solanaceae (Angiospermae)[279]	Chronic toxicity (vitamin D intoxication—calcinoses)
Ranunculin	In *Ranunculus* and *Caratocephalus* spp., Ranunculaceae (Angiospermae)[180,195]	Acutely toxic; irritant to mucous membranes; Upon absorption, it affects several organs such as the heart, the lungs, etc.
	C-Glycosides (Some Also Occurring as O-Glycosides) In medicinal plants	
Anthraquinone, anthrone, and dianthrone glycosides	Sennosides in *Cassia angustifolia,* Fabaceae (= Leguminosae; Angiospermae)[152,280]	Laxative effect; some compounds are drastica
	S (Thio)–Glycosides In food and feed resources	
Glucosinolates	In many species within the families of Capparales (Angiospermae)[23]	Chronic toxicity due to release of thiocyanate and other compounds; sharp (burning) taste

cyanogenic epoxides.[31,182] Cyanogenesis has been detected in prokaryotes, fungi, plants, and animals. Cyanogens have been isolated from a great number of organisms; the glycosides, however, have been isolated only from plants and insects.[182]

The release of cyanide from a cyanogen implies the degradation of the compound, a reaction that may be either spontaneous or enzyme catalyzed.[52,182] Cyanogenic lipids and cyanogenic glycosides are broken down to cyanohydrins (hydroxynitriles); these are cyanogens in themselves (Figure 4–2). The cyanogenesis starts on crushing of the tissue, the cyanogens, and the degradative enzymes being compartmentalized either at the subcellular level or at tissue level in the intact plant.[211]

The glycosides are the most common cyanogens, and comprise more than 60 structures.[146,241] They were recognized early as substances that are poisonous to animals.[86] Cyanogens are of some systematic importance at the level of higher plant taxa,[182,265] and within certain families and genera.[182,194,242] The ingestion of cyanide and cyanogenic compounds may lead to acute[172] as well as chronic intoxications, the latter including the central nervous system (CNS) syndrome, konzo.[269,270]

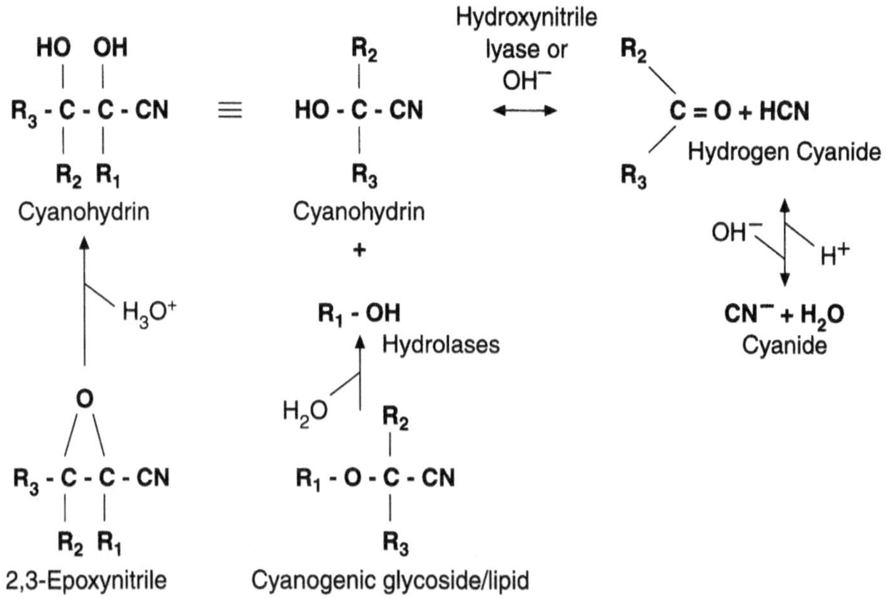

Figure 4–2 The interrelationship between cyanogenic compounds and cyanide/hydrogen cyanide. In a cyanogenic glycoside, R_1 is a saccharide moiety; in a cyanogenic lipid, an acyl moity. Hydrolases: glycosidase(s)—Refer also to Figure 4–3—or lipase. Note that the cyanohydrin formed upon hydrolysis of one of the three types of cyanogens (epoxynitriles, glycosides, or lipids) is a cyanogen itself.

Glycoalkaloids

Steroidal alkaloids and alkaloid glycosides occur throughout the genus *Solanum* (Solanaceae). The common potato (*S. tuberosum)* contains in its edible tuber the two compounds α-chaconin and α-solanin.[140] The total content may vary from 10 mg/kg to 390 mg/kg, with a mean of 73 mg/kg.[96] In other species of *Solanum* and closely related genera, different glycosides and free genins may dominate.[92,206] Gastrointestinal absorption of steroidal alkaloid glycosides varies between animal species. Some hydrolysis of the glycosidic bond and further metabolism seem to occur in different animal species, as judged from analyses comparing the serum level of α-chaconin and/or α-solanin to that of total alkaloids at different times after ingestion.[140] The toxicity of the potato glycosides to humans includes gastrointestinal upset with diarrhea, vomiting, and abdominal pain. In severe cases, neurological symptoms, some of which are clearly a result of the acetylcholinesterase inhibitor activity of these glycosides,[216] are seen.[140] Both α-chaconin and α-solanin, together with their aglycones, are teratogenic in one or more animal models.[96] However, Kuiper-Goodman & Nawrot[140] did not find the suggested association of the consumption of blighted potatoes during pregnancy with increasing incidences of spina bifida substantiated.

Methylazoxymethanol Glycosides

Glycosides of methylazoxymethanol (MAM) have been found only in cycads (*Macrozamia* and *Cycas* spp). The concentration is high in seeds; smaller quantities are found in stems and leaves.[154] Extensive losses of sheep have occurred in Australia due to consumption of *Macrozamia* and *Cycas* spp.[113] The first isolation of a MAM glycoside, macrozamin (the β-primeverosid of MAM), was from seeds of *M. spiralis* Miq., an Australian cycad. Today, other MAM glycosides are known, among these cycasin (the β-D-glucopyranosid of MAM), which was shown to be characteristic of, and exclusive to, all the genera of cycads.[154] The relative concentrations of cycasin to macrozamin in ripe seeds differ within the cycad genera.[177] The glycosides release MAM on the hydrolysis, which is catalyzed by β-glycosidases. MAM is a mutagenic and carcinogenic alkylating agent.[163,178]

Oligosaccharides

Flatulence is a common phenomenon that is associated with the ingestion of legumes, among others, and caused by the microbial fermentation of low molecular weight sugars. Many of these sugars are α-galactosides because humans do not have α-galactosidase in their digestive tract.[104] The legume oligosaccharides, raffinose, stachyose, and verbascose,[215] are of particular interest. Soybeans contain (by weight) approximately 1% of raffinose and 2.5% of stachyose;[215] the winged bean contains 1–2% of raffinose, 2–4% of stachyose, and 0.2–1% of verbascose.[94]

Ptaquiloside

Bracken fern(s) (*Pteridium* spp.) found throughout the world causes cancer of the urinary bladder in ruminants and is the only higher plant shown to cause cancer naturally in animals.[249] Enzootic hematuria, the clinical name given to the urinary bladder neoplasia of ruminants, tends to occur persistently in localized bracken-infested regions. The major carcinogen of bracken is the mutagenic and clastogenic norsesquiterpene glucoside, ptaquiloside, which in laboratory animals has been shown to be carcinogenic *per os*.[111,190,249] Bracken has further been associated with carcinoma of the upper digestive tract of cattle, where it is believed to transform the bovine papilloma (type 4) to a malign tumor. After hydrolysis of the glucoside, the genin is partly converted under alkaline conditions to a dienone, which can then undergo further reactions to form adducts with DNA bases. A preliminary investigation of the alkylation patterns produced has been presented.[250] Bracken fern is acutely toxic to several farm animals such as horses, cattle, and sheep, the syndromes being different for the different animal species.[82] The administration of pure ptaquiloside to a calf resulted in the same symptoms as known for the bracken intoxications of this species, thus demonstrating that the causative principle of cattle bracken poisoning is ptaquiloside.[112]

Saponins

A great number of food and feed plants contain saponins. Saponins may belong either to the group of pentacyclic triterpenoid saponins or to the steroidal saponins. The latter include in a broad sense the steroidal alkaloid glycosides that are found, for example, in potatoes. Although certain saponins such as the medicinally used quillaja saponin have been known for centuries to damage mucous membranes,[125] most saponins are considered quite unproblematic when they are administered orally. Saponin fractions from certain *Yucca* spp. have even been used as a feed additive to promote growth of, for example, turkeys.[71] However, concerns have been raised recently that saponins in food or feed may promote oral sensitisation to allergens through their membranolytic action in the gastrointestinal tract, resulting in enhanced uptake of the allergens.[128] This concern is based on the fact that saponins have been shown to act as oral adjuvants.[59,132,158] Food plants that may contain considerable amounts of saponins include the seeds of *Quinoa* spp., fruits of *Borassus flabellifer* (palmyrah) and *Balanites* spp., and roots and stolons of *Glycyrrhizae glabra* (licorice) (Table 4–1). The palmyrah fruits are fermented to wine (palm wine) in Sri Lanka, whereas experimental solid-state tempeh fermentations have been described for quinoa. However, no information is available concerning the fate of the saponins in any of these products.

Vicine and Convicine

Vicia faba (faba/fava bean), *V. harbonensis,* and *V. sativa* contain the two glycosides, vicine and convicine,[219] which after hydrolysis in the intestine and uptake of the genins (divicine and isouramil) cause hemolytic anemia (favism) in glucose-6-phosphate dehydrogenase-deficient individuals.[166,284] Together with condensed tannins, these two glycosides limit the use of the proteinaceous raw faba beans as feed for monogastric animals.[167,276] Vicine and convicine have not been detected in significant concentrations in other plant species.

Cucurbitacins

Cucurbitacins were first characterized as the bitter compounds of cucumbers, marrows, and squashes (Cucurbitaceae). The cucurbitacins as a group are thought to be among the most bitter substances known to man. Cucurbitacin B can be detected in dilutions as low as 1 ppb, and the glycosides of cucurbitacin E at 10 ppb.[175] Cucurbitacins make up a group of oxygenated tetracyclic triterpenes, some of which occur as glycosides.[103] Some cucurbitacins are not only bitter, but also toxic. Thus, the lethal dose for 10% of a test group of mice (LD_{10} orally mice) of cucurbitacin B is approximately 5 mg/kg b.w.[103] This is quite strong toxicity, as seen from the fact that it is equal to the lowest dose used in the International Organization for Economic Cooperation and Development Guidelines test for acute oral toxicity (Fixed Dose procedure, guideline no. 420).

Glucosinolates

In 1990, more than 100 glucosinolates were already known.[252] They occur in Capparales, Salvadorales, Violales, Euphorbiales, and Tropaeolales within Violiflorae sensu Dahlgren.[57,58] Reasons for interest in glucosinolates or glucosinolate-containing plants are the various antinutritional and toxic effects, the flavors, and the positive physiological effects associated with these constituents and their byproducts.[23] Rape (*Brassica napus, B. campestris,* and *B. juncea)* is among the most important crop containing glucosinolates. Seeds of these species contain approximately 400 g of oil and approximately 250 g of protein per kg. However, the use of rapeseed meals as a protein source in livestock rations and human diets is limited because of compounds associated with the protein fractions. These include phytic acid, phenolic compounds, and glucosinolates. Rapeseed that is bred to contain less than 2% erucic acid in its oil and less than 30 μg/g of aliphatic glucosinolates is termed "double low" or "canola." All pure glucosinolates tested in animal diets have caused antinutritional or toxic effects even when they were in concentrations relevant to levels based on double-low rapeseed as the protein source.[243]

RISKS ASSOCIATED WITH THE OCCURRENCE OF TOXIC GLYCOSIDES IN DIFFERENT COMMODITIES

Regarding toxins in food, the compounds that call for discussions in further detail are the cyanogenic glycosides, but also the MAM glycosides, ptaquiloside, the saponins, the favism agents (vicine and convicine), and the glucosinolates.

Cyanogenic Glycosides

Although discussions concerning a toxicity of intact cyanogenic glycosides may be found, the literature at present concludes that known intoxication syndromes, whether acute or chronic, are mainly due to HCN that is formed from the compounds.[235,236]

A plant containing cyanogenic glycosides may or may not contain enzymes that catalyze their breakdown (i.e., hydrolases [β-glycosidases] and cyanohydrin lyases). These are stored separately from the glycosides.[211] When a tissue containing both cyanogenic glycosides and these enzymes is crushed, enzyme(s) and substrate(s) are brought together and hydrolysis and further lysis (i.e., cyanogenesis) starts. Thus, the intake of raw or processed cyanogenic material normally will mean an intake of a mixture of the genuine glycoside(s) and accompanying hydrolysis products. Tissues that only contain the glycosides (and not the enzymes) will only give rise to exposure to the genuine glycoside(s).

Although cyanogenic glycosides may undergo acid hydrolysis,[30,83] the conditions in the stomach of a nonruminant, together with the very short residence time, will let the main fraction pass to the intestine. In the intestine, the glycosides will be absorbed, as shown for linamarin in a number of animal species and in humans,[19,37,110,208] and for prunasin and amygdalin in different animal species,[44,220,221,235] or it will be hydrolyzed by microorganisms.[28,44,210] In humans, Carlson *et al.*[43] very recently found that approximately 25% of linamarin ingested in a stiff porridge prepared from cassava flour was absorbed and excreted unchanged in the urine, whereas a little less than 50% was converted to cyanohydrin or cyanide and absorbed as such. The rest could not be accounted for. Most or all of the absorbed glycosides will be excreted in the urine, as shown for both linamarin and amygdalin in animals and humans.[7,37,110,173] HCN as well as cyanohydrins will give rise to cyanide exposure through absorption and nonenzymatic lyses of the cyanohydrins.

Whereas the absorbed glycosides will be excreted unchanged in the urine, the HCN will be totally or partly metabolized, the main metabolite being the goitrogenic compound, thiocyanate. The rate of this conversion will depend on the nutritional status of the individual. Current knowledge concerning the known biomarkers for cyanide exposure (acute and long term) and their use in clinical and experimental toxicology was reviewed by Rosling.[231] The detoxification processes (metabolization) and methods for the estimation of the sulphane sulphur pools available for this were reviewed by Westley.[281] Acute human intoxications have been described as a result of the intake of cassava products and almonds, whereas sorghum and cyanogenic acacia leaves and pods have caused veterinary intoxications.

Acute Intoxication

Acute intoxications in humans caused by the intake of insufficiently processed cassava meals have been reported from nearly all parts of the cassava consuming area, although it must be emphasized that the published reports are very scarce in relation to the extensive use of cassava as human food.[5,76,78,172] The symptoms of acute intoxication include vomiting, nausea, headache, dizziness, difficulty with vision, and collapse.[172]

Chronic Intoxication Syndromes

Evidence has accumulated that cyanide exposure from the diet is a causative factor in konzo,[115,269] and may aggravate iodine deficiency disorders.[60] The influence, if any, on the development of special types of diabetes remains a matter of discussion.[3,263] Symptoms and diagnosis of konzo have been described by Rosling & Tylleskär.[232]

Based on the knowledge available concerning the toxicity of cyanide and cyanogenic glycosides, the Joint World Health Organization (WHO)/Food and Agricultural Organization (FAO) Expert Commitee on Food Additives and Contaminants (JECFA) tried to estimate a safe level for the intake of cyanogenic glycosides by humans. The committee concluded that "because of lack of quantitative toxicological and epidemiological information, a safe level of intake of cyanogenic glycosides could not be estimated." However, the committee also concluded that "a level of up to 10 mg HCN/kg of product is not associated with acute toxicity."[253(p332)] Thus, no authority has yet felt confident to set scientifically based safe levels for the intake of one or more of the known cyanogenic glycosides (or their products of degradation), that is, levels that take the risk(s) for the development of chronic intoxications into consideration. In acknowledgement of this, the "International Workshop on Cassava Safety," held in Ibadan, Nigeria in 1994, concentrated on making recommendations concerning steps to be taken in research; in breeding programs; and in information to extension workers in the agricultural, food, and nutrition sectors.[10]

Long before humans knew the identity of cyanide, they did know that bitter cassava is a good starch crop, but that it must be detoxified before consumption.[67,69] Today, we know that this is because of its content of the cyanogenic glucosides, linamarin and lotaustralin. Overviews of the occurrence of cyanogenic glycosides in plants used for human or animal consumption are provided in Conn[51] and Jones.[129] Some of the important species of plants have been subjected to selection/breeding for a low total cyanogenic potential (TCP). Examples of the constituents and the TCP of economically important crops are provided in the following sections, together with some remarks concerning their importance as food or feed commodities.

Phaseolus lunatus (Seeds) and Other Beans. Seeds from several species of legumes are used for human consumption, many of which contain toxic and antinutritional substances. Thus, seeds from, for example, *P. lunatus, P. aureus, Cajanus cajan, Canavalia gladiata,* and *Vigna unguiculata* have been examined due to concerns about the possibility for cyanide intoxications.[63,192] All of these species are known to be cyanogenic in one or more tissues.[242] *P. lunatus* contains linamarin as its main cyanogenic constituent; the cyanogens have not been identified in the other species.[242] Only *P. lunatus* has been subjected to investigations concerning the variation in the cyanogenic potential.[21] However, several of the other species certainly may contain toxic amounts of cyanogens in the seeds.[192]

Prunus Species (Seeds). Peach, plum, cherry, apricot, and almond (family Amygdalaceae sensu Dahlgren) are all drupes (stone fruits) of great importance to man. Cyanogenic glycosides typical for Amygdalaceae are phenylalanine derived.[182] Thus, the ripe seeds of *P. persica* (peach), *P. domestica* (plum), *P. avium/cerasus* (cherry), *P. dulcis (P. amygdalus)* (almond), and *P. armeniaca* (apricot) all contain amygdalin as the major cyanogenic constituent. The total cyanogenic potential per gram dry weight of whole fruit rises during the early development, and the relative composition of cyanogens changes from 100% prunasin in the beginning to nearly 100% of amygdalin in the ripe seed.[90,170,189] Amygdalin and different *Prunus* seeds have, in spite of their ineffectivity, been commercially promoted for years as medicines to treat different cancers.[108]

- *Almond*—This tree is very widely cultivated around the Mediterranean. The naming of the species and its varieties/cultivars has changed through time.[99] The tree comes in two varieties, var. *dulcis* and var. *amara*, of which var. *amara* contains high concentrations of amygdalin in its ripe seeds (also denoted "bitter almonds").[39,50,99] The seeds are used in confectionary and bakery.[99] They contain approximately 50% of lipids, the oil being used in cosmetics and dermatology.[39,99] Bitter almonds (but also, e.g., apricot seeds) are also used to produce an essential (volatile) oil called "oil of almonds." This competes with synthetic benzaldehyde as a source of flavor.[39] Only few references exist concerning the content of cyanogenic glycosides in almonds.[50,90] Conn[50] found bitter almond seeds to release 290 mg of HCN/100 g of seed. According to Sturm,[260] commercial sweet almonds from California in general contain less bitter seeds (approximately 1%) than the 2–3% that is normally found in the Mediterranean ones.
- *Apricot*—Apricots have considerable economic importance for several countries such as Italy, the production of which was approximately 200,000 tons in 1988.[259] Different products are marketed from apricots, including fresh, dried, and canned fruits; nectar; jam; and distilled liqueur.[176] The number of varieties and hybrids of apricots are numerous.[174] Thus, Audergon *et al.*[14] tested more than 400 varieties as part of a physicochemical characterization program. Several marketed products of apricots require destoning,[55] leaving the stones as a byproduct from which oil can be extracted. The use of the seed/presscake is, however, restricted by the toxicity.[266] Depending on the variety and type of apricot, the apricot stone is relatively small, representing 6–8% of the fruit weight, even if it can sometimes reach 10%.[176] To the best knowledge of the present author, no investigations have been published concerning the variation in content of amygdalin in seeds of different cultivars. However, as part of investigations concerning the microbial degradation of cyanogens in such

seeds, Tuncel *et al.* [266,267] analyzed two batches of bitter and one of sweet Turkish apricot seeds, obtained on the commercial market. The bitter ones were found to contain approximately 52 and 92 µmol/g d.w., respectively; the sweet ones contained approximately 2.5 µmol/g.

- *Peach*—Much of the same that has been said for apricot can be said for peach. Seeds from *P. persica* Batsch (peach) also contain amygdalin as their major cyanogenic constituent.[193] Kupchella & Syty[141] analyzed the total cyanogenic potential of the seeds from an undefined cultivar and found it to correspond to a content of amygdalin of approximately 2.45% w/w.

Linum usitatissimum (Seeds = Linseed/Flaxseed). Flax is grown for two main purposes, fibers and seeds. Different cultivars are used for the two products. Whole seeds are used as a laxative due to the swelling seed coat polysaccharides.[200] Both full-fat flaxseed flour and defatted meal from the oil extraction are on the commercial market, the latter in two qualities, with 30% and 40% protein, respectively.[198] Flax is one of the major industrial oilseeds traded in world markets. Global production for crop year 1994–1995 was 2.44 million metric tons, with Canada contributing a major share. Flaxseed oil is used for a multitude of purposes, the oil being priced up to four times that of the whole seed.[198] The extraction cake (linseed meal) is traditionally used for fodder purposes. Recently, research into the refinement of flax products has accelarated. Thus, two patents have been issued for the use of flaxseed polysaccharide (gum) for cosmetic and medical preparations,[13,196] and an optimization of protein extraction from defatted flaxseed meal has been presented.[199]

Until 1980, linamarin was thought to be the main cyanogen in linseed. However, looking for the factor(s) in linseed meal responsible for its protective effect against selenium toxicity, Smith *et al.*[251] isolated two new cyanogenic glycosides (linustatin and neolinustatin). A later TLC-based investigation concerning the concentrations of different cyanogenic glycosides in a linseed sample gave the following µmol/g: linustatin+neolinustatin 4.6, linamarin 0.46, and lotaustralin 0.36,[32] pointing to linustatin and neolinustatin as the major cyanogenic constituents. This was further confirmed by a high performance liquid chromatography (HPLC) analysis of 48 samples, which on the other hand only found traces of linamarin and lotaustralin.[240] However, a recent investigation showed quite some variation between 10 cultivars. Two contained no linamarin, whereas in the cultivar Vimy, 7.8% of the weight of the total cyanogenic glycosides were linamarin.[201] This is close to the findings of Brimer *et al.*[32] for an unspecified sample. Frehner *et al.*[90] analyzed both the cyanogenic potential and the relative cyanogen composition during fruit development—one cultivar. As in *Prunus* seeds, the monoglucosides predominated at anthesis, shifting toward diglycosides during maturation. Rosling[230] found the cyanogenic potential of a nonspecified number of commercially sold linseed in Sweden to range from 4 mmol/kg to 12 mmol/kg (112–336 mg kg^{-1} HCN). The acute lethal dose is less than 2 mmol in 24 hours in sick and malnourished patients.[270]

Manihot esculenta (Roots and Leaves). The genus *Manihot* (Euphorbiaceae) incorporates more than 200 species, all originating in tropical America, from where several have been spread to other continents. Thus, *M. esculenta* Crantz (cassava) is today grown as a major source of starch in tropical Africa, India, Indochina, Indonesia, and Polynesia.[184] As early as 1605, Clusius reported that cassava could be toxic to man. The two cyanogenic glucosides, linamarin and lotaustralin, are responsible for this.[70,183] The cyanogenic potential (CNp) of several cassava germplasm collections has been investigated. Thus, Aalbersberg & Limalevu[1] analyzed 28 cultivars grown in Fiji, and found a variation from approximately 15 mg to 120 mg HCN equivalents/kg f.w. Dufour[67,69] looked at 14 cultivars of the Tukanoan Indians in northwest Amazonia and found very high levels (310–561 mg HCN eq./kg f.w.) in *Kii* (toxic cultivars) and 171 mg HCN eq./kg f.w. in the only *Makasera* (nontoxic/safe cultivar) grown. The Tukanoan Indians clearly expressed that they preferred toxic varieties as the main staple (70% of calorie intake) component of their diet. In this connection, it should be noted that the so-called "safe" (*Makasera*) cultivar had a higher CNp than the 100 ppm (f.w.) that was proposed as the upper limit by Koch[67,69] based on acute toxicity. Finally, Bokanga[24] examined 1,768 different cassava collections and found that the content of the central pith of the root varied from approximately 1 mg to more than 530 mg HCN equivalent/ kg d.w. The peel surrounding the pith has a much greater content, as have the leaves.[24] No acyanogenic cassava was found. While discussing the cyanogenic potential in this precise way, it should be born in mind that variations of up to 100% may be recorded between roots of the same plant.[24] It has also been shown that age, agricultural practices,[73] and environment may have a strong influence on its cyanogenic potential.[24,26]

The leaves of *M. esculenta* also serve as food and feed.[25] The cyanogenic potential of leaves from the same plant is less variable than that of the roots,[24] and is usually 5 to 20 times higher on a fresh weight basis.[25] The high content in the leaves normally does not present a problem for their use in food, given the methods generally used in their preparation.[25] In contrast, the roots of many cultivars, if not properly processed, have actually caused both acute and chronic intoxications worldwide. However, it should be emphasized again that the cassava root (even highly cyanogenic types) is a very valuable and irreplaceable crop. To ensure its safe use in every community under all conditions, the effectiveness of the different processing techniques (under rural as well as industrialized conditions) needs to be verified and the knowledge spread.[10]

Sorghum Species (Leaves and Seeds). Seedlings of *S. bicolor* (Poaceae) and other *Sorghum* species synthesize the cyanogenic glucoside dhurrin that is localized in the ariel shoots of the plant.[101] Thus, three-day-old etiolated seedlings of *S. vulgare* (i.e., the name for any cultivated grain sorghum) was found to contain up to 15 μmoles/g.[4] The content in mature leaves is much lower. The concentration depends on species, subspecies, and race/cultivar, and is also influenced by ecological factors.[65] Although most intoxications are seen in cattle browsing a newly sprouted field, forage may not be totally safe.[65,282] Grain sorghums constitute an important part of human nutrition in several semi-arid areas of the world.[61,88] Generally, the grains are considered completely safe for human consumption,[88,136] although the digestibility and biological value are not always high as a result of the occurrence of quite high concentrations of phytate and polyphenolics in several cultivated types.[88,120] Especially in Sudan, sorghum is irreplaceable, being the traditional stable food.[64] Although sorghum seeds in general are safe, germinated seeds are not. In certain African countries, germinated sorghum seeds are used traditionally for the production of malt,[88] which in turn is used for the brewing of alcoholic beverages[64] and for the production of the baked products called *Hulu-mur*.[64] According to FAO,[88] the traditional methods of preparation of these products remove the dhurrin effectively; however, it is stressed that the existence of these products must not be seen as an indication of sprouted sorghums being safe—they are not.[88]

MAM Glycosides

A metabolic fate and mechanism of toxicity, including the same alkylating end product as with dimethylnitrosamine, has been proposed for the MAM that is released from the MAM glycosides.[209] Thus, cycasin has been shown to be toxic to a number of animals, causing hepatic lesions and demyelination with axonal swelling in the spinal cord.[22,246] Cow's milk may be a vector of transmission of plant toxins. Thus, Mickelsen *et al.*[168] showed that MAM can pass into the milk of lactating rats, causing tumors in the offspring. The seeds of several *Cycas* spp. are traditionally eaten in Australia[22] and on certain islands.[150,254] A special neurological syndrome occurring on the island of Guam, and termed *Guam ALS-PDC,* has been hypothesized to be due to the intake of seeds of *C. circinalis*.[150,254] In 1987, Spencer *et al.*[254] proposed that the causative factor of this syndrome was the neuroexcitotoxic amino acid β-N-methylamino-L-alanine (BMAA). However, a number of subsequent investigations doubted this, as reviewed by Stone[258] in an article on the gradual disappearance of this disease. Thus, it may never be known whether the MAM glycosides could have a role in this disease, though it remains a possibililty given the spinal cord lesions reported in goats as a result of chronic intake of cycasin.[246]

Ptaquiloside

The carcinogenicity of ptaquiloside demonstrated in feeding experiments with rats, mice, hamsters, guinea pigs, and cattle, among others, is alarming because the young shoots of bracken fern are highly regarded as a tasty dish in Japan.[111] Hence, this intake of bracken has been linked to high incidences of stomach cancer in Japan,[111] and in Costa Rica among people who have been exposed to milk that was produced in bracken-infested grasslands.[6] The theory has been supported by the finding of a high tumor incidence in rats and mice that were fed milk from cows that had been fed with dietary complements of bracken, and by the subsequent demonstration of ptaquiloside in bovine milk.[6]

Saponins

Food and feed containing saponins include soybean, guar, quinoa, balanites fruits, and others. Besides the membranolytic action of many saponins, certain of these compounds exert special effects due to the structure of their aglycone.[286] Such effects include (1) lowering of blood cholesterol;[144] (2) reversible sodium retention and potassium loss leading to hypertension, water retention, and electrolyte imbalance (e.g., glycyrrhizinic acid found in licorice root, the roots and stolons from *Glycyrrhiza glabra,* and for products to which licorice root extract, or glycyrrhizinic acid, has been added);[100,133,239,256] and (3) crystal formation in the liver and biliary system, which may inhibit the excretion of phylloerythrin (from chlorophyll degradation), causing a subsequent photosensitation as seen in "Geeldikkop" (a *Tribulus terristris* intoxication).[131] A number of saponins are bitter. The occurrence of bitter saponins in palmyrah (*Borassus flabellifer* L) fruit pulp thus reduces the use of juices based on this fruit.[74] Likewise, seeds of *Chenopodium* spp. used for human consumption (*C. quinoa* [quinoa], *C. pallidicaule* [canihua], and *C. berlandieri* ssp. *nuttaliae* [Safford] Wilson and Heiser [huauzontle]) contain bitter saponins,[95,107,226] most of which are concentrated in the outer layers of the grain.[40,41,223]

Vicine and Convicine

Favism is characterized by anemia, jaundice, and hemoglobinuria, and may develop in subjects with glucose-6-phosphate dehydrogenase (G6PD) deficiency as a consequence of faba bean intake. Favism has also been reported in breast-fed infants whose mothers had eaten faba beans, and in newborn infants.[54] More than 300 variants of G6PD are known.[273] In addition, an association between the genotype of ACP_1 (human red cell acid phosphatase) and favism has been shown, and a possible biochemical mechanism has been proposed.[27] Most cases of G6PD deficiency described in the past were from Italy and other countries around the Mediterranean, that is, patients with the common Mediterranean B-form of G6PD, rather than the common African A (-) form.[273] However, recent investigations have shown that subjects with variants that result in a relatively mild G6PD deficiency may also develop favism.[93,181] Preventive measures and treatments have been described elsewhere.[102,162,188]

Glucosinolates

The most prominent toxic manifestation of glucosinolates in humans is the occurrence of goiter.[214] In animal experiments, this and other effects were generally more pronounced when myrosinases were included in the diet.[252] The effects seen were related to differences in the side chains and to chirality.[252] The fact that there are several mechanisms behind the toxic and antinutritional effects has also been very recently stressed by the results of the most detailed studies on the degradation products of various glucosinolates.[23] These authors presented an overview of the different degradation products formed from glucosinolates, which also include, for example, oligomers. From the degradation of glucobrassicin (an indole glucosinolate), indolyl-3-methanol is formed in considerable amounts, but it disappears very quickly, giving rise to, among others, appreciable amounts of thiocyanate ion. No organic isothiocyanates and thiocyanates are formed. In contrast, the degradation of various aliphatic glucosinolates results in the formation of nitriles as well as isothiocyanates and thiocyanates.[23] Toxic effects of glucosinolates in *B. oleracea* have been reviewed by Stoewsand[257] and those of crambe (*Crambe abyssinica*) meal fed to broiler chicks by Kloss *et al.*[137] The mechanism behind the observed decrease in cancer risk for people on diets with a high content of cruciferous vegetables has been investigated by Wallig *et al.*[274]

VARIATION IN TOXIN CONCENTRATION AMONG VARIETIES AND CULTIVARS: THE INFLUENCE OF TRADITIONAL DOMESTICATION AND MODERN BREEDING

Several toxic glycosides (including various saponins and cyanogenic glycosides, etc.) are known to be bitter tasting in addition to toxic. Hence, the term "bitter," as opposed to "sweet," has been used traditionally to designate naturally occurring or selected groups within a plant species that contain high amounts of the toxic (and bitter) substance. Depending on the view of the botanical author, the groups in question may be divided on the level of variety, form, or cultivar. Examples of plant species for which the division bitter/sweet has been used are *P. dulcis* and other *Prunus* spp. (containing amygdalin), as well as *M. esculenta* (cassava, containing linamarin), and in quinoa.[139] In most such cases, a correlation between the toxicity (content of glycoside) and the degree of bitterness of the plant part has been established. However, it is only seldom that a proper investigation concerning the degree to which this correlation holds has been performed. Thus, a positive correlation, but with exceptions, was found in a number of smaller studies on cassava roots.[247] Hence, King & Bradbury[135] took up the challenge of investigating in more detail the bitter-tasting substances in cassava parenchyma and cortex. Linamarin was found to be the sole contributer to bitterness present in the parenchyma; a new structure (isopropyl-β-D-apiofuranosyl-(1–6)-β-D-glucopyranoside) contributing in the cortex of some cultivars. This is in agreement with a very recent study from Malawi,[234] which compared the content of cyanogenic glucosides in the cortex of 492 cassava roots with their taste as estimated by a taste panel. The correlation had an $r^2 = 0.96$ when looking at the cultivar level.

It is well documented, at least for a number of cyanogenic plant species, that the concentration of both the glycosides and the enzymes degrading them can show a discrete variation (polymorphism) as well as a continous one. The polymorphism is genetically based, whereas the continous variations observed may be both genetically and environmentally influenced.[24,26,116–119,130,159,186] The genetic polymorphism (discrete variation, chemical races) with respect to the occurrence of both cyanogenic constituents and hydrolytic enzymes makes it difficult to define what is meant by a "cyanogenic species." Furthermore, it should be noted that the cyanohydrin lyase, which cleaves the cyanohydrins formed after the hydrolysis of the glycoside(s), may be expressed in certain organs and not in others. Thus, White *et al.*[283] recently showed that this enzyme, although present in the leaves, is not expressed in the roots of cassava. This observation explains why very high intermediate concentrations of cyanohydrins are formed during the processing of cassava roots. The environmental influences mentioned above may furthermore mean that certain plants will be found positive at some times of the year and negative at others.

Increased use of more highly cyanogenic cultivars of cassava among small farmers has been reported from several places. Thus, Dufour reported on a clear preference for Kii (toxic varieties) for most purposes by the Tukanoan Indians,[67,69] whereas Aalbersberg & Limalevu[1] stated that planting of the toxic (bitter) cultivars increased in New Guinea. Also, Onabolu *et al.*[197] found that the three most commonly grown cultivars in Ososa (a semi-urban farming community approximately 80 km east of Lagos, Nigeria), where cassava has been the main staple for decades, were all stated to be poisonous and to need processing. However, this was not regarded as a disadvantage. Farmers' reasons for preferentially growing cassava cultivars providing bitter roots were studied in Malawi.[48] In many traditional agricultural communities, the farmers (often the women) judge the "safeness" of the roots by chewing a small piece. According to Dufour,[67,69] the Tukanoan Indians appear to be able to distinguish accurately less from more poisonous cultivars by the taste. Very recent studies from Malawi prove such a procedure to exist, and to be very effective (Chiwona-Karltun, personal communication, October 2000).

Several of the species within the family Cucurbitaceae, which are used as human food, naturally contain cucurbitacins in amounts that are unacceptable to the market. However, intense domestication and breeding have resulted in cultivars low in bitter compounds.[126,127] Breeding programs for curcurbits are constantly aware of the bitterness.[84]

Great variations (0–13000 μg/g) may also be found in the content of ptaquiloside in bracken fern as a result of both ecological and genetic variation, a tendency for higher contents being reported when originating in relatively colder climates.[249] In addition, *P. esculentum* contains the cyanogenic glucoside, prunasin, the concentration of which similarly has been related to climatic conditions.[153]

Also, quinoa cultivars vary concerning the quantitative content of saponins, and the tradition has, as for other crops, been working with so-called sweet and bitter varieties.[139]

For *V. faba,* it should be mentioned that, although Duc *et al.*[66] gave the first report of a gene that codes for nearly a zero content of vicine and convicine, present-day cultivars contain approximately 7 and 2.5 mg g^{-1} respectively.[276]

REMOVAL OF TOXINS THROUGH PROCESSING

Traditional Household versus Modern Industrial Processing

When discussing the removal of toxic and antinutritional constituents, one must distinguish between traditional household processing and industrial processing. The two procedures may use different starting materials and will have different means of analyzing these and different methods available for processing. The priorities may indeed be very different when choosing between slow versus fast processing methods, and between processes that require low input of water and/or energy as compared to methods requiring a high input. The general trend of a greater number of traditional food fermentation processes being industrialized has been discussed recently by Rombouts & Nout.[228]

A number of industrial processing methods or laboratory methods meant for industrial development were investigated quite early on for soybean (e.g., removal of oligosaccharides and proteinase inhibitors)[47,56,91,109,224,245,248,261,264] and for cruciferous plants (glucosinolates).[20,72,85,151,255] Later, linseed (flax, cyanogenic glycosides),[149,156,157,275] jojoba meal (organic nitrile glycosides),[2] citrus juices (limonoids and naringin),[105,145,212,229,244] cotton seeds (gossypol),[179] and quinoa seeds (saponins)[95,226] were focused on. For commodities such as cassava roots and leaves, lima beans, cycas seeds, bracken leaves, and yam tubers (alkaloids as well as terpenes), most methods described and investigated scientifically are actually household processing methods. A look at the processing of cassava roots will help illustrate the important characteristics for each of the two sectors.

Household Processing (Cassava Roots)

Both in South America and in sub-Saharan Africa, sweet and bitter roots are generally regarded as two well-known different crops, and most traditional methods of processing that have been studied have proven very effective in removing the cyanogens.[17,68,197] Likewise, studies in Mozambique,[37,169] Tanzania,[172] and Zaire[18] have demonstated that problems with intoxications, whether accute or chronic, are seen only when shortcuts are made in the traditional methods of processing. Shortcuts may be introduced due to shortage of food or to a wish to produce products for the commercial market faster.

Industrial Processing (Cassava Roots)

Although industrial processing of cassava roots for starch has existed for decades in Brazil,[45] cassava processing industries are coming into existence only gradually in Africa. These include small-scale producers of flours, biscuits, and fermented products such as gari,[16,237] as well as bigger industries making the same products or products such as dry snacks. If industrial processing starts from bitter cultivars, precautions must be taken to ensure that methods are followed that give safe products. Often, traditional household methods will be scaled up because they are well known and give products that are well accepted by the local population. However, because these methods generally require prolonged processing,[15,202] it may be tempting to make shortcuts in order to increase production and/or reduce the requirements for storage/processing capacity, water, or energy. In such cases, it will be crucial to investigate the new process and the products.

Unit Operations that May Influence the Concentration or the Identity of Toxins

A reduction in the level of a toxin or an antinutritional constituent in a product may be achieved in different ways. Analyzing the reductions reported in a number of studies on one or more of the commodities such as soybean,[42,47,56,151,233] rapeseed,[20,72,85,255] yam tubers,[123,278] jojoba meal,[2] *Cycas* seeds,[22] citrus juices,[75,105,145] quinoa seeds,[95] faba beans,[12,106,167,287] cassava roots and leaves,[8,9,11,15,77,142,143,202] flaxseeds,[156,157,275] lima beans and other beans,[124,192,218] *Prunus* seeds,[185,238,266,267] and *Sorghum* green parts,[282] one can conclude that substances may (1) leach during soaking (steeping/retting), (2) be deliberately extracted using aqueous or nonaqueous solvents or two-phase systems, or (3) be degraded. The rate and extent of leaching/extraction will depend on the degree of comminution (particle size) or, for whole seeds or roots, on whether dehulling or peeling has taken place. Tissue softening/maceration and cell disruption by the action of growing microorganisms or added enzymes will also play a role. Degradation may be the result of a chemical reaction catalyzed by endogenous and/or added enzymes. It may also be the result of a chemical or heat treatment. Finally, the compound may be degraded by microorganisms during a fermentation. In all cases, the degradation will normally depend on parameters such as moisture content, temperature, and pH. In the case of enzyme-catalyzed reactions and reactions with added chemicals (e.g., oxidations), the substrate (the toxin) must be released so as to come into contact. The method of grating, how effective cell rupture is, and the resulting particle size will have an influence on the end result.[266] Contact may be facilitated further by the action of microorganisms or of added pectinolytic and cellulolytic enzymes. As mentioned previously, the level of endogenous enzyme(s) present may vary from cultivar to cultivar.

Again, one can learn from the production of a cassava product, namely, the fermented product, gari. Early investigations into the hydrolysis of linamarin during the production of gari reported that hydrolysis took days,[121,134,191] whereas Vasconcelos *et al.*[271] found that 95% of initial linamarin was hydrolyzed three hours after grating the roots. A fast degradation was also observed by Giraud *et al.*,[97(p81)] which led this group to conclude: "The observed differences may be explained by the use of non-traditional means for the preparation of gari, particularly during the grating of the cassava roots or as Mkpong *et al.* (1990) [see ref. 171] noticed, by the utilisation of cassava varieties showing different levels of endogenous linamarase." Both of these variables have already been touched on; the example thus further underlining the importance of chemical control whenever a change is made, either in starting materials, processing equipment, or the process. However, there is more in this quotation that calls for a discussion, specifically, the words "non-traditional means for the preparation." From this, one could get the impression that because the two most recent papers find a "fast" degradation, this is most probably what one will get using "traditional equipments/methods for the grating." But, what are "traditional equipments/methods"? Obviously, a person from Nigeria, the world's largest cassava producer[89] (where the crop is mainly produced for human consumption by subsistence farmers) and the home country of gari, would think of the equipment/methods used in his or her region as "traditional." Hence, it is interesting to note that a very recent investigation where local female Nigerian processors prepared gari using their own utensils could report a period of hydrolysis of approximately four days. A control study at the research station International Institute of Tropical Agriculture (IITA) in Ibadan, Nigeria, using the same method gave the same result, even though two different cultivars were used in the two examples.[197]

Degradation Products

The degradation of a toxic constituent may lead to one or more nontoxic reaction products; however, it may also lead to the formation of new toxins. For glycosides, the most common degradation processes will be those that start with the hydrolysis of the glycosidic bond(s) giving the aglycone and one or more saccharides. However, other routes of degradation have been described. Thus, the cyanogenic glucoside, linamarin, is metabolized to the glycoside of the corresponding amide by a bacterium, *Brevibacterium sp.* strain R312,[147,148] whereas Verbiscar *et al.*[272] reported that a number of strains of *Lactobacillus acidophilus* and *L. bulgaricus* seem to degrade the toxic (noncyanogenic) cyanoglycosides in jojoba meal by acting on the cyanogroup. Returning to the route starting with the hydrolysis of the glycosidic bond(s), one should still be aware that several different possibilities often exist, depending on the glycoside and the enzyme/microorganism in question. Again, the degradation of cyanogenic glycosides will be used as an example.

A cyanogenic β-bis-glycoside such as amygdalin or vicianin may in principle be hydrolyzed through a sequential (two-step) mechanism or a simultanous (one-step) mechanism (Figure 4–3). Both systems have been found in cyanogenic plants.[114] For degrading microoganisms, however, Brimer and coworkers[34,36] demonstrated that most seem to cleave such glycosides in two steps, although important differences in the overall process have been observed. Thus, a considerable concentration of the intermediate product prunasin, another (toxic) cyanogenic glycoside, is formed during the hydrolysis of amygdalin by a yeast such as *Endomycopsis fibuliger* and by filamentous fungi such as *Mucor circinelloides* and *Penicillium aurantiogriseum*. In contrast, only negligible concentrations were detected using a strain of *L. plantarum*.[149] In all four cases, the end product was the cyanohydrin mandelonitrile, a toxin in itself, and in general more toxic than the parent glycoside[270] because the latter may be absorbed and excreted unchanged.[37]

Depending on the product/process in which the cyanohydrin is formed, it may have different fates. The commodity itself may contain a cyanohydrin lyase (α–hydroxynitrilelyase) that catalyzes the cleavage to form HCN and an oxocompound (Figure 4–2). This is true for cassava leaves, almond, and linseed,[165] but apparently not for cassava root, where the enzyme seems not to be expressed.[283] Also, if the starting material has been heated (e.g., an oil extraction cake), this process may not happen because enzymes may have been inactivated. If the resulting pH is higher than around six, the cyanohydrin will disintegrate spontaneously to form the same two products. Exogenous enzymes or microoganisms can also metabolize the cyanohydrin;[122,138] however, described examples on the latter using food-acceptable processes are few. In an alkaline medium, any cyanide that is formed will remain present if it is not extracted.

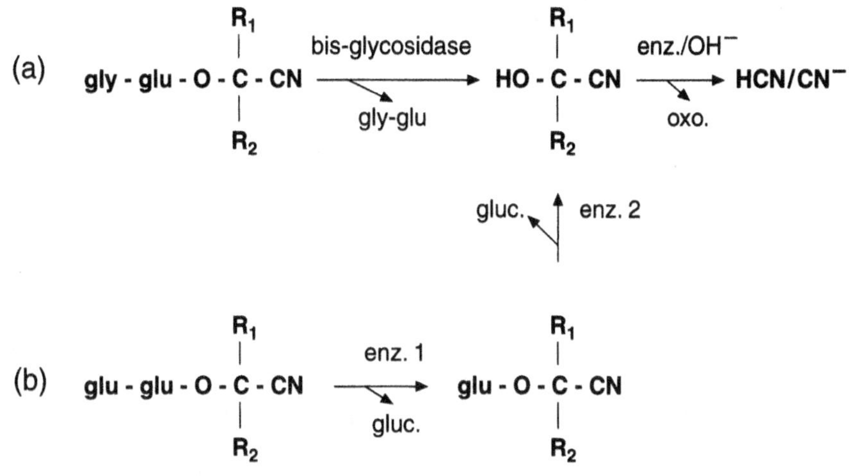

Figure 4–3 Different mechanisms for the enzymatic degradation (hydrolysis) of a cyanogenic glycoside (containing a disaccharide moiety) to a cyanohydrin, which in turn may release HCN either spontaneously or by enzyme catalysis. (a) Cleavage of a glycoside by a bis-glycosidase (simultaneous mechanism, e.g., Vicianin in a *Davallia* species). (b) Cleavage by the concerted action of two hydrolytic reactions (sequential mechanism, e.g., the hydrolysis of the bis-glucoside amygdalin by glucosidase A and B in almond). *Note:* Glu = Glucose Moiety, Gluc. = Glucose, Gly = a monosaccharide moiety (i.e., glucose or another).

In moist acid media, the cyanohydrin, if not further metabolized, will prove quite stable.[87,97] However, it may be driven out by drying or by heating/cooking.[197,266] Looking at the consequences of these examples, the following should be noted.

1. Degradation of a toxin may result in the formation of another. This means:
   a. When new processes/toxic raw materials are taken into a production, the whole degradation process should be clarified or the end product should be tested toxicologically.
   b. If a process that results in the formation of other toxins is used, the chemical analysis used for end-product control must quantify all possible toxins (starting, intermediate, and end products).
2. New toxins formed may be removed by the same type of process(es) that led to degradation of the parent compound(s) or may need other measures for their removal.

Microorganisms and Enzymes that Can Facilitate the Removal of Toxins through Degradation

Recently, Reddy & Pierson[222] published a short review on the reduction of antinutritional and toxic compounds in plant foods by fermentation. Chemical entities discussed included phytates, tannins, cyanogenic glycosides, oxalates, saponins, lectins, and inhibitors of enzymes such as α-amylase, trypsin, and chymotrypsin. Phytates may be hydrolyzed by endogenous as well as microbial phytases. The reduction of phytates during breadmaking and tempeh fermentations were discussed in particular. Most of the parameters pointed to earlier in this text are shown to be of importance, depending on the product and raw materials. Among the glycosides, only cyanogens are discussed and only briefly. Hence, more details concerning current knowledge about enzymes and microorganisms that can degrade toxic glycosides will be provided next. A review of all processes and all organisms shown to be able to degrade one or another toxin is beyond the scope of this article. Thus, reference will only be made to a limited number of review articles and original papers.

Microorganisms

A number of microorganisms isolated from fermenting food and feed products have been investigated for their ability to degrade toxins, especially glycosides. Likewise, a number of screenings have been performed on other species/strains from different collections of microorganisms. The glycosides used as substrates have mainly been the cyanogenic glycosides, amygdalin,[33,38] linamarin,[35,285] and linustatin;[149] the jojoba cyanoglucosides;[272] and vicine/convicine.[164] The resulting overall picture is that high β-glycosidase activities are most often found within filamentous fungi, followed by yeasts and, finally, bacteria. Among the bacteria, lactic acid bacteria such as *L. plantarum* tend to show relatively high activities. The level of activity is strongly strain dependent. Compared to plant enzymes,[114] most crude as well as purified microbial β-glycosidases seem to possess quite a broad substrate specificity,[149,207] a factor that may be important to remember and analyze in further detail, particularly when screening using an artificial substrate.[149] In addition to the ability to degrade the glycosides, it is also crucial that the organism grows well on the vegetable substrate.[185] Degradation of toxic glycosides by means other than hydrolysis of the glycosidic link has only been described in a few cases: The case of the bacteria *Brevibacterium sp.*, and strains of *L. acidophilus* and *L. bulgaricus,* have already been mentioned.

Whether the vegetable toxin itself can inhibit the growth of microorganisms has been addressed in only a few studies. The study by Brabban & Edwards[29] on the effect of glucosinolates on microbial growth showed an antimicrobial effect of both the parent compounds and the aglycones formed as a result of microbial hydrolysis of the glucosinolates. The paper is stongly recommended for its detailed and illustrative discussions.

Enzymes

Studies have been published regarding the use of added plant enzymes for the hydrolysis of toxic glycosides.[12,268,282] In the case of the addition of β-glycosidase from almond (as powdered almond) to faba bean dishes, quite effective treatments have been demonstrated at relatively low cost.[12] However, most other studies have shown a need for such high amounts of partly purified enzyme(s) that it seems unfeasible from a cost-benefit point of view. Crude microbial enzymes with known glycosidase activities from microorganisms have also been investigated. Such studies have demonstrated effective acceleration of detoxification processes in cassava[207] or hydrolysis of pure glycosides from faba bean.[164] However, most such crude enzymes have been derived from filamentous fungi, and the relative role of the glycosidase(s) compared to other activities present, such as cellulase and pectinases, is still to be clarified. A significant role for cellulase in particular, in both the softening of tissue and the acceleration of natural toxin degradation, has been demonstated for linamarin degradation in cassava root.[8,9,11,77,79–81]

CONCLUSION

A number of food plant species contain toxic or antinutritional substances. In cases where cultivars/varieties are grown that have a content that is too high for consumption, the level must be reduced through processing. A number of such methods of processing have been developed succesfully for a wide variety of crops throughout the world. The reduction of glycosides is especially well documented in literature, safe products being produced by the fermentation of roots from even highly toxic cultivars of cassava. However, incidents of both acute and chronic intoxications have been described, but in general only when the usual methods of fermentation were not adhered to.

Several vegetables studied show great variation among varieties/cultivars with respect to the content of both toxins and endogenous enzymes that can cause changes in the level or nature of the toxins during processing. Likewise, the activity of enzymes such as β-glycosidases, pectinases, and cellulases are strongly strain dependent when looking at the fermenting microorganisms. Although glycosidases may contribute directly to the degradation of glycosidic toxins, the tissue and cell wall degrading enzymes may facilitate the detoxification process, releasing both the toxins and endogenous plant enzymes.

In conclusion, the safety of the end product depends on a number of factors such as the (1) overall process scheme, (2) overall time of processing, (3) equipment used (e.g., type of grinders), (4) variety/cultivar of vegetable commodity used, and (5) identity (genetic stability) of the microorganisms used in fermentations. Any change in any of these factors, such as when production is industrialized, should trigger a check of the safety of the end product through chemical analysis or by other means.

REFERENCES

1. Aalbersberg, W. G. L. & Limalevu, L. (1991). Cyanide content in fresh and processed Fijian cassava (*Manihot esculenta*) cultivars. *Trop Sci* 31, 249–256.
2. Abbott, T. P., Nakamura, L. K., Buchholz, G., Wolf, W. J., Palmer, D. M., Gasdorf, H. J., Nelsen, T. C. & Kleiman, R. (1991). Processes for making animal feed and protein isolates from jojoba meal. *J Agric Food Chem* 39, 1488–1493.
3. Akanji, A. O. (1994). Cassava intake and risk of diabetes in humans. *Acta Hortic* 375, 349–359.
4. Akazawa, T., Miljanaich, P. & Conn, E. E. (1960). Studies on cyanogenic glycoside of *Sorghum vulgare*. *Plant Physiol* 35, 535–538.
5. Akintonwa, A., Tunwashe, O. & Onifade, A. (1994). Fatal and non-fatal acute poisoning attributed to cassava-based meal. *Acta Hortic* 375, 285–288.
6. Alonso-Amelot, M. E., Pérez-Mena, M., Calcagno, M. P. & Jaimes-Espinoza, R. (1992). Quantitation of Pterosins A and B, and Ptaquiloside, the main carcinogen of *Pteridium aquilinum* (L. Kuhn), by high pressure liquid chromatography. *Phytochem Anal* 3, 160–164.
7. Ames, M. M., Kovach, J. S. & Flora, K. P. (1978). Initial pharmacologic studies of amygdalin (laetrile) in man. *Res Commun Chem Pathol Pharmacol* 22, 175–185.
8. Amoa-Awua, W. K. A., Appoh, F. E. & Jakobsen, M. (1996). Lactic acid fermentation of cassava dough into agbelima. *Int J Food Microbiol* 31, 87–98.
9. Amoa-Awua, W. K. A., Frisvad, J. C., Sefa-Dedeh, S. & Jakobsen, M. (1997). The contribution of moulds and yeasts to the fermentation of "Agbelima" cassava dough. *J Appl Microbiol* 83, 288–296.
10. Summary and recommendations: proceedings of the International Workshop on Cassava Safety, Ibadan, Nigeria, March 1–4, 1994. (1994). *Acta Hortic* 375, 11–19.
11. Amoa-Awua, W. K. A. & Jakobsen, M. (1995). The role of *Bacillus* species in the fermentation of cassava. *J Appl Bacteriol* 79, 250–256.
12. Arbid, M. S. S. & Marquardt, R. R. (1985). Hydrolysis of the toxic constituents (vicine and convicine) in Faba bean (*Vicia faba L.*) food preparations following treatments with β-glucosidase. *J Sci Food Agric* 36, 839–846.
13. Attström, R., Glantz, P. O., Hakansson, H. & Larsson, K. (1993). U.S. Patent 5,260,282.
14. Audergon, J. M., Duffillol, J. M., Souty, M., Breuils, L. & Reich, M. (1991). Biochemical and physicochemical caracterisation of 400 apricot varieties: consequences in the apricot selection and improvement process. *Acta Hortic* 293, 111–119.
15. Balagopalan, C., Padmaja, G., Nanda, G. & Moorthy, S. N. (1988). *Cassava in Food, Feed and Industry*. Boca Raton, FL: CRC Press.
16. Banalya, J. (1998). Ghana hi-tech to process cassava. *Post Harvest Systems (IITA)* 3, 12.
17. Banea, M., J. -P., Nkiabungu, B., Tylleskär, T. & Rosling, H. (1998). High cassava consumption without cyanide exposure in Kinshase, in former Zaire. *Ecol Food Nutr* 37, 363–377.
18. Banea, M., Poulter, N. & Rosling, H. (1992). Shortcuts in cassava processing and risk of dietary cyanide exposure in Zaire. *Food Nutr Bull* 14, 137–143.
19. Barrett, M. D. P., Hill, D. C., Alexander, J. C. & Zitnak, A. (1977). Fate of orally dosed linamarin in rats. *Can J Physiol Pharmacol* 55, 134–136.
20. Bau, H. -M., Villaume, C., Lin, C. -F., Evrard, J., Quemener, B., Nicolas, J. -P. & Mejean, L. (1994). Effect of a solid-state fermentation using *Rhizopus oligosporus* sp.T-3 on elimination of antinutritional substances and modification of biochemical constituents of defatted rapeseed meal. *J Sci Food Agric* 65, 315–322.

21. Baudoin, J. P., Barthelemy, J. P. & Ndungo, V. (1991). Cyanide production of the lima bean, *Phaseolus lunatus* L. genetic variability in the primary and secondary gene pools and in some intraspecific hybride populations. *Bull Rech Agron Gembloux* 26, 367–388.
22. Beck, W. (1992). Aboriginal preparation of *Cycas* seeds in Australia. *Econ Bot* 46, 133–147.
23. Bjergegaard, C., Li, P. W., Moeller, P., Otte, J. & Soerensen, H. (1994). Glucosinolates and their transformation products—compounds with a broad biological activity. In *Bioactive Substances in Food of Plant Origin.* Vol. 1, *Low Molecular Weight N-Containing Compounds, Phenolic Compounds and Antioxidants, Enzyme Inhibitors,* pp. 1–15. Edited by H. Kozlowska, J. Fornal & Z. Zdunczyk. Olztyn, Poland: Centrum Agrotechnologii i Weterynarii - Polska Akademia Nauk.
24. Bokanga, M. (1994). Distribution of cyanogenic potential in cassava germplasm. *Acta Hortic* 375, 117–123.
25. Bokanga, M. (1994). Processing of cassava leaves for human consumption. *Acta Hortic* 375, 203–207.
26. Bokanga, M., Ekanayake, I. J., Dixon, A. G. O. & Porto, M. C. M. (1994). Genotype-environment interactions for cyanogenic potential in cassava. *Acta Hortic* 375, 131–139.
27. Bottini, E., Bottini, F. G., Borgiani, P. & Businco, L. (1997). Association between ACP_1 and Favism: a possible biochemical mechanism. *Blood* 89, 2613–2615.
28. Bourdoux, P., Mafuta, M., Hanson, A. & Ermans, A. M. (1982). Cassava toxicity: the role of linamarin. In *Nutritional Factors Involved in the Goitrogenic Action of Cassava,* pp. 15–27. Edited by F. Delange, F. B. Iteke & A. M. Ehrmans. Ottawa, Canada: IDRC.
29. Brabban, A. D. & Edwards, C. (1995). The effect of glucosinolates and their hydrolysis on microbial growth. *J Appl Bacteriol* 79, 171–177.
30. Bradbury, J. H., Egan, S. V. & Lynch, M. J. (1991). Analysis of cyanide in cassava using acid hydrolysis of cyanogenic glucosides. *J Sci Food Agric* 55, 277–290.
31. Brimer, L. (1998). Determination of cyanide and cyanogenic compounds in biological systems. In *Cyanide Compounds in Biology; CIBA Foundation Symposium No. 140,* pp. 177–200. Edited by David Evered and Sarah Harnett. Chichester, England: John Wiley & Sons.
32. Brimer, L., Broegger Christensen, S., Moelgaard, P. & Nartey, F. (1983). Determination of cyanogenic compounds by thin-layer chromatography. *J Agric Food Chem* 31, 789–793.
33. Brimer, L., Cicalini, A. R., Federici, F. & Petruccioli, M. (1994). Production of β-glycosidases (linamarase and amygdalase) and pectolytic enzymes by *Penicillium* spp. *World J Microbiol Biotechnol* 10, 203–206.
34. Brimer, L., Cicalini, A. R., Federici, F. & Petruccioli, M. (1998). Amygdalin degradation by *Mucor circinelloides* and *Penicillium aurantiogriseum*: mechanism of hydrolysis. *Arch Microbiol* 169, 106–112.
35. Brimer, L., Cicalini, A. R., Federici, F. & Petruccioli, M. (1998). Microbial linamarases for the detoxification of vegetable products. *Trop Agric* 75, 166–168.
36. Brimer, L., Nout, M. J. R. & Tuncel, G. (1998). β-glycosidase (amygdalase and linamarase) from *Endomyces fibuliger* (LU677): formation and crude enzyme properties. *Appl Microbiol Biotechnol* 49, 182–188.
37. Brimer, L. & Rosling, H. (1993). Microdiffusion method with solid state detection for determination of cyanogenic glycosides from cassava in human urine. *Food Chem Toxicol* 31, 599–603.
38. Brimer, L., Tuncel, G. & Nout, M. J. R. (1993). Simple screening procedure for microorganisms to degrade amygdalin. *Biotechnol Tech* 7, 683–687.
39. Bruneton, J. (1995). *Pharmacognosy, Phytochemistry, Medicinal Plants.* Andover, United Kingdom: Lavoisier/Intercept.
40. Burnouf-Radosevich, M. & Delfel, N. E. (1984). High-performance liquid chromatography of oleane-type triterpenes. *J Chromatogr* 292, 403–409.
41. Burnouf-Radosevich, M. & Paupardin, C. (1983). Elaboration de saponines triterpéniques par de tissus de *Chenopodium quinoa* Willd. cultivés in vitro. *C.R. Seances Acad Sci Ser III Sci Vie (Physiol Végét)* 296, 429–432.
42. Calloway, D. H., Hickey, C. A., & Murphy, E. L. (1971). Reduction of intestinal gas-forming properties of legumes by traditional and experimental food processing methods. *J Food Sci* 36, 251–255.
43. Carlson, L., Mlingi, N., Juma, A., Ronquist, G. & Rosling, H. (1999). Metabolic fates in humans of linamarin in cassava flour ingested as stiff porridge. *Food Chem Toxicol* 37, 307–312.
44. Carter, J. H., McLafferty, M. A. & Goldman, P. (1980). Role of the gastrointestinal microflora in amygdalin (laetrile)-induced cyanide toxicity. *Biochem Pharmacol* 29, 301–304.
45. Cereda, M. P. (1994). Processing of cassava roots in Brazil: safety implications. *Acta Hortic* 375, 225–226.
46. Cheeke, P. R. & Shull, L. R. (1985). Glycosides. In *Natural Toxicants in Feeds and Poisonous Plants,* pp. 173–234. Westport, CT: AVI Publishing Company.
47. Chen, S. (1989). Preparation of fluid soymilk. In *Proceedings of the World Congress on Vegetable Protein Utilization in Human Foods and Animal Feedstuffs,* pp. 341–352. Edited by T. H. Applewhite. Champaign, IL: American Oil Chemists Society.
48. Chiwona-Karltun, L., Mkumbira, J., Saka, J., Bovin, M., Mahungu, N. M. & Rosling, H. (1998). The importance of being bitter—a qualitative study on cassava cultivar preference in Malawi. *Ecol Food Nutr* 37, 219–245.
49. Cole, R. J., Cutler, H. G. & Stuart, B. P. (1989). Carboxyatractyloside. In *Toxicants of Plant Origin.* Vol. 2, *Glycosides,* pp. 253–263. Edited by P. R. Cheeke. Boca Raton, FL: CRC Press.

50. Conn, E. E. (1979). Cyanogenic glycosides. In *International Review of Biochemistry, Biochemistry of Nutrition IA*, Vol. 27, pp. 21–43. Edited by A. Neuberger & T. H. Jukes. Baltimore, MD: University Park Press.
51. Conn, E. E. (1981). Unwanted biological substances in foods: cyanogenic glycosides. In *Impact of Toxicology on Food Processing (IFT Basic Symposium Series)*. Westport, CT: AVI Publishing Company.
52. Conn, E. E. (1991). The metabolism of a natural product: lessons learned from cyanogenic glycosides. *Planta Med* 57 (Suppl. 1), S1–S9.
53. Cooper, M. R. & Johnson, A. W. (1998). *Poisonous Plants and Fungi in Britain (Animal and Human Poisoning)*, 2nd edn. London: The Stationary Office.
54. Corchia, C., Balata, A., Meloni, G. F. & Meloni, T. (1995). Favism in a female newborn infant whose mother ingested fava beans before delivery. *J Pediatr* 127, 807–808.
55. Crivelli, G., Torreggiani, D., Senesi, E., Forni, E., Bertolo, G. & Maestrelli, A. (1991). Researches on the osmotic dehydration of apricots. *Acta Hortic* 293, 657–666.
56. Cruz, R., Batistela, J. C. & Wosiacki, G. (1982). Microbial α-galactosidase for soymilk processing. *J Food Sci* 46, 1196–1200.
57. Dahlgren, R. M. T. (1980). A revised system of classification of the angiosperms. *Bot J Linn Soc* 80, 91–124.
58. Dahlgren, R., Rosendal-Jensen, S. & Nielsen, B. J. (1981). A revised classification of the angiosperms with comments on correlation between chemical and other characters. In *Phytochemistry and Angiosperm Phylogeny*, pp. 149–204. Edited by D. A. Young & S. Seigler. New York: Praeger Publishers.
59. Dalsgaard, K., Hilgers, L. & Trouve, G. (1990). Classical and new approaches to adjuvant use in domestic food animals. *Adv Vet Sci Comp Med* 35, 121–160.
60. Delange, F., Ekpechi, L. O. & Rosling, H. (1994). Cassava cyanogenesis and iodine deficiency disorders. *Acta Hortic* 375, 289–293.
61. Dendy, D. A. V. (1995). Sorghum and the millets: production and importance. In *Sorghum and Millets, Chemistry and Technology*, pp. 11–26. Edited by D. A. V. Dendy. St. Paul, MN: American Association of Cereal Chemists.
62. Deshpande, S. S. (1992). Food legumes in human nutrition: a personal perspective. *CRC Crit Rev Food Sci Nutr* 32, 333–363.
63. Dibofori, A. N., Okoh, P. N. & Onigbinde, A. O. (1994). Effect of germination on the cyanide and oligosaccharide content of lima beans (*Phaseolus lunatus*). *Food Chem* 51, 133–136.
64. Dirar, H. A. (1993). *The Indigenous Fermented Foods of the Sudan, a Study in African Food and Nutrition.* Wallingford, United Kingdom: CAB International.
65. Doggett, H. (1988). *Sorghum.* London: Longman.
66. Duc, G., Sixdenier, G., Lila, M. & Furstoss, V. (1989). Search of genetic variability for vicine and convicine content in *Vicia faba* L.: a first report of a gene which codes for nearly zero-vicine and zero-convicine contents. In *Recent Advances of Research in Antinutritional Factors in Legume Seeds*, pp. 305–313. Edited by J. Huisman, T. F. B. van der Poel & I. E. Liener. Wageningen, Netherlands: Pudoc.
67. Dufour, D. L. (1988). Cyanide content of cassava (*Manihot esculenta*, Euphorbiaceae) cultivars used by Tukanoan Indians in Northwest Amazonia. *Econ Bot* 42, 255–266.
68. Dufour, D. (1989). Effectiveness of cassava detoxification techniques used by indigenous peoples in north west Amazonia. *Interciencia* 14, 86–91.
69. Dufour, D. L. (1993). The bitter is sweet: a case study of bitter cassava (*Manihot esculenta)* in Amazonia. In *Tropical Forests, People and Food, Biocultural Interactions and Applications to Development*, pp. 575–588. Edited by C. M. Hladik, A. Hladik, O. F. Linares, H. Pagezy, A. Semple & M. Hadley. Paris/Carnforth: Unesco/The Parthenon Publishing Group.
70. Dunstan, W. R., Henry, T. A. & Auld, S. J. M. (1906). Cyanogenesis in plants V. the occurrence of phaseolutin in cassava (*Manihot aipi* and *Manihot utilissima*). *Proc R Soc Lond SER B Biol SCI* 78, 152–158.
71. Dziuk, H. E., Duke, G. E., Buck, R. J. & Janni, K. A. (1985). Digestive parameters in young turkeys fed yucca saponin. *Poult Sci* 64, 1143–1147.
72. Eggum, B. O., Jensen, S. K. & Soerensen, H. (1993). The nutritive value of dehulled protein rich rapeseed meal produced by aqueous enzymatic extraction. *Bull GCIRC* 9, 37–41.
73. Ekanayake, I. J. & Bokanga, M. (1995). *A Review on Production, Agronomy and Cyanogenesis. The Cassava Biotechnology Network: Proceedings of the Second International Scientific Meeting. Bogor, Indonesia, August 22–26, 1994.* Working Document no. 150. Cali, Columbia: Centro International de Agricultura Tropical.
74. Errol, R. J., Nikawela, J. K., Gooneratne, J. & Theivendirarajah, K. (1994). Studies on the bitter principle and debittering of Palmyrah fruit pulp. *J Sci Food Agric* 65, 185–189.
75. Esaki, S., Ohishi, A., Katsumata, A., Sugiyama, N. & Kamiya, S. (1993). Synthesis of δ-mannopyranosyl-containing disaccharides and phenols as substrates for the α-L-mannosidase activity of commercial naringinase. *Biosci Biotechnol Biochem* 57, 2009–2103.
76. Espinoza, O. B., Perez, M. & Ramirez, M. S. (1992). Bitter cassava poisoning in eight children: a case report. *Vet Hum Toxicol* 34, 65.

77. Essers, A. J. A. (1995). Removal of cyanogens from cassava roots; studies on domestic sun-drying and solid-substrate fermentation in rural Africa. PhD diss., Agricultural University Wageningen, Wageningen, the Netherlands.

78. Essers, A. J. A., Alsen, P. & Rosling, H. (1992). Insufficient processing of cassava induced acute intoxications and the paralytic disease konzo in a rural area of Mozambique. *Ecol Food Nutr* 27, 17–27.

79. Essers, A. J. A., Bennik, M. H. J. & Nout, M. J. R. (1995). Mechanism of increased linamarin degradation during solid-substrate fermentation of cassava. *World J Microbiol Biotechnol* 11, 266–270.

80. Essers, A. J. A., Ebong, C., Van der Grift, R. M., Nout, M. J. R., Otim-Nape, W. & Rosling, H. (1995). Reducing cassava toxicity by heap-fermentation in Uganda. *Int J Food Sci Nutr* 46, 125–136.

81. Essers, A. J. A., Jurgens, C. M. G. A. & Nout, M. J. R. (1995). Contribution of selected fungi to the reduction of cyanogen levels during solid-substrate fermentation of cassava. *Int J Food Microbiol* 26, 251–257.

82. Evans, W. C., Patel, M. C. & Koohy, Y. (1982). Acute bracken poisoning in homogastric and ruminant animals. *Proc R. Soc Edinb Sect B* 81, 29–64.

83. Eyjolfsson, R. (1970). Recent advances in the chemistry of cyanogenic glycosides. *Fortschr Chem Org Naturst* 27, 74–108.

84. Fanourakis, N. E. & Tzifaki, E. E. (1993). Correlated inheritance of fruit neck with fruit length and linkage relations with 10 other characteristics of cucumber. *Euphytica* 65, 71–77.

85. Finnegan, T. J. A. & Lewis, M. J. (1985). Nitrogen extraction from defatted rapeseed, with particular reference to United Kingdom commercial rapeseed meal. *J Sci Food Agric* 36, 520–530.

86. Finnemore, H. & Gledhill, W. C. (1928). The presence of cyanogenetic glucosides in certain species of Acacia. *Aust J Pharm* 9, 174–178.

87. Fomunyam, R. T., Adegbola, A. A. & Oke, O. L. (1985). The stability of cyanohydrins. *Food Chem* 17, 221–225.

88. Food and Agricultural Organization. (1995). *Sorghum and Millets in Human Nutrition (FAO Food and Nutrition Series, No. 27).* Rome: Author.

89. Food and Agricultural Organization. (1996). *Year Book of Food Production.* Rome: Food and Agricultural Organization.

90. Frehner, M., Scalet, M. & Conn, E. E. (1990). Pattern of the cyanide-potential in developing fruits. *Plant Physiol* 94, 28–34.

91. Fulmer, R. W. (1989). The preparation and properties of defatted soy flours and their products. In *Proceedings of the World Congress on Vegetable Protein Utilization in Human Foods and Animal Feedstuffs,* pp. 55–61. Edited by T. H. Applewhite. Champaign, IL: American Oil Chemists Society.

92. Gaffield, W. & Keeler, R. F. (1994). Plant steroidal alkaloid teratogens: structure-activity relations and implications. In *Plant-Associated Toxins, Agricultural, Phytochemical and Ecological Aspects,* pp. 333–338. Edited by S. M. Colegate & P. R. Dorling. Wallingford, United Kingdom: CAB International.

93. Galiano, S., Gaetani, G. F., Barabino, A., Cottafava, F., Zeitlin, H., Town, M. & Luzzatto, L. (1990). Favism in the African type of glucose-6-phosphate dehydrogenase deficiency (A-). *Bri Med J* 300, 236.

94. Garcia, V. V. & Palmer, J. K. (1980). Carbohydrates of winged beans, *Psophocarpus tetragonologus* (L.). *J Food Technol* 15, 477–484.

95. Gee, J. M., Price, K. R., Ridout, C. L., Wortley, G. M., Hurrell, R. F. & Johnson, I. T. (1993). Saponins of Quinoa (*Chenopodium quinoa*): effect of processing on their abundance in quinoa products and their biological effects on the intestinal mucosal tissue. *J Sci Food Agric* 63, 201–209.

96. Gelder, van W. M. J. (1991). Chemistry, toxicology, and occurrence of steroidal glycoalkaloids: potential contaminants of the potato (*Solanum tuberosum* L.). In *Poisonous Plant Contamination of Edible Plants,* pp. 117–156. Edited by Abdel-Fattah M. Rizk. Boca Raton, FL: CRC Press.

97. Giraud, E., Gosselin, L. & Raimbault, M. (1993). Production of a *Lactobacillus plantarum* starter with linamarase and amylase activities for cassava fermentation. *J Sci Food Agric* 62, 77–82.

98. González, J. A., Estévez-Braun, A., Estévez-Reyes, R., Bazzocchi, I. L., Moujir, L., Jimenez, I. A., Ravelo, A. G. & González, A. G. (1995). Biological activity of secondary metabolites from *Bupleurum salicifolium* (Umbelliferae). *Experientia* 51, 35–39.

99. Grasselly, C. & Crossa-Raynaud, P. (1980). *L'amandier, Techniques Agricoles et Productions Méditerranéennes.* Paris: G. -P. Maisonneuve et Larose.

100. Haberer, J. P., Jouve, P., Bedock, B. & Bazin, P. E. (1984). Severe hypokalaemia secondary to overindulgence in alcohol-free pastis. *Lancet* 8376 (March 25), 575–576.

101. Halkier, B. A. & Moeller, B. L. (1990). The biosynthesis of cyanogenic glucosides in higher plants. *J Biol Chem* 265, 21114–21121.

102. Hampl, J. S., Holland, K. A., Marple, J. T., Hutchins, M. R. & Brockman, K. K. (1997). Acute hemolysis related to consumption of fava beans: a case study and medical nutition therapy approach. *J Am Diet Assoc* 97, 182–183.

103. Harborne, J. B. & Baxter, H., eds. (1995). *Phytochemical Dictionary, a Handbook of Bioactive Compounds from Plants.* London: Taylor and Francis.

104. Hartemink, R., ed. (1998). *Non-Digestible Oligosaccharides: Healthy Food for the Colon? Proceedings of the International Symposium, Wageningen, the Netherlands, 4–5 Dec., 1997.* Wageningen, Netherlands: Wageningen Pers.
105. Hasegawa, S. & Maier, V. P. (1990). Biochemistry of limonoid citrus juice bitter principles and biochemical debittering processes. In *Bitterness in Foods and Beverages (Developments in Food Science 25)*, pp. 293–308. Edited by R. L. Rouseff. Amsterdam: Elsevier.
106. Hegazy, M. I. & Marquardt, R. R. (1983). Development of a simple procedure for the complete extraction of vicine and convicine from faba beans (*Vicia faba* L.). *J Sci Food Agric* 34, 100–108.
107. Heiser, C. B. & Nelson, D. C. (1974). On the origin of the cultivated chenopods (*Chenopodium*). *Genetics* 78, 503–505.
108. Herbert, V. (1979). Laetrile: the cult of cyanide: promoting poison for profit. *Am J Clin Nutr* 32, 1121–1158.
109. Hernandez, J. L., Adris, J., de Rank, E. F., Farias, R. & Samman, N. (1981). Predigested soybeans. *J Am Oil Chem Soc* 58, 510–511.
110. Hernandez, T., Lundquist, P., Oliveira, L., Crista, R. P., Rodriguez, E. & Rosling, H. (1995). Fate in humans of dietary intake of cyanogenic glycosides from roots of sweet cassava consumed in Cuba. *Nat Tox* 3, 114–117.
111. Hirono, I. (1989). Carcinogenic bracken glycosides. In *Toxicants of Plant Origin.* Vol. 2, *Glycosides*, pp. 239–251. Edited by P. R. Cheeke. Boca Raton, FL: CRC Press.
112. Hirono, I., Kono, Y., Takahashi, K., Yamada, K., Niwa, H., Ojika, M., Kigoshi, H., Niiyama, K. & Uosaki, Y. (1984). Reproduction of acute bracken poisoning in a calf with ptaquiloside, a bracken constituent. *Vet Rec* 115, 375–378.
113. Hooper, P. T. (1978). Cycad poisoning in Australia—etiology and pathology. In *Effects of Poisonous Plants on Livestock*, pp. 337–347. Edited by R. F. Keeler, K. R.Van Kampen & L. F. James. New York: Academic Press.
114. Hösel, W. (1981). Glycosylation and glycosidases. In *The Biochemistry of Plants, a Comprehensive Treatise.* Vol. 7, *Secondary Plant Products*, pp. 725–753. Edited by P. K. Stumpf & E. E. Conn. London: Academic Press.
115. Howlett, W. P. (1994). Konzo: a new human disease entity. *Acta Hortic* 375, 323–329.
116. Hughes, A. (1981). The genetic control of plant cyanogenesis. In *Cyanide in Biology*, pp. 495–508. Edited by B. Vennesland, E. E. Conn, C. J. Knowles, J. Westley & F. Wissing. New York: Academic Press.
117. Hughes, A. & Conn, E. E. (1976). Cyanoglucoside biosynthesis in white clover (*Trifolium repens*). *Phytochemistry* 15, 697–701.
118. Hughes, A. & Maher, E. P. (1973). Studies on the nature of the Li locus in *Trifolium repens* L. I. purification and properties of the enzyme component. *Biochem Genet* 8, 1–12.
119. Hughes, M. A., Sharif, A. L., Dunn, M. A. & Oxtoby, E. (1988). The molecular biology of cyanogenesis. In *Cyanide Compounds in Biology; CIBA Foundation Symposium No. 140*, pp. 111–130. Edited by David Evered and Sarah Harnett. Chichester, England: John Wiley & Sons.
120. Hulse, J. H., Laing, E. M. & Pearson, O. E. (1980). Nutritional inhibitors and toxic factors. In *Sorghum and the Millets: Their Composition and Nutritive Value*, pp. 297–395. London: Academic Press.
121. Ikediobi, C. O. & Onyike, E. (1982). The use of linamarase in gari production. *Process Biochem* 17, 2–5.
122. Ingvorsen, K., Yde, B., Godtfredsen, S. E. & Tsuchiya, R. T. (1988). Microbial hydrolysis of organic nitriles and amides. In *Cyanide Compounds in Biology; CIBA Foundation Symposium No. 140*, pp. 16–31. Edited by David Evered and Sarah Harnett. Chichester, England: John Wiley & Sons.
123. Jadhav, S. J., Sharma, R. P. & Salunke, D. K. (1981). Naturally occurring alkaloids in foods. *CRC Crit Rev Toxicol* 9, 21–104.
124. Jansz, E. R. & Pieris, N. (1978). Studies on some local legumes. II. Cyanogenic glucosides. *J Nat Sci Counc Sri Lanka* 6, 1–9.
125. Joint FAO/WHO Expert Committee on Food Additives. (1986). Evaluation of certain food additives and contaminants: twenty-ninth report of the joint FAO/WHO Expert Committee on Food Additives. *WHO Tech Rep Ser* 733, 41.
126. Johns, T. (1990). *With Bitter Herbs They Shall Eat It: Chemical Ecology and the Origins of Human Diet and Medicine.* Tucson, AZ: The University of Arizona Press.
127. Johns, T. (1994). Defence of nitrogen-rich seeds constrains selection for reduced toxicity during the domestication of the grain legumes. In *Advances in Legume Systematics.* Part 5, *The nitrogen Factor*, pp. 151–167. Edited by J. I. Sprent & D. M. McKey. London: The Royal Botanic Gardens.
128. Johnson, I. T., Gee, J. M., Price, K. R. & Fenwick, G. R. (1989). Gastrointestinal effects of some membranolytic plant constituents. In *Recent Advances of Research in Antinutritional Factors in Legume Seeds*, pp. 206–209. Edited by J. Huisman, T. F. B. van der Poel, & I. E. Liener. Wageningen, Netherlands: Pudoc.
129. Jones, D. A. (1998). Why are so many food plants cyanogenic? *Phytochemistry* 47, 155–162.
130. Kaplan, M. A. C., Figueiredo, M. R. & Gottlieb, O. (1983). Variation in cyanogenesis in plants with season and insect pressure. *Biochem Syst Ecol* 11, 367–370.

131. Kellerman, T. S., Miles, C. O., Erasmus, G. L., Wilkins, A. L. & Coetzer, J. A. W. (1994). The possible role of steroidal saponins in the pathogenesis of Geeldikkop, a major hepatogenous photosensitation of small stock in South Africa. In *Plant-Associated Toxins, Agricultural, Phytochemical and Ecological Aspects*, pp. 287–292. Edited by S. M. Colegate & P. R. Dorling. Wallingford, United Kingdom: CAB International.

132. Kensil, C. R. (1996). Saponins as vaccine adjuvants. *Cr Rev Therap Drug Carr Syst* 13, 1–55.

133. Kerlan, V. Ogor, C. & Bercovici, J. P. (1994). Intoxication á la glycyrrhizine après un sevrage tabagique. *Presse Med* 23(1), 50.

134. Ketiku, A. O., Akinyele, I. O., Keshiro, O. O. & Akinnawo, O. O. (1978). Changes in the hydrocyanic acid concentration during traditional processing of cassava into "gari" and "lafun." *Food Chem* 3, 221–228.

135. King, N. L. R. & Bradbury, J. H. (1995). Bitterness of cassava: identification of a new apiosyl glucoside and other compounds that affect its bitter taste. *J Sci Food Agric* 68, 223–230.

136. Klopfenstein, C. F. & Hoseney, R. C. (1995). Nutritional properties of sorghum and the millets. In *Sorghum and Millets, Chemistry and Technology*, pp. 125–168. Edited by D. A. V. Dendy. St. Paul, MN: American Association of Cereal Chemists.

137. Kloss, P., Jeffery, E., Tumbleson, M., Zhang, Y., Parsons, C. & Wallig, M. (1996). Studies on the toxic effects of crambe meal and two of its constituents, 1-cyano-2-hydroxy-3-butene (CHB) and epi-progoitrin, in broiler chick diets. *Br Poult Sci* 37, 971–986.

138. Knowles, C. J. (1988). Cyanide utilization and degradation by microorganisms In *Cyanide Compounds in Biology; CIBA Foundation Symposium No. 140*, pp. 3–15. Edited by David Evered and Sarah Harnett. Chichester, England: John Wiley & Sons.

139. Koziol, M. J. (1991). Afrosimetric estimation of threshold saponin concentration for bitterness in quinoa (*Chenopodium quinoa* Willd). *J Sci Food Agric* 54, 211–219.

140. Kuiper-Goodman, T. & Nawrot, P. S. (1993). Solanine and chaconine. *WHO Food Addit Ser* 30, 339–372.

141. Kupchella, L. & Syty, A. (1984). Determination of cyanogenic glycosides in seeds by molecular absorption spectrometry in the gas phase. *J Assoc Off Anal Chem* 67, 188–191.

142. Lancaster, P. A. & Brooks, J. E. (1983). Cassava leaves as human food. *Econ Bot* 37, 331–348.

143. Lancaster, P. A., Ingham, J. S., Lim, M. Y. & Coursey, D. G. (1982). Traditional cassava-based foods: survey of processing techniques. *Econ Bot* 36, 12–45.

144. Lansky, P. S. (1993). Plants that lower cholesterol. *Acta Hortic* 332, 131–136.

145. Lea, A. G. H. (1991). Enzymes in the production of beverages and fruit juices. In *Enzymes in Food Processing*, pp. 194–220. Edited by G. A. Tucker & L. F. J. Woods. Glasgow, Scotland: Blackie Academic and Professional.

146. Lechtenberg, M. & Nahrstedt, A. (1999). Cyanogenic glycosides. In *Naturally Occurring Glycosides*, pp. 147–191. Edited by R. Ikan. Chichester, England: John Wiley & Sons.

147. Legras, J. L., Jory, M., Arnaud, A. & Galzy, P. (1990). Detoxification of cassava pulp using *Brevibacterium* sp. R312. *Appl Microbiol Biotechnol* 33, 529–533.

148. Legras, J. L., Kaakeh, M. R., Arnaud, A. & Galzy, P. (1989). Degradation of cyanoglucosides by *Brevibacterium* sp. R312 strain. *J Gen Appl Microbiol* 35, 451–461.

149. Lei, V., Amoa-Awua, W. K. A. & Brimer, L. (1999). Degradation of cyanogenic glycosides by *Lactobacillus plantarum* strains from spontaneous cassava fermentation and other microorganisms. *Int J Food Microbiol* 53, 169–184.

150. Lewin, R. (1987). Environmental hypothesis for brain diseases strengthened by new data. *Science* 237, 483–484.

151. Liener, I. E. (1987). Detoxifying enzymes. In *Food Biotechnology*. Vol. 1, pp. 249–271. Edited by R. D. King & P. S. J. Cheetham. London: Elsevier Applied Science.

152. Loew, D., von Bergmann, U., Schmidt, M. & Überla, K. H. (1994). Anthranoidlaxanzien. Ursache für kolonkarzinom? *Dtsch Apoth Ztg* 134, 3180–3183.

153. Low, V. H. K. & Thomson, J. A. (1990). Cyanogenesis in Australian bracken (*Pteridium esculentum*): distribution of cyanogenic phenotypes and factors influencing activity of the cyanogenic glucosidase. In *Bracken Biology and Management*, pp. 105–111. Edited by J. A. Thomson & R. T. Smith. Wahroonga, NSW Australia: The Australian Institute of Agricultural Science.

154. Luca, P., de Moretti, A., Sabato, S. & Gigliano, S. S. (1980). The ubiquity of cycasin in cycads. *Phytochemistry* 19, 2230–2231.

155. Luckner, M. (1990). *Secondary Metabolism in Microorganisms, Plants, and Animals*, 3rd edn. Verlag, Jena: VEB Gustav Fischer.

156. Madhusudhan, K. T. & Singh, N. (1985). Effect of detoxification on the physicochemical properties of linseed. *J Agric Food Chem* 33, 12–19.

157. Madhusudhan, K. T. & Singh, N. (1985). Effect of heat treatment on the functional properties of linseed meal. *J Agric Food Chem* 33, 19–22.

158. Maharaj, I., Froh, K. J., & Campbell, J. B. (1986). Immune responses of mice to inactivated rabies vaccine administered orally: potentiation by Quillaja saponin. *Can J Microbiol* 32, 414–420.

159. Maher, E. P. & Hughes, A. (1973). Studies on the nature of the Li locus in *Trifolium repens* L. II. the effect of genotype on enzyme activity and properties. *Biochem Genet* 8, 13–26.

160. Majak, W. & Pass, M. A. (1989). Aliphatic nitrocompounds. In *Toxicants of Plant Origin.* Vol. 2, *Glycosides,* pp. 143–159. Edited by P. R. Cheeke. Boca Raton, FL: CRC Press.

161. Malcolm, S. B. (1991). Cardenolide-mediated interactions between plants and herbivores. In *Herbivores: Their Interactions with Secondary Plant Metabolites, 2nd edn.* Vol. 1, *The Chemical Participants,* pp. 251–296. Edited by G. A. Rosenthal & M. R. Berenbaum. San Diego: Academic Press.

162. Marquardt, R. R. & Arbid, M. S. S. (1988). Protection against the toxic effects of the favism factor (divicine) in rats by vitamins E, A and C and iron chelating agents. *J Sci Food Agric* 43, 155–166.

163. Matsushima, T., Matsumoto, H., Shirai, A., Sawamura, M. & Sugimura, T. (1979). Mutagenicity of the naturally occurring carcinogen cycasin and synthetic methylazoxymethanol conjugates in *Salmonella typhimurium. Cancer Res* 39, 3780–3782.

164. McKay, A. M. (1992). Hydrolysis of vicine and convicine from faba beans by microbial β-glucosidase enzymes. *J Appl Bacteriol* 72, 475–478.

165. McMahon, J. M., White, W. L. B. & Sayre, R. T. (1995). Cyanogenesis in cassava (*Manihot esculenta* Crantz). *J Exp Bot* 46, 731–741.

166. McMillan, D. C., Schey, K. L., Meier, G. P. & Jollow, D. J. (1993). Chemical analysis and hemolytic activity of the fava bean aglycone divicine. *Chem Res Toxicol* 6, 439–444.

167. Meijer, M. M. T. & Muuse, B. G. (1989). Optimalization of dehulling technique and enzymatic hydrolysis of vicine/convicine to eliminate ANF's of Faba bean. In *Recent Advances of Research in Antinutritional Factors in Legume Seeds,* pp. 268–271. Edited by J. Huisman, T. F. B. van der Poel, & I. E. Liener. Wageningen, Netherlands: Pudoc.

168. Mickelsen, O., Campbell, M. E., Yang, M., Mugera, G. M. & Whitehair, C. K. (1964). Studies with cycad. *Fed Proc* 23, 1363–1365.

169. Ministry of Health Mozambique. (1984). Mantakassa: an epidemic of spastic paraparesis associated with chronic cyanide intoxication in a cassava stable area in Mozambique. *Bull WHO* 62, 477–492.

170. Mizutani, F., Hirota, R., Amano, S., Hino, A. & Kadoya, K. (1991). Changes in cyanogenic glycoside content and β-cyanoalanine synthetase activity in flesh and seeds of Japanese plum (*Prunus salicina* Lindl.) during development. *J Jpn Soc Hortic Sci* 59, 863–867.

171. Mkpong, O. E., Yan, H., Chism, G. & Sayre, R. T. (1990). Purification, characterization, and localization of linamarase in cassava. *Plant Physiol* 93, 176–181.

172. Mlingi, N., Poulter, N. H. & Rosling, H. (1992). An outbreak of acute intoxications from human consumption of insufficiently processed cassava in Tanzania. *Nutr Res* 12, 677–687.

173. Moertel, C. G., Ames, M. M., Kovach, J. S., Mayer, T. P., Rubin, J. R. & Tinker, J. A. (1981). A pharmacologic and toxicological study of amygdalin. *J Am Med Assoc* 245, 48–50.

174. Monastra, F. & Fideghelli, C. (1980). L'albicocco. *Agricoltura e Ricerca* 10, 24–33.

175. Moncrieff, R. W. (1967). *The Chemical Senses.* Cleveland, OH: CRC Press.

176. Monzini, A. & Gorini, F. (1991). Pre-freezing, storage and industrial use of apricot with relationship to quality. *Acta Hortic* 293, 599–655.

177. Moretti, A., Sabato, S. & Gigliano, G. S. (1981). Distribution of macrozamin in australasian cycads. *Phytochemistry* 20, 1415–1416.

178. Morgan, R. W. & Hoffman, G. R. (1983). Cycasin and its mutagenic metabolites. *Mutat Res* 114, 19–58.

179. Munro, J. M. (1987). *Cotton,* 2nd edn. Harlow, United Kingdom: Longman Scientific and Technical.

180. Nachman, R. J. & Olson, J. D. (1983). Ranunculin. a toxic constituent of the poisonous range plant bur buttercup (*Ceratocephalus testiculatus*). *J Agric Food Chem* 31, 1358–1360.

181. Nafa, K., Reghis, A., Osmani, N., Baghli, L., Aït-Abbes, H., Benabadji, M., Kaplan, J. -C., Vulliamy, T. & Luzzatto, L. (1994). At least five polymorphic mutants account for the prevalence of glucose-6-phosphate dehydrogenase deficiency in Algeria. *Hum Genet* 94, 513–517.

182. Nahrstedt, A. (1987). Recent developments in chemistry, distribution and biology of the cyanogenic glycosides. In *Biologically Active Natural Products,* pp. 213–234. Edited by K. Hostettmann & P. J. Lea. Oxford: Oxford University Press.

183. Nartey, F. (1968). Studies on cassava, *Manihot utilissima* Pohl. I. cyanogenesis: the biosynthesis of linamarin and lotaustralin in etiolated seedlings. *Phytochemistry* 7, 1307–1312.

184. Nartey, F. (1978). Manihot esculenta *(Cassava), cyanogenesis, ultrastructure and seed germination.* Copenhagen: Munksgaard.

185. Nout, M. J. R., Tuncel, G. & Brimer, L. (1995). Microbial degradation of amygdalin of bitter apricot seeds (*Punus armeniaca*). *Int J Food Microbiol* 24, 407–412.

186. O'Brien, G. M., Wheatley, C. C., Iglesias, C. & Poulter, N. H. (1994). Evaluation, modification and comparison of two rapid assays for cyanogens in cassava. *J Sci Food Agric* 65, 391–399.

187. Oelrichs, P. B., Pearce, C. M., Kudo, K. & Kelly, W. R. (1994). The isolation, structure elucidation and toxicity of the kaurene glycosides parquin and carboxyparquin

in *Cestrum parqui.*. In *Plant-Associated Toxins, Agricultural, Phytochemical and Ecological Aspects,* pp. 251–255. Edited by S. M. Colegate & P. R. Dorling. Wallingford, United Kingdom: CAB International.

188. Ohga, S., Higashi, E., Nomura, A., Matsuzaki, A., Hirono, A., Miwa, S. & Fujii, H. (1995). Haptoglobin therapy for acute favism: a Japanese boy with glucose-6-phosphate dehydrogenase Guadalajara. *Br J Haematol* 89, 421–423.
189. Ohtsubo, T. & Ikeda, F. (1994). Seasonal changes of cyanogenic glycosides in Mume (*Prunus mume* Sieb. et Zucc.) seeds. *J Jpn Soc Hortic Sci* 62, 695–700.
190. Ojika, M., Sugimoto, K., Okazaki, T. & Yamada, K. (1989). Modification and cleavage of DNA by ptaquiloside: a new potent carcinogen isolated from bracken fern. *J Chem Soc Chem Commun* 22, 1775–1777.
191. Okafor, N. & Ejiofor, A. O. (1990). Rapid detoxification of cassava mash fermenting for gari production following the inoculation with a yeast simultaneously producing linamarase and amylase. *Process Biochem* 25, 82–86.
192. Okolie, N. P. & Ugochukwu, E. N. (1989). Cyanide contents of some Nigerian legumes and the effect of simple processing. *Food Chem* 32, 209–216.
193. Okuyama, T., Takata, M., Takahashi, K., Ishikawa, T., Miyasaka, K. & Kaneyama, N. (1989). High-performance liquid chromatographic analysis of naturally occurring glycosides and saponins. *J Chromatogr* 466, 390–398.
194. Olafsdottir, E. S., Andersen, J. V. & Jaroszewski, J. W. (1989). Cyanohydrin glycosides of Passifloraceae. *Phytochemistry* 28, 127–132.
195. Olson, J. D., Anderson, T. E., Murphy, J. C. & Madsen, G. (1983). Bur buttercup poisoning of sheep. *J Am Vet Med Assoc* 183, 538–543.
196. O'Mullane, J. E. & Hayter, I. P. (1993). International Patent PCT/GB93/00343.
197. Onabolu, A. O., Bokanga, M. & Rosling, H. (1999). Cassava processing in a Nigerian community affected by a neuropathy attributed to dietary cyanide exposure. *Trop Sci* 39, 129–135.
198. Oomah, B. D. & Mazza, G. (1995). Functional properties, uses of flaxseed protein. *Inform* 6, 1246–1252.
199. Oomah, B. D., Mazza, G. & Cui, W. (1994). Optimization of protein extraction from flaxseed meal. *Food Res Int* 27, 355–361.
200. Oomah, B. D., Kenaschuk, E. O., Cui, W. & Mazza, G. (1995). Variation in the composition of water-soluble polysaccharides in flaxseed. *J Agric Food Chem* 43, 1484–1488.
201. Oomah, B. D., Mazza, G. & Kenaschuk, E. O. (1992). Cyanogenic compounds in flaxseed. *J Agric Food Chem* 40, 1346–1348.
202. Padmaja, G. (1995). Cyanide detoxification in cassava for food and feed uses. *CRC Cri Rev Food Sci Technol* 35, 299–339.
203. Palter, R., Lundin, R. E. & Haddon, W. F. (1972). A cathartic lignan glycoside isolated from *Carthamus Tinctorus* L., Compositae. *Phytochemistry* 11, 2871–2874.
204. Pass, M. A. (1994). Toxicity of plant-derived aliphatic notrotoxins. In *Plant-Associated Toxins, Agricultural, Phytochemical and Ecological Aspects,* pp. 541–545. Edited by S. M. Colegate & P. R. Dorling. Wallingford, United Kingdom: CAB International.
205. Pederson, G. A. & Brink, G. E. (1998). Cyanogenesis effect on insect damage to seedling white clover in a Bermuda grass sod. *Agron J* 90, 208–210.
206. Petersen, H. W., Moelgaard, P., Nyman, U. & Olsen, C. E. (1993). Chemotaxonomy of the tuber-bearing *Solanum* species, subsection *Potatoe* (Solanaceae). *Biochem Biophys Ecol* 21, 629–644.
207. Petruccioli, M., Bimer, L., Cicalii, A. R. & Federici, F. (1999). The linamarase of *Mucor circinelloides* LU M40 and its detoxifying activity on cassava. *J Appl Microbiol* 86, 302–310.
208. Philbrick, D. J., Hill, D. C. & Alexander, J. C. (1977). Physiological and biochemical changes associated with linamarin administration to rats. *Appl Pharmacol* 42, 539–551.
209. Pieters, L. A. C. & Vlietinck, A. J. (1991). Naturally occurring carcinogens in plant foodstufs. In *Poisonous Plant Contamination of Edible Plants,* pp. 1–25. Edited by Abdel-Fattah M. Rizk. Boca Raton, FL: CRC Press.
210. Poulton, J. E. (1983). Cyanogenic compounds in plants and their toxic effects. In *Handbook of Natural Toxins,* pp. 118–157. Edited by R. F. Keeler & A. T. Tu. New York: Marcel Dekker.
211. Poulton, J. E. & Li, C. P. (1994). Tissue level compartmentation of (R)-amygdalin and amygdalin hydrolase prevents large-scale cyanogenesis in undamaged *Prunus* seeds. *Plant Physiol* 104, 29–35.
212. Puri, A. (1990). Removal of bitter compounds from citrus products by adsorption techniques. In *Bitterness in Foods and Beverages (Developments in Food Science 25),* pp. 325–336. Edited by R. L. Rouseff. Amsterdam: Elsevier.
213. Pusztai, A., Ewen, S. W. B., Grant, G., Brown, D. S., Stewart, J. C., Peumans, W. J., van Damme, E. J. M. & Bardicz, S. (1993). Antinutritive effects of wheat-germ agglutinin and other N-acetylglucosamine-specific lectins. *Br J Nutr* 70, 313–321.
214. Rabot, S., Nugon-Baudon, L., Raibaud, P. & Szylit, O. (1993). Rape-seed meal toxicity in gnotobiotic rats: influence of a whole human faecal flora or single human strains of *Escherichia coli* and *Bacteriodes vulgatus*. *Br J Nutr* 70, 323–331.

215. Rachis, J. J. (1975). Oligosaccharides of food legumes: alpha-galactosidase activity and the flatus problem. In *Physiological Effects of Food Carbohydrates,* pp. 207–222. Edited by A. Jeanes & J. Hodge. Washington, DC: American Chemical Society.

216. Raddick, J. G. (1989). The acetylcholinesterase-inhibitory activity of steroidal glycoalkaloids and their aglycones. *Phytochemistry* 28, 2631–2634.

217. Raffauf, R. F. (1996). *Plant Alkaloids: A Guide to Their Discovery and Distribution.* New York: Food Products Press.

218. Rahman, S. & Subrahmanyan, V. (1947). Effect of different treatments on the removal of hydrocyanic acid from the Burma bean (*P. lunatus* Linn.). *Curr Sci* 11, 351–352.

219. Ramsay, G. & Griffiths, D. W. (1996). Accumulation of vicine and convicine in *Vicia faba* and *V. narbonensis*. *Phytochemistry* 42, 63–67.

220. Rauws, A. G., Olling, M. & Timmerman, A. (1982). The pharmacokinetics of amygdalin. *Arch Toxicol* 49, 311–319.

221. Rauws, A. G., Olling, M. & Timmerman, A. (1983). The pharmakokinetics of prunasin, a metabolite of amygdalin. *J Toxicol Clin Toxicol* 19, 851–856.

222. Reddy, N. R. & Pierson, M. D. (1994). Reduction in antinutritional and toxic components in plant foods by fermentation. *Food Res Int* 27, 281–290.

223. Reichert, R. D., Tatarynovich, J. E. & Tyler, R. T. (1986). Abrasive dehulling of quinoa (*Chenopodium quinoa*): effect on saponin content as determined by an adapted haemolytic assay. *Cereal Chem* 63, 471–475.

224. Reynolds, J. H. (1974). An immobilized α-galactosidase continous flow reactor. *Biotechnol Bioeng* 16, 135–147.

225. Rezakhani, A. & Maham, M. (1994). Cardiac manifestations of oleander poisoning in cattle and donkeys. In *Plant-Associated Toxins, Agricultural, Phytochemical and Ecological Aspects,* pp. 534–537. Edited by S. M. Colegate & P. R. Dorling. Wallingford, United Kingdom: CAB International.

226. Ridout, C. L., Price, K. R., DuPont, M. S., Parker, M. L. & Fenwick, G. R. (1991). Quinoa saponins—analysis and preliminary investigations into the effect of reduction by processing. *J Sci Agric Food* 54, 165–176.

227. Robins, R. J., Rhodes, M. J. C., Parr, A. J. & Walton, N. J. (1990). Biosynthesis of bitter compounds. In *Bitterness in Foods and Beverages (Developments in Food Science 25),* pp. 49–79. Edited by R. L. Rouseff. Amsterdam: Elsevier.

228. Rombouts, F. M. & Nout, M. J. R. (1995). Microbial fermentation in the production of plant foods. *J Appl Bacteriol* 79 (Symp. Suppl.), 108S–117S.

229. Romero, C., Manjón, A., Bastida, J. & Iborra, J. L. (1985). A method for assaying the rhamnosidase activity of naringinase. *Anal Biochem* 149, 566–571.

230. Rosling, H. (1993). Cyanide exposure from linseed. *Lancet* 341, 177.

231. Rosling, H. (1994). Measuring effects in humans of dietary cyanide exposure from cassava. *Acta Hortic* 375, 271–283.

232. Rosling, H. & Tylleskär, T. (1996). Konzo. In *Tropical Neurology,* pp. 353–364. Edited by R. A. Shakin, P. K. Newman & C. M. Poser. London: W.B. Saunders Company.

233. Ruiz-Terán, F. & Owens, J. D. (1999). Fate of oligosaccharides during production of soya bean tempe. *J Sci Food Agric* 79, 249–252.

234. Saka, J. D. K., Mhone, A. R. K., Mkambira, J., Brimer, L., Bokanga, M., Mahungu, N. M., Chiwona-Karltun, L. & Rosling, H. (1998). Correlation between cyanogenic glucoside content and taste of fresh cassava roots. *Trop Agric* 75, 169–173.

235. Sakata, M., Yoshida, A., Yuasa, C., Sakata, K. & Haga, M. (1987). Toxicity of D,L-mandelonitrile-β-D-glucoside "prulaurasin" in rat. *J Toxicol Sci* 12, 47–55.

236. Salkowski, A. A. & Penney, D. G. (1994). Cyanide poisoning in animals and humans: a review. *Vet Hum Toxicol* 36, 455–466.

237. Sanni, M. O., Sobaminwa, A. O., Eyinla, C. M. & Rosling, H. (1994). Safety aspects of processing cassava to gari in Nigeria. *Acta Hortic* 375, 227–231.

238. Schab, R. & Yannai, S. (1973). An improved method for debittering apricot kernels. *J Food Sci Technol* 10, 57–59.

239. Schambelan, M. (1994). Licorice ingestion and blood pressure regulating hormones. *Steroids* 59, 127–130.

240. Schilcher, H. & Wilkens-Sauter, M. (1986). Quantitative determination of cyanogenic glycosides in *Linum usitatissimum* using HPLC. *Fette, Seifen, Anstrichm* 8, 287–290.

241. Seigler, D. S. (1991). Cyanide and cyanogenic glycosides. In *Herbivores: Their Interactions with Secondary Plant Metabolites,* 2nd edn, Vol. 1., pp. 35–77. Edited by G. A. Rosenthal & M. R. Berenbaum. San Diego: Academic Press.

242. Seigler, D. S., Maslin, B. R. & Conn, E. E. (1989). Cyanogenesis in the Leguminosae. In *Advances in Legume Biology, pp.* 645–672. Edited by C. H. Stirton & J. L. Zarucchi. St. Louis, MO: Missouri Botanical Garden.

243. Shahidi, F. & Naczk, M. (1990). Removal of glucosinolates and other antinutrients from canola and rapeseed by methanol/ammonia processing. In *Canola and Rapeseed; Production, Chemistry, Nutrition and Processing Technology,* pp. 291–306. Edited by F. Shahidi. New York: Van Nostrand Reinhold.

244. Shaw, P. E. (1990). Cyclodextrin polymers in the removal of bitter compounds in citrus juices. In *Bitterness in Foods and Beverages (Developments in Food Science 25),* pp. 309–324. Edited by R. L. Rouseff. Amsterdam: Elsevier.

245. Sherba, S. E. (1970). South African Patent 6,902,503.

246. Shimizu, T., Yasuda, N., Kono, I., Yagi, F., Tadera, K. & Kobayashi, A. (1986). Hepatic and spinal lesions in goats chronically intoxicated with cycasin. *Jpn J Vet Sci* 48, 1291–1295.

247. Sinha, S. K. & Nair, T. V. R. (1968). Studies on the variability of cyanogenic glucoside content in cassava tubers. *Ind J Agric Sci* 38, 958–963.

248. Smiley, K. L., Hensley, D. E. & Gasdorf, H. J. (1976). Alpha-galactosidase production and use in a hollow-fiber reactor. *Appl Environ Microbiol* 31, 615–617.

249. Smith, B. L., Seawright, A. A., Ng, J. C., Hertle, A. T., Thomson, J. A., & Bostock, P. D. (1994). Concentration of ptaquiloside, a major carcinogen in bracken fern (*Pteridium* spp.), from eastern Australia and from a cultivated worldwide collection held in Sydney, Australia. *Nat Toxins* 2, 347–353.

250. Smith, B. L., Shaw, G., Prakash, A. & Seawright, A. A. (1994). Studies on DNA adduct formation by ptaquiloside, the carcinogen of bracken ferns (*Pteridium* spp.). In *Plant-Associated Toxins, Agricultural, Phytochemical and Ecological Aspects,* pp.167–172. Edited by S. M. Colegate & P. R. Dorling. Wallingford, United Kingdom: CAB International.

251. Smith, C. R., Weisleder, D., Miller, R., Palmer, I. S. & Olson, O. E. (1980). Linustatin and Neolinustatin: cyanogenic glycosides of linseed meal that protect animals against selenium toxicity. *J Org Chem* 45, 507–510.

252. Soerensen, H. (1990). Glucosinolates: structure-properties-function. In *Canola and Rapeseed: Production, Chemistry, Nutrition and Processing Technology,* pp. 149–172. Edited by F. Shahidi. New York: Van Nostrand Reinhold.

253. Speijers, G. (1993). Cyanogenic glycosides. *WHO Food Addit Ser* 30, 299–337.

254. Spencer, P. S., Nunn, P. B., Hugon, J., Ludolph, A. C., Ross, S. M., Roy, D. N. & Robertson, R. C. (1987). Guam amyotrophic lateral sclerosis-parkinsonism-dementia linked to a plant exitant neurotoxin. *Science* 237, 517–522.

255. Staron, T. (1974). La détoxification des tourteaux de colza par voie biologique. *L' alimentation et la Vie*, 62, 165–179.

256. Stoermer, F. C., Reistadt, R. & Alexander, J. (1993). Glycyrrhizic acid in liquorice: evaluation of health hazard. *Food Chem Toxicol* 31, 303–313.

257. Stoewsand, G. S. (1995). Bioactive organosulfur phytochemicals in *Brassica oleracea* vegetables: a review. *Food Chem Toxic* 33, 537–543.

258. Stone, R. (1993). Guam: deadly disease dying out. *Science* 261, 424–426.

259. Strada, G. D., Fideghelli, C., Monastrata, F., Pennone, F. & Quarta, R. (1991). Apricot breeding at the Instituto Sperimentale per la Frutticoltura (ISF) of Rome. *Acta Hortic* 293, 121–132.

260. Sturm, W. (1982). Mandeln aus dem Mittelmeerraum und/oder Kalifornien. Vorkommen und geschmacklische auswirkungen bitterer mandeln und die bedeutung rein sü(er mandeln. *Süsswaren* 26, 196–198.

261. Sugimoto, H. & Van Buren, J. P. (1970). Removal of oligosaccharides from soy milk by an enzyme from *Aspergillus saitoi*. *J Food Sci* 35, 655–660.

262. Sunnerheim-Sjöberg, K. (1991). *Chemical studies of secondary metabolites in betula and pinus—with emphasis on defence against mammalian herbivores.* PhD diss., Swedish University of Agricultural Sciences, Uppsala, Sweden.

263. Swai, A. B. M., Mclarty, D. G., Mtinangi, B. L., Tatala, S., Kitange, H. M., Mlingi, N., Rosling, H., Howlett, W. P., Brubaker, G. R. & Alberti, K. G. M. M. (1992). Diabetes is not caused by cassava toxicity. *Diabetes Care* 15, 1378–1385.

264. Thananukul, D., Tanaka, N., Chichester, C. O. & Lee, T. -C. (1976). Degradation of raffinose and stachyose in soybean milk by α-galactosidase from *Mortierelle vinecea*: entrapment of galactosidase within polyacrylamide gel. *J Food Sci* 41, 173–175.

265. Tjon Sie Fat, L. A. (1979). *Contribution to the knowledge of cyanogenesis in angiosperms (Cyanogene Verbindingen bij Poaceae, Commelinaceae, Ranunculaceae en Campanulaceae).* PhD diss., University of Leiden, Leiden.

266. Tuncel, G., Nout, M. J. R. & Brimer, L. (1995). The effect of grinding, soaking and cooking on the degradation of amygdalin of bitter apricot seeds. *Food Chem* 53, 447–451.

267. Tuncel, G., Nout, M. J. R. & Brimer, L. (1998). Degradation of cyanogenic glycosides of bitter apricot seeds (*Prunus armeniaca*) by endogenous and added enzymes as affected by heat treatments and particle size. *Food Chem* 63, 65–69.

268. Tuncel, G., Nout, M. J. R., Brimer, L., & Göktan, D. (1990). Toxicological, nutritional and microbiological evaluation of tempe fermentation with *Rhizopus oligosporus* of bitter and sweet apricot seeds. *Int J Food Microbiol*, 11, 337–344.

269. Tylleskär, T. (1994). The association between cassava and the paralytic disease konzo. *Acta Hortic* 375, 331–339.

270. Tylleskär, T., Banea, M., Bikangi, N., Cooke, R., Poulter, N. & Rosling, H. (1992). Cassava cyanogens and konzo, an upper motoneuron disease found in Africa. *Lancet* 339, 208–211.

271. Vasconcelos, A. T., Twiddy, D. R., Westby, A. & Reilly, P. J. A. (1990). Detoxification of cassava during gari preparation. *Int J Food Sci Tech* 25, 198–203.

272. Verbiscar, A. J., Baniga, T. F., Weber, C. W., Reid, B. L., Swingle, R. S., Tei, J. E. & Nelson, E. A. (1981). Detoxification of jojoba meal by lactobacilli. *J Agric Food Chem* 29, 296–302.

273. Verjee, Z. H. (1993). Glucose-6-phosphate dehydrogenase deficiency in Africa—a review. *East Afr Med J* 70(4) supplement, 40–47.

274. Wallig, M. A., Kingston, S., Staack, R. & Jeffery, E. H. (1998). Induction of rat pancreatic glutathione S- transferase and quinone reductase activities by a mixture of glucosiolate breakdown derivatives found in brussels sprouts. *Food Chem Toxicol* 36, 365–373.

275. Wanasundara, P. K. J. P. D., Amarowicz, R., Kara, M. T. & Shahidi, F. (1993). Removal of cyanogenic glycosides of flaxseed meal. *Food Chem* 48, 263–266.

276. Wang, P. -X. & Ueberschär, K. H. (1990). The estimation of vicine, convicine and condensed tannins in 22 varieties of faba beans (*Vicia faba* L.). *Anim Feed Sci Technol* 31, 157–165.

277. Weber, C. W., Berry, J. W. & Cooke, E. M. (1983). Influence of jojoba meal upon growth and reproduction in mice. In *Jojoba and Its Uses Through 1982, Proceedings of the Fifth International Conference on Jojoba and Its Uses*, pp. 93–99. Edited by A. Elias-Cesnik. Tucson, AZ: Office of Arid Lands Studies, University of Arizona.

278. Webster, J., Beck, W. & Ternai, B. (1984). Toxicity and bitterness in Australian *Dioscoria bulbifera* L. and *Dioscoria hispida* Dennst. from Thailand. *J Agric Food Chem* 32, 1087–1090.

279. Weissenberg, M. (1989). Calcinogenic glycosides. In *Toxicants of Plant Origin*. Vol. 2, *Glycosides*, pp. 201–238. Edited by P. R. Cheeke. Boca Raton, FL: CRC Press.

280. Westendorf, J. (1993). Anthranoid derivatives—general discussion. In *Adverse Effects of Herbal Drugs*. Vol. 2, pp. 105–118. Edited by P. A. G. M. De Smet, K. Keller, R. Hänsel & R. F. Chandler. Berlin: Springer-Verlag.

281. Westley, J. (1988). Mammalian cyanide detoxification with sulphane sulphur. In *Cyanide Compounds in Biology; CIBA Foundation Symposium No. 140*, pp. 201–218. Edited by David Evered and Sarah Harnett. Chichester, England: John Wiley & Sons.

282. Wheeler, J. L. & Mulcahy, C. (1989). Consequences for animal production of cyanogenesis in sorghum forage and hay: a review. *Trop Grass* 23, 193–202.

283. White, W. L. B., Arias-Garzon, D. I., McMahon, J. M. & Sayer, R. T. (1998). Cyanogenesis in cassava: the role of hydroxynitrile lyase in root cyanide production. *Plant Physiol* 116, 1219–1225.

284. Yannai, S. & Marquardt, R. R. (1985). Induction of favism-like symptoms in the rat: effects of vicine and divicine in normal and buthionide sulphoxime-treated rats. *J Sci Food Agric* 36, 1161–1168.

285. Yeoh, H. H., Tan, T. K. & Loh, C. M. (1995). Sources of fungal linamarases. *World J Microbiol Biotechnol* 11, 678–680.

286. Yoshiki, Y., Kudou, S. & Ukubo, K. (1998). Relationship between chemical structures and biological activities of triterpenoid saponins from soybean. *Biosci Biotechnol Biochem* 62, 2291–2299.

287. Zee, J. A., Boudreau, A., Bourgeois, M. & Breton, R. (1988). Chemical composition and nutritional quality of faba bean (*Vicia faba* L. Minor) based tofu. *J Food Sci* 53, 1772–1774.

Chapter 5

Chemical Hazards and Their Control: Toxins

Maurice O. Moss

MICROBIAL TOXINS

Many food commodities are susceptible to contamination by microorganisms, the subsequent growth of which can lead to the presence of toxic compounds in food. Some of the bacterial toxins, such as those of *Staphylococcus aureus*, *Clostridium botulinum*, and many Gram-negative species associated with food poisoning are macromolecules such as proteins. However, some bacteria are able to produce low molecular weight toxic metabolites, which may contaminate foods and cause serious poisoning. One example is *Burkholderia* (*Pseudomonas*) *cocovenenans*, which produces toxins such as bonkrekic acid and toxoflavin (see bongkrek poisoning on page 112). Another example is the macrocyclic depsipeptide, cereulide, which is recognized as the emetic toxin of *Bacillus cereus*.

Several species of cyanobacteria can also produce very toxic metabolites, but rarely are these directly associated with foods. An exception may be the contamination of species, such as *Spirulina*, grown as food or as components of "health food" products, with toxigenic species. More directly associated with foods are the toxins that are produced by groups of eukaryotic algae such as dinoflagellates and diatoms. These may be ingested by shellfish or fish and pass through the food chain to humans.

A number of species of fungi can produce relatively low molecular weight secondary metabolites, which are toxic to humans and domesticated animals and are referred to as mycotoxins.[49] Mycotoxin biosynthesis may be associated with the preharvest stage of crop production by fungi that are obligate endophytes of plants, plant pathogens, or members of the flora responsible for the decay of senescent plant material.[47] However, the highest concentrations of many of these toxic metabolites are produced by fungi growing on postharvest commodities that are stored under inappropriate conditions.[31] The majority of mycotoxins are especially important in the context of animal husbandry, but several are also significant as contaminants of human foods and will be dealt with in the following sections. Table 5–1 summarizes the toxins considered, their most common sources, and the food commodities most frequently implicated.

A number of fermented foods involve a mold-ripening stage, usually with species of *Penicillium*. Some of these molds are known to be potentially toxigenic; thus, strains of *P. roqueforti* can produce PR-toxin but not during the production of the blue cheeses. Many strains of *P. camemberti* produce cyclopiazonic acid, and this compound has been detected in the crusts of Camembert-type cheese but not in the interior of the cheeses.[35,36] Cyclopiazonic acid has also been found associated with mold-ripened fermented sausage.[76] *Aspergillus versicolor* is a common member of the surface flora of hard cheeses stored for long periods, and this species is known to produce sterigmatocystin, a mycotoxin that is a biosynthetic precursor of the aflatoxins and is both acutely toxic and carcinogenic but very much less so than aflatoxin B_1.

Table 5–1 Toxic Microbial Metabolites that May Be Associated with Foods

Toxin	*Sources*	*Commodities*	*LD_{50} (mg/kg)*
Aflatoxin B_1	*Aspergillus flavus* *A. parasiticus* *A. nomius*	Maize, groundnuts, treenuts, dried figs, spices	0.5 (dog) 9.0 (mouse)
Ochratoxin A	*Penicillium verrucosum* *A. ochraceus*	Cereals, coffee, spices, dried vine fruits	28 (rat)
Patulin	*P. expansum* *A. clavatus*	Apple juice, other fruit juices, malted barley residues	35 (mouse)
T-2 toxin	*Fusarium sporotrichioides*	Overwintered cereals	5.2 (rat)
Deoxynivalenol	*F. graminearum*	Cereals	46 (mouse)
Fumonisin B_1	*F. moniliforme*	Maize	?
Cyanoginosin	*Microcystis aeruginosa*	Water	0.05 (mouse)
Bonkrekic acid	*Burkholderia cocovenenans*	Tempeh bonkrek	6.84 (oral LD_{100} in mouse)
Toxoflavin	*B. cocovenenans*	Tempeh bonkrek	8.4 (oral in mouse)
Saxitoxin	*Alexandrium catenella*	Mussels	0.012 (ip in mouse)
Okadaic acid	*Dinophysis fortii*	Mussels	0.2 (LD_{99} in mouse)
Domoic acid	*Pseudonitzschia pungens*	Mussels	3.6 (mouse)

This mycotoxin can be found in cheeses but usually only at the surface.[52] A recent study has demonstrated the occurrence of sterigmatocystin in Ras cheese that was purchased in local markets in Egypt.[43]

AFLATOXINS

Although aflatoxins are produced by a small number of species of the genus *Aspergillus*, they are especially widespread because there are essentially three routes to the contamination of food commodities.

1. Direct contamination through the mold spoilage of stored products by species such as *A. flavus*, *A. parasiticus*, and *A. nomius*
2. Preharvest production in the field by the establishment of an endophytic association of one of these species with plants such as maize and groundnuts, followed by some form of stress on the growing crop, such as drought
3. Passage through the food chain into animal products, such as milk, following the consumption by farm animals of contaminated animal feeds

The aflatoxins are a family of metabolites; the most important of which in terms of both toxicity and prevalence is aflatoxin B_1 (Figure 5–1). Aflatoxins are acutely toxic, carcinogenic, and immunosuppressive. Aflatoxins have been found in a wide range of tropical and subtropical products such as figs, pistachio and Brazil nuts, spices, peanuts, and maize. Pittet[59] provided a useful update on the natural occurrence of aflatoxins and other mycotoxins with detailed information concerning the incidence and range of concentrations found. The most important of those commodities that might be used as a raw material for fermented foods is maize, although there have been reports of low concentrations of aflatoxins in both rice[50] and wheat.[56] A diverse range of products associated with foods and beverages of the Far East, such as rice wine, soy sauce, and miso, involves the use of a source of amylolytic, lipolytic, and proteolytic enzymes. This material, known as *koji*, traditionally is produced by growing appropriate strains of *A. oryzae* on substrates such as rice, wheat, and soya beans. This mold is closely related to *A. flavus*, and may indeed be a "domesticated" form of this species.[15] Some koji fungi are indistinguishable from *A. flavus* but are nontoxigenic. It seems that the selection of strains producing increased levels of secreted hydrolytic enzymes has selected strains that do not produce aflatoxins.

Aflatoxins are susceptible to both microbial and mammalian metabolism. Indeed, it is a consequence of metabolism in the mammalian liver that the aflatoxins are toxic (Figure 5–2). The wide range in both acute and carcinogenic toxicity (Table 5–2) in different animal species results from differences in the metabolic activities of these animals. Carcinogenicity is associated with the formation of aflatoxin epoxide and its subsequent reaction with guanine residues in DNA, whereas the acute toxicity requires the hydroxylation of this epoxide to form dihydroxyaflatoxin, which can react with the lysine residues of proteins. The possibility that microorganisms could be used to degrade aflatoxins, and hence detoxify contaminated foods, has been reviewed.[5] One of the earliest reports of the successful removal of aflatoxins by microorganisms was that of Lillehoj and his colleagues at the Northern Regional Research Laboratories in Peoria, Illinois.[37,38] Arising from a wide screen of both eukaryotic and prokaryotic microorganisms, the bacterium *Flavobacterium aurantiacum* (NRRL B-184) was the most effective, but the phenomenon has not been converted into a practical detoxification process.

When cows are fed on feed contaminated with aflatoxin B_1, they secrete a proportion of the contaminant in their milk as the metabolite, aflatoxin M_1. This compound is less toxic than its precursor, but, because it is uniformly distributed in a liquid food such as milk, and very young and elderly people may be exposed to a significant extent, the European Commission (EC) has set a particularly stringent maximum permissible level for aflatoxin M_1 in milk and dairy products of 0.05 μg/kg. The comparable level for aflatoxin B_1 in groundnuts, nuts, dried fruit, and cereals for direct human consumption is 2 μg/kg (EC No 1525/98), although less stringent levels are set in other parts of the world.[8] Unfortunately, there is no consistent evidence that aflatoxin M_1 is removed during any of the most commonly used processes in the milk industry, such as heat treatment, cold storage, or spray drying.[79] Neither is aflatoxin completely degraded during the fermentation processes used in the manufacture of cheese, cream, or butter, although it may be partitioned between the different components of each process. Thus,

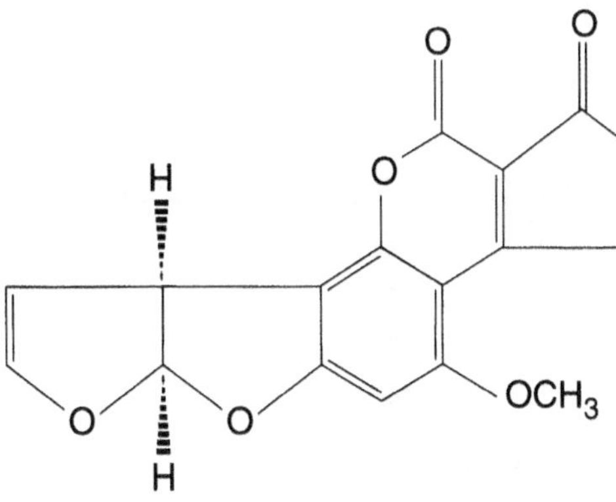

Figure 5–1 Aflatoxin B_1.

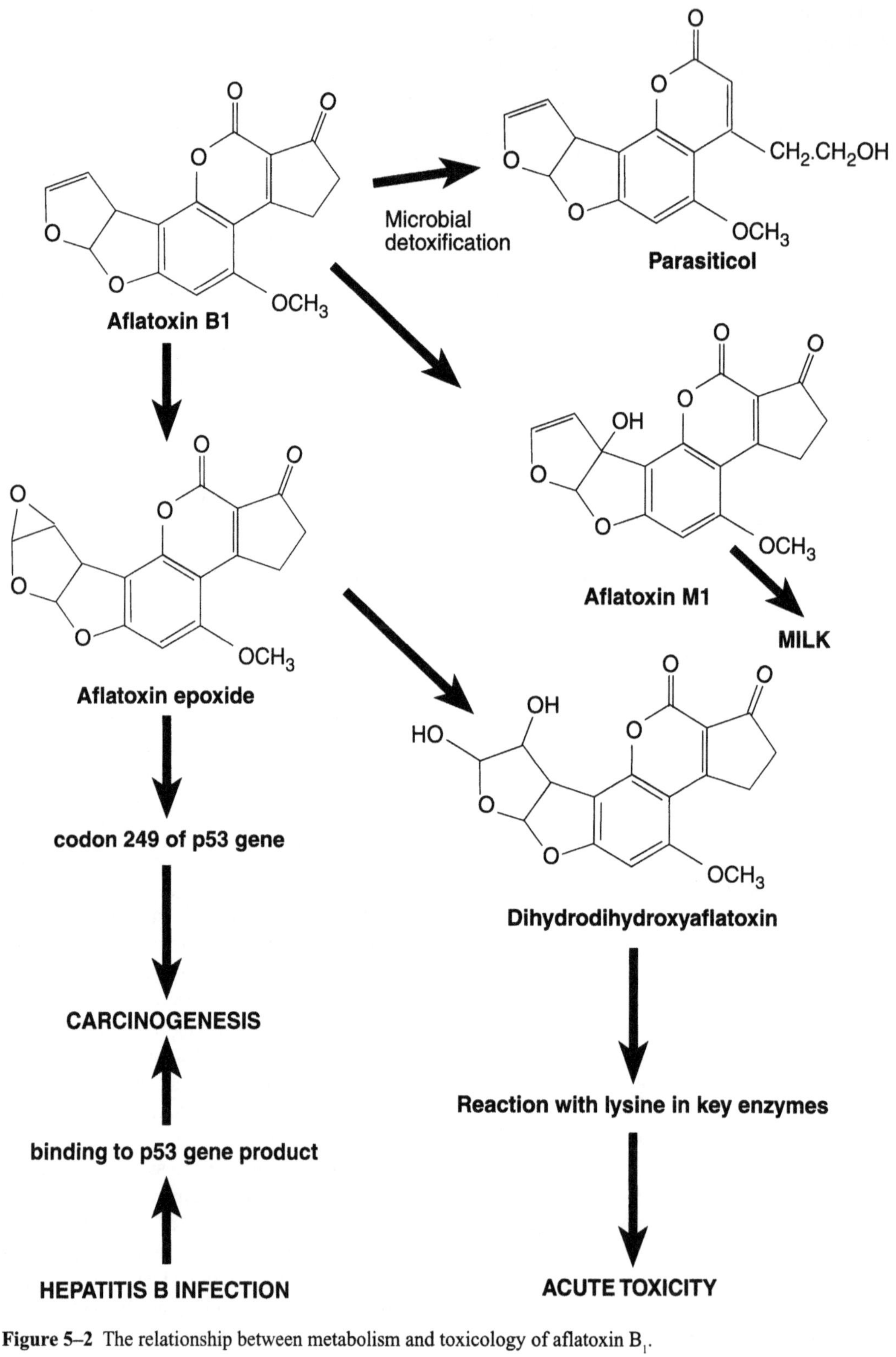

Figure 5–2 The relationship between metabolism and toxicology of aflatoxin B_1.

Table 5–2 Oral Acute LD_{50} Values and TD_{50} Values for Carcinogenesis of Aflatoxin B_1

Animal Species	*LD_{50} (mg/kg body wt)*	*TD_{50} (μg/kg body wt/day)*
Rabbit	0.3	–
Cat	0.6	–
Dog	0.5–1.0	–
Pig	0.6	–
Baboon	2.0	–
Rat (male)	5.5	1.3–5.8
Rat (female)	17.9	6.9–12.5
Macaque monkey	7.8	–
Rhesus monkey		156
Cynamolgus monkey		848
Mouse	9.0	>5,300
Hamster	10.2	–
Humans	5.0*	132†

*Based on epidemiological evidence from cases of acute poisoning[31]
†Based on an analysis of intake versus incidence data[78]

cheese may retain as much as 60% of the aflatoxin M_1 in the milk that is used in its manufacture, whereas cream may retain only approximately 10%, and butter as little as 2%.

Aflatoxin B_1 may be detoxified during the fermentation of milk because *Lactococcus lactis* (previously known as *Streptococcus lactis*) is able to convert it into aflatoxin B_{2a} and aflatoxicol (Figure 5–3).[40,41] In fact, the formation of aflatoxin B_{2a} may simply be the result of the reduced pH; there is the possibility of it being reconverted to the parent compound. However, it has been claimed that the formation of aflatoxin B_{2a} is not simply a result of acid-catalyzed hydration of aflatoxin B_1, and also that viability is not a prerequisite for the removal of aflatoxin B_1 by probiotic strains of *Lactobacillus rhamnosus*.[20,21] There have been searches for other microbial systems for the removal of aflatoxins from foods and a possibility actively being investigated is the use of enzymes from a nontoxic, edible species of fungus, *Armillaria tabescens*, which is valued already in China for its medicinal properties in alleviating a number of disorders.[39]

An alternative strategy to detoxifying aflatoxin directly in foods is the possibility of removal from the gastrointestinal tract by probiotic bacteria. Such a study has been carried out using probiotic strains of *Lactobacillus* and *Propionibacterium*.[2] The results from these investigations suggest that such probiotic bacteria have a role in reducing the bioavailability of foodborne carcinogens such as aflatoxin B_1. However, the most effective strategies for limiting human exposure to aflatoxins are to avoid contamination in the first place or a chemical process, such as ammoniation, to degrade aflatoxin irreversibly in animal feeds.[63]

OCHRATOXIN A

Ochratoxin A (Figure 5–4) is produced by *P. verrucosum* in temperate climates and by a number of *Aspergillus* species, especially *A. ochraceus*, in warmer parts of the world.[48] Ochratoxin A is most common in cereals of temperate countries, such as barley, oats, rye, and wheat,[30,45] but has also been found in maize,[32] coffee, cocoa, dried vine fruits, wine,[59] and beer.[65] Although ochratoxin A can survive fermentation processes, it normally does not survive the malting process used in the production of beer.[33] Its presence in beer at very low concentrations (ca 0.2 ng/ml) may be due to its pres-

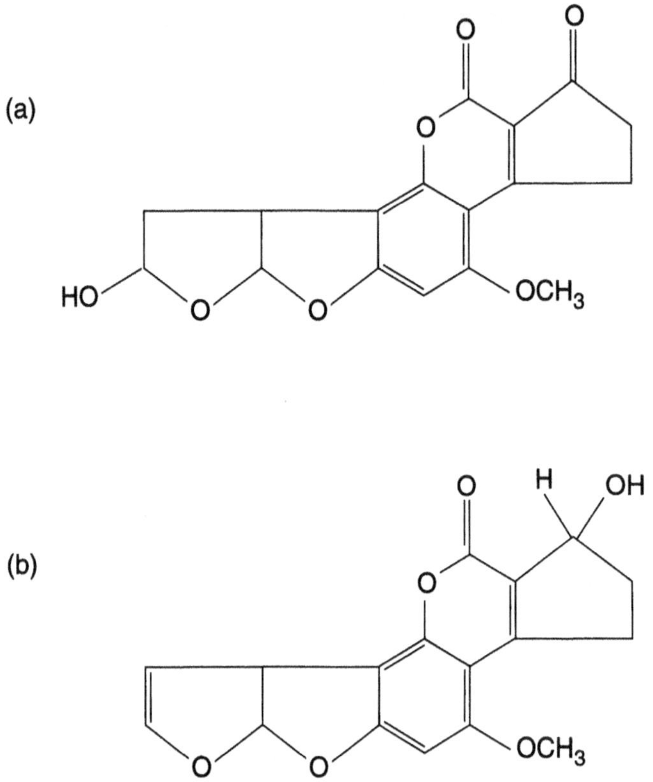

Figure 5–3 (a) Aflatoxin B_{2a} (b) Aflatoxicol.

ence in adjuncts that are used in commercial beer production. A survey of the occurrence of ochratoxins in a range of wines from the Swiss retail market showed them to be present at very low concentrations and to be more frequent, and at higher concentrations, in red wines from more southerly regions of Europe.[82] It is probable that contamination precedes the fermentation stage. These results confirm that ochratoxins are not removed by an alcoholic fermentation.

Ochratoxin A is relatively thermostable, having a half life at 100 °C of more than 10 hours in

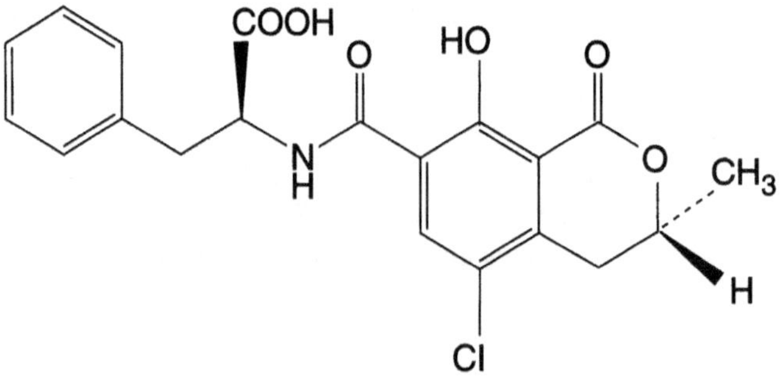

Figure 5–4 Ochratoxin A.

dry wheat and nearly 2.5 hours in moist wheat.[7] Several studies have confirmed that ochratoxin A will survive in most food processes involving a heating stage. Like aflatoxin, ochratoxin A can also pass through the food chain and may be found in meat products, especially of the pig,[32] but it does not seem to be secreted effectively into cow's milk.[75] However, surveys in Scandinavian countries have shown the occurrence of ochratoxin A in cow's milk in Sweden[9] and Norway.[72] Because of the low biotransfer of ochratoxin A from animal feeds to milk, Skaug[72] speculated that the inhalation of contaminated airborne particles may be the route into cow's milk. If ochratoxin is not normally present in milk, it would not be expected to occur in dairy products. However, some cheeses are readily contaminated with molds, and some of the molds isolated from cheeses have been shown to produce ochratoxin in the laboratory. Scott,[64] in his detailed review of the occurrence of mycotoxins in dairy products, concluded that "cheese is generally a good substrate for fungal growth but a poor substrate for experimental mycotoxin production."[(p221)] There are a few reports of the natural occurrence of ochratoxin A in moldy cheese, and these are referenced in this review, but the molds used deliberately for the production of mold-ripened cheeses (i.e., *P. roqueforti* and *P. camemberti*) do not produce this mycotoxin.

The transfer of ochratoxin from animal feeds to animal tissues, such as muscle, liver, and kidneys, combined with the extended residence time for ochratoxin A in animal tissues, leads to the possibility of its presence in meat products. There are many reports of the occurrence of ochratoxin A in kidneys, liver, and even sausages, and these have been documented extensively.[34] The contamination of meat products by transfer from animal feeds should be distinguished from the occurrence of ochratoxin A in moldy meat products such as smoked pork, other smoked meats, and sausages, in which much higher levels of ochratoxin A can be found.[34]

At the acute level, ochratoxin A is a nephrotoxin, which is certainly responsible for most cases of porcine nephropathy and has been suspected to be an etiological agent in Balkan endemic nephropathy. A detailed risk assessment of this mycotoxin has been carried out[34] and it seems prudent to assume that it is also carcinogenic.[17] For this reason, and because there is no doubt regarding human exposure to ochratoxin A from a range of foods, the member countries of the European Union are presently seeking to agree to maximum levels in foods for human consumption; these are likely to include maximum acceptable levels in fermented beverages such as beer and wine.

PATULIN

Patulin (Figure 5–5) is produced by a number of species of *Penicillium*, *Aspergillus,* and *Byssochlamys,* but, in the context of human foods, the most important species is *P. expansum*. This mold is associated especially with a soft rot of apples, but may also occur on a wide range of other fruits. *P. expansum* is able to form a rapid brown soft rot in apples, once infection has been established, and it becomes immediately recognizable once the conidiophores bearing large numbers of blue green spores form as pustules on the surface of the rot. Although infection is usually through a wound and the rot is usually very evident, some apple varieties can be infected from within, and, although the apple looks superficially sound, the core may be infected and contaminated with patulin. It is unlikely that fresh fruit will be a hazard because as soon as mold contamination is apparent, it will normally be discarded because of the obvious visual and taste defects. However, an organolep-

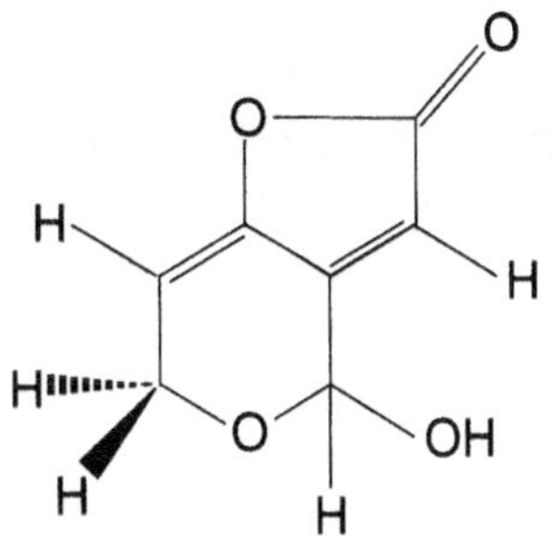

Figure 5–5 Patulin.

tically acceptable fruit juice can be expressed from fruit containing some rot, and patulin is stable at the low pH values of most fruit juices.

The natural occurrence of patulin in commercial apple juice was reported as long ago as 1972.[69] Surveys since then have demonstrated that the contamination of fruit juices with patulin is a continuing nuisance.[10,44,61] Patulin was discovered originally as a potentially useful broad spectrum antibiotic, but the acute toxicity precluded its use. The report of Dickens & Jones in 1961[19] suggested that, at high enough doses, patulin could induce sarcomas in experimental animals at the site of injection and hence may be carcinogenic. However, several studies since then have failed to provide conclusive evidence that patulin is carcinogenic,[26] but the United Kingdom has set an advisory maximum level of 50 µg/l in apple juice for human consumption, and there is not yet any statutory limit for patulin in the European Community. A report of the U.K. Ministry of Agriculture, Fisheries and Food[44] demonstrated that a few samples of fresh apple juice taken from retail outlets in the United Kingdom were contaminated with patulin at levels that were above the advisory limit. A detailed assessment of this report and others, as well as the analytical difficulties associated with the determination of low concentrations of patulin in apple juice, is available.[71]

It has been known for some time that patulin disappears during the fermentation of apple juice to cider using the yeast *Saccharomyces cerevisiae*.[23] The microbial decomposition of patulin requires the yeast to be viable; it is an inducible phenomenon, and occurs during fermentative metabolism rather than respiratory metabolism. More recently, it has been shown that patulin is metabolized to a number of products including ascladiol during a yeast fermentation.[71] If due diligence is paid to the quality of apple juice used in the manufacture of cider, it seems unlikely that patulin will be a problem in cider.

FUSARIUM TOXINS

The genus *Fusarium* contains several species that are important plant pathogens causing serious losses in agriculture and horticulture. They are thus of special importance in the field during the growth and development of a crop. Some species are also capable of continuing growth and metabolic activity postharvest, but they require higher water activities (A_w) than most species of *Penicillium* and *Aspergillus*. *Fusarium* is associated with a very wide range of secondary metabolites, many of which are toxic to farm animals and humans. The three groups that are particularly important in human foods are the polyketide-derived zearalenone, the sesquiterpenoid trichothecenes, and the recently discovered fumonisins. Bennett & Richard[4] reviewed the effects of processing on these toxins in contaminated grains and it is clear that they are stable during wet and dry milling and ethanol fermentations with the exception of deoxynivalenol, for which there is conflicting evidence concerning the effect of fermentation. Although these processes may not destroy the common fusarial toxins, they do influence their segregation among the various fractions of the process. Thus, for example, wet milling will produce toxin-free starch from maize, but the other products, used in animal feeds, will have higher levels of zearalenone than the starting material.

Zearalenone

Zearalenone (Figure 5–6) has little or no acute toxicity but has potent estrogenic activity and is responsible for a serious disorder in pigs known as vulvovaginitis. Zearalenone is produced by several species of *Fusarium*, but the most important are *F. graminearum* and *F. culmorum*, both of which may be responsible for head rot (often characterized by a pink or red discoloration of the de-

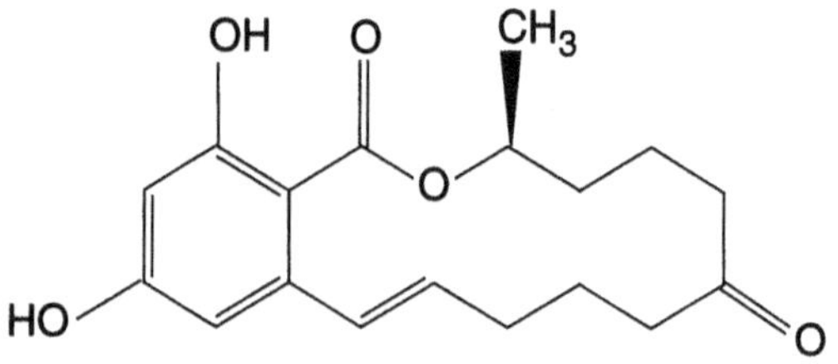

Figure 5–6 Zearalenone.

veloping grain) in cereals such as maize, wheat, and barley. The ecology of *Fusarium* in cereals has been reviewed in detail by Chelkowski,[13] and there are many reports of the occurrence of zearalenone, with or without other *Fusarium* toxins, in cereals.[25, 59, 73] Zearalenone can be formed in the field (up to 21 mg/kg),[55] but the largest concentrations are usually associated with poor storage postharvest. Concentrations as high as 2.9 g/kg have been reported.[46] A survey of Canadian beers for *Fusarium* mycotoxins failed to demonstrate the occurrence of zearalenone,[66] and it seems unlikely that this metabolite will occur at significant levels in fermented milk products.

Trichothecenes

The trichothecenes form a very large family of sesquiterpene metabolites, the most important of which, in the context of foods, are produced by species of *Fusarium*. The terrible outbreaks of alimentary toxic aleukia in humans, and of hemorrhagic moldy corn toxicosis in farm animals, are especially associated with T-2 toxin (Figure 5–7a), which is produced primarily by *F. sporotrichioides* and *F. poae*. This is one of a group of very toxic trichothecenes that fortunately are relatively rare in crops grown for human consumption. The conditions leading to outbreaks of alimentary toxic aleukia are hopefully avoidable (i.e., widespread famine following a war). A much more common trichothecene is deoxynivalenol (Figure 5–7b), which is formed by *F. graminearum*, *F. culmorum*, and related species. Indeed, there are some years when samples from crops such as barley may show 100% incidence of deoxynivalenol (DON) because these species of *Fusarium* are patho-

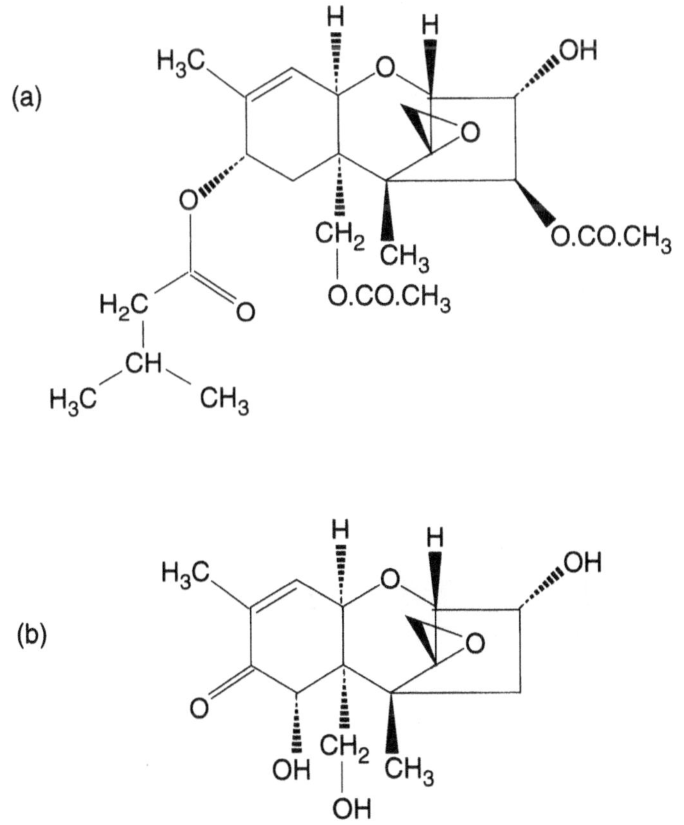

Figure 5–7 (a) T-2 toxin (b) Deoxynivalenol.

gens that may establish themselves in the field, causing red scab diseases, and continue their activity postharvest.

During the period 1996–1998, DON was widespread in cereals throughout the temperate parts of the world, with incidences from 43% to 100% and concentrations from 2 μg/kg to 62,050 μg/kg.[59] There is a clear account of the type of weather conditions that occurred during the growing season for maize in Maryland and Delaware during 1994, which led to DON contamination prior to harvest.[77] Although not as toxic as T-2 toxin, DON, like all the trichothecenes, is immunosuppressive and biologically active at concentrations very much less than the LD_{50} of 50–70 mg/kg. DON is a stable molecule and may pass into fermented cereal products, but it does not appear to be a problem in fermented dairy products. It would be surprising if DON were not found in beers and, indeed, it has been reported that 29 of 50 samples of Canadian and imported beer analyzed contained DON.[66] Although the majority of beers had very low concentrations of DON, nine had more than 5 ng/ml, and a single sample had as much as 50.3 ng/ml.

Fumonisins

The fumonisins are produced by *F. moniliforme* and related fusaria that do not produce trichothecenes. Fumonisin B_1 (Figure 5–8) is known to cause equine encephalomalacia, pulmonary edema in pigs, and a number of other illnesses in a range of animal species. It may be associated with esophageal carcinoma in humans[62] and has recently been the subject of a monograph.[28] An assessment of human exposure to the fumonisins for people in the Netherlands with a special emphasis on those people with gluten intolerance who would be especially at risk has been carried out.[18] The fumonisins are associated with maize and maize products, reflecting the host specificity of the molds producing it. In those commodities, it is remarkably widespread and can occur in relatively high concentrations. In his extensive review, Pittet[59] provided documentation of surveys in many countries during the period 1995–1998 with incidence of occurrence ranging from 20% to 100% and concentrations from 10 μg/kg to 37,650 μg/kg. These surveys included maize products such as polenta, corn flakes, and popcorn.

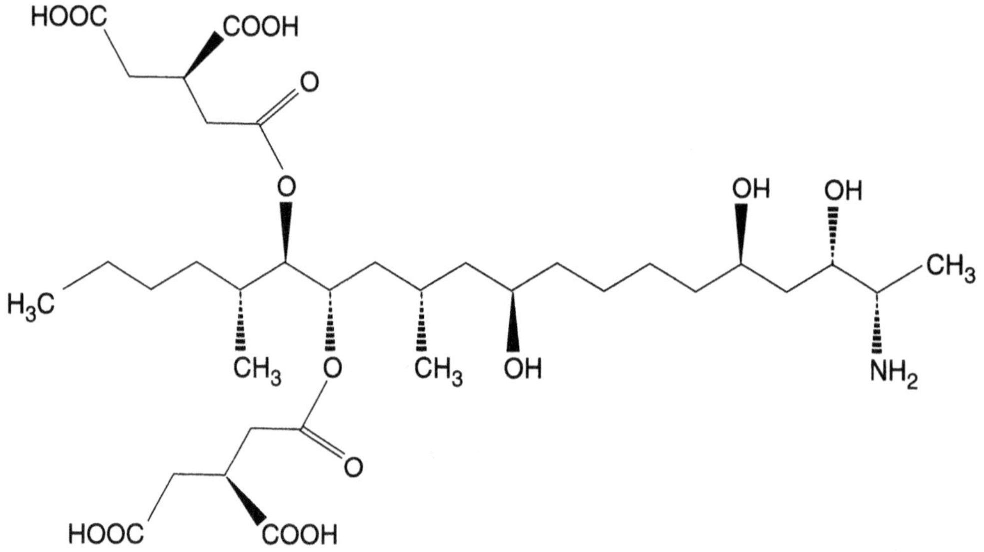

Figure 5–8 Fumonisin B_1.

As with DON, it would seem likely that fumonisins will be found in beers in which maize products are used at some stage during the manufacture, but there is no evidence of transmission from animal feeds to milk and hence to milk products. By using immunoaffinity columns for extraction and cleanup, it has been shown that 86% of samples of beer that was purchased in retail outlets in Lincoln, Nebraska, contained fumonisin B_1.[24] The Lincoln study followed an earlier report of fumonisin contamination of beer in Canada that had included beers that were imported from the United States.[67,70] These authors agreed that the most likely source of fumonisin in beer is the maize grits that are used as a brewing adjunct.

The fumonisins are relatively stable to elevated temperatures and survive a range of cooking, baking, and frying processes. Thus, after baking corn muffins spiked with fumonisin B_1 for 20 minutes at 175 or 200 °C, as much as 84% and 72%, respectively, survived.[29] A significant reduction when spiked corn masa was fried at 140–170 °C for up to six minutes was also found. Even frying contaminated chips for 15 minutes at 190 °C only reduced the fumonisin level by 67%. However, it has also been reported that, during extrusion cooking of corn grits, the reduction of fumonisin B_1 depended on several factors, including moisture content and whether or not mixing screws were used.[12] One of the problems in assessing the significance of the breakdown of the fumonisins is that they are esters of an aminopentol compound, and some food processes lead to the hydrolysis of the ester groups to the parent compound, which is still toxic.

The preparation of tortillas in Central and South America requires the treatment of maize with lime to produce nixtamal before cooking. Scott & Lawrence[68] validated the methodology for measuring the aminopentol formed from fumonisin B_1 in calcium hydroxide-processed foods such as tortilla and nacho chips, taco shells, and air-dried corn tortillas. They found significant levels of the aminopentol, but always at lower levels than the fumonisin B_1 from which it was derived. This implies that the latter had partly survived the alkaline process. Tortillas from villages in Guatemala may have as much as 185 mg kg^{-1} of AP_1 (i.e., the aminopentol derived from the hydrolysis of fumonisin B_1) and also still contain up to10 mg kg^{-1} of fumonisin B_1 itself.[42] AP_1 may not always be formed during cooking and food processing of contaminated maize, even if fumonisin levels are reduced.[57]

ALTERNARIA TOXINS

Species of *Alternaria* cause dry and soft black rots of a range of commercially important crops and are often associated with the postharvest spoilage of fruits and vegetables.[58] The most common species is *A. alternata,* which can produce a number of toxic metabolites, the most important of which is tenuazonic acid (Figure 5–9). Other toxins include alternariol and its monomethyl ether, altenuene, and the altertoxins. The analysis of such a complex mixture requires a multitoxin method.[51] One or more of these toxins have been isolated from a number of cereals, oilseed rape, tomatoes, sunflower seed, and olives. There is the potential for these toxins to occur in the products of cereals, oilseeds such as sunflower, fruit juices, and tomato products, but there is no information regarding the effects of food processing on the survival of *Alternaria* toxins into fermented foods. Although isolates of *Alternaria* have been reported as part of the mold flora of some cheeses,[64] it is unlikely that *Alternaria* toxins will have any significance for human health from fermented foods. A detailed account of *Alternaria* species and their toxins is given by Bottalico & Logrieco.[6]

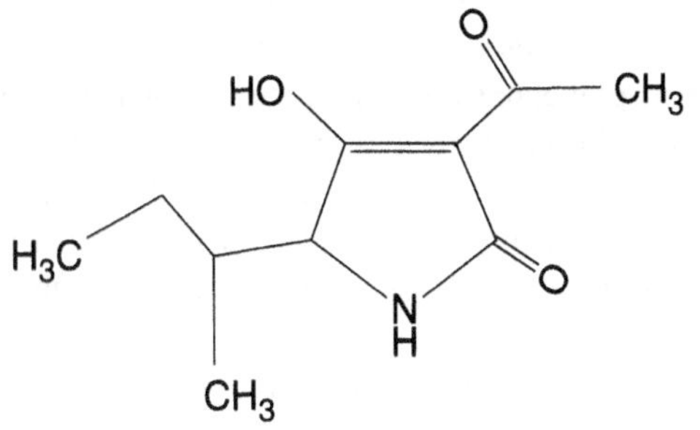

Figure 5–9 Tenuazonic acid.

BACTERIAL AND ALGAL TOXINS

The conditions for the growth of most food poisoning bacteria are generally well understood, and the control of growth can be effected by heat treatment, refrigeration, reduced A_w and low or high pH, alone or in combination. One or more of these parameters can also be combined with the presence of both synthetic or natural growth inhibitors. However, once bacterial toxins have been formed and released into the food, they are generally resistant to the conditions that inhibit or kill the producing organism.

Botulinum toxins are relatively large polypeptides (i.e., molecular weights of approximately 150 kDa) that are usually complexed with proteins as a progenitor toxin when they are released into the food. Such complexes are more stable at elevated temperatures and low pH than the pure neurotoxins. Of 14 outbreaks of botulism associated with dairy products between 1912 and 1997, the majority involved cheese or cheese spreads.[3,14,60] An especially serious outbreak in the United Kingdom during 1989 involving 27 cases and one death was a result of the consumption of hazelnut yogurt that was contaminated with type B neurotoxin produced by a proteolytic strain of *C. botulinum*.[54] The problem was not with the yogurt itself, which has a pH too low for growth and toxin production, but with the canned hazelnut conserve that was added to it. This was occasioned by a change in production from a product with a high sucrose content receiving a mild heat treatment to a product in which sucrose was replaced with aspartame, in response to a perceived public demand for low sugar products. The heat treatment remained unchanged and was insufficient to kill *C. botulinum* spores, but the high sucrose content prevented growth in the original formulation. The elevated A_w of the new formulation allowed growth and toxin formation in the hazelnut conserve and, although the pH of the yogurt to which it was added prevented further growth of the organism, it did not inactivate the toxin that had already been produced.

The enterotoxins of *S. aureus* are small single polypeptide chain proteins (i.e., molecular weights 26 to 30 kDa) that are more heat resistant than the producing organism. Although fermentation processes would normally prevent growth of this organism, there have been instances of enterotoxin production in cheese when the starter culture failed to grow fast enough,[80] and in salami sausage when a change in production conditions again allowed *S. aureus* to outcompete the organisms used for the fermentation.[53,74]

The emetic toxin of *B. cereus* known as cereulide is a cyclic dodecadepsipeptide[1] that is stable to proteolytic enzymes and remarkably heat stable (i.e., no loss of activity after 90 minutes at 121 °C). Cereulide is usually produced in cooked rice, pasta, and noodles, and is very unlikely to occur in fermented foods.

Bongkrek Poisoning

An Indonesian food known as tempeh bongkrek is made by inoculating coconut presscake with the mold *Rhizopus oligosporus*. Contamination of this food by the pseudomonad *B. cocovenenans*, previously known as *P. cocovenenans*,[81] unfortunately is not uncommon and has been responsible for quite a number of deaths. *B. cocovenenans* produces two toxic metabolites, toxoflavin (Figure 5–10) and bongkrekic acid (Figure 5–11), which has been shown to be a heptenetrioic acid.[16] Normally, the rapid growth of *R. oligosporus* over and into the solid substrate effectively inhibits bacterial growth, although bonkrekic acid formation is not necessarily directly correlated with bacterial numbers.

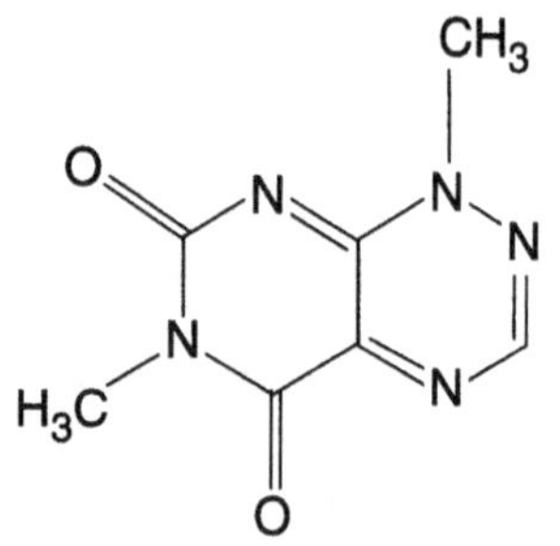

Figure 5–10 Toxoflavin.

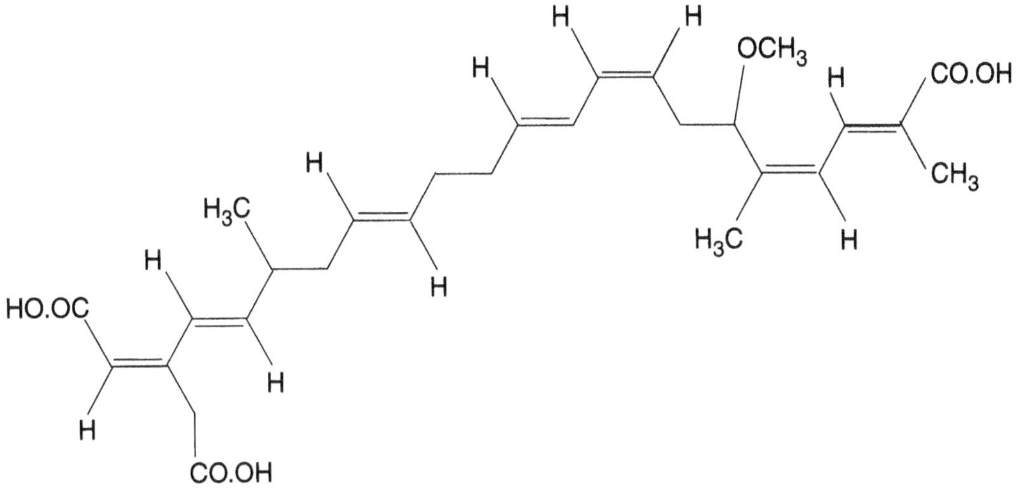

Figure 5–11 Bongkrekic acid.

Outbreaks of bonkrek poisoning seem to occur most often when coconut presscake is mixed with excessive amounts of coconut milk. The presence of both an aqueous environment and elevated levels of lipids seems to encourage bacterial growth and bongkrekic acid formation. It seems that temperature, pH, moisture, and salt concentration themselves do not fully explain why some tempeh bonkrek becomes toxic and another factor, such as the presence of appropriate lipids, may be implicated. In a useful review of the conditions influencing the formation of these toxins, it was specifically demonstrated that fatty acids have an important role in the formation of bongkrekic acid.[22] Although comparable in acute toxicity, bongkrekic acid is usually present at much higher concentrations than toxoflavin and is most likely to be of greatest significance in causing illness. A few hours after consumption of toxic tempeh bongkrek, people complain of malaise, abdominal pains, dizziness, sweating, and fatigue. Coma and death can follow within 20 hours of the onset of symptoms. The edible jelly fungus (*Tremella fuciformis*), which is cultivated in several countries in southeast Asia, is contaminated frequently with *B. cocovenenans*, and has also been implicated in bongkrek poisoning.

Cyanobacterial Toxins

A number of species of freshwater cyanobacteria belonging to the genera *Microcystis*, *Anabaena,* and *Aphanizomenon* can form extensive blooms in standing water and may cause deaths of animals drinking the contaminated water.[11] Cyanoginosin (Figure 5–12), a toxic metabolite of *Microcystis aeruginosa*, is an hepatotoxin. It is unlikely that these toxins will find their way into fermented foods, although there may be some concern that they may contaminate samples of *Spirulina* that are collected from the wild and used as food and health products.

Algal Toxins

Among the eukaryotic algae, it is the dinoflagellates and a small group of diatoms that cause concern because of their ability to produce very potent toxins. A number of toxic responses to contamination of seafoods occur, of which paralytic shellfish poisoning, diarrheal shellfish poisoning, and neurotoxic shellfish poisoning are all associated with marine dinoflagellates, and amnesic shellfish poisoning with species of marine diatom. Table 5–3 lists a selection of the

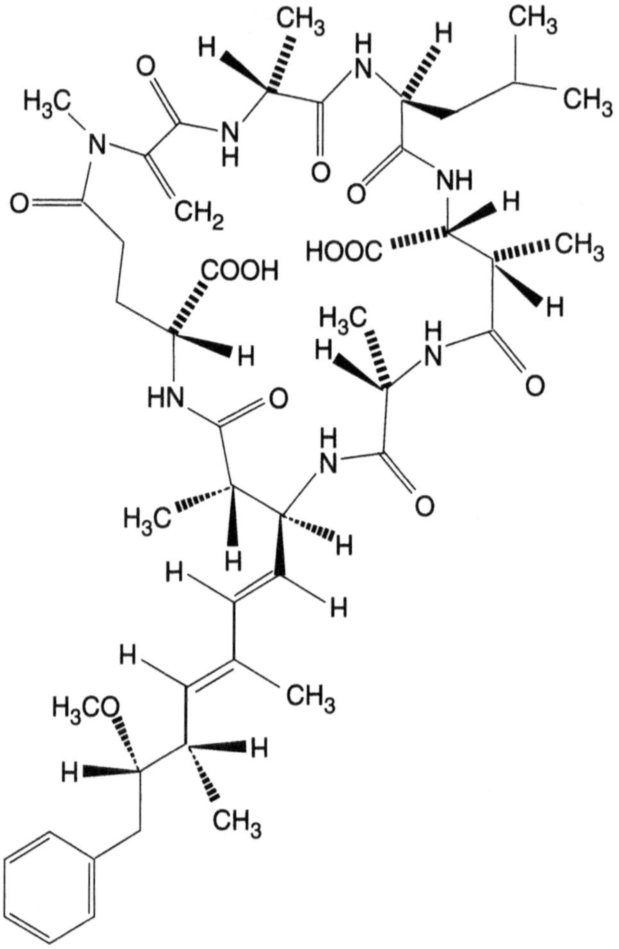

Figure 5–12 Cyanoginosin.

Table 5–3 A Selection of Eukaryote Algal Toxicoses

Toxicosis	*Associated Species*	*Toxins*
	Dinoflagellates	
Paralytic shellfish poisoning (PSP)	*Alexandrium catanella*	Saxitoxin (Figure 5–13)
Diarrheal shellfish poisoning (DSP)	*Dinophysis fortii*	Okadaic acid (Figure 5–14)
Neurotoxic shellfish poisoning (NSP)	*Gymnodium breve*	Brevitoxin
Ciguatera poisoning	*Gambierdiscus toxicus*	Ciguatoxin
	Diatoms	
Amnesic shellfish poisoning (ASP)	*Pseudonitzschia pungens*	Domoic acid (Figure 5–15)

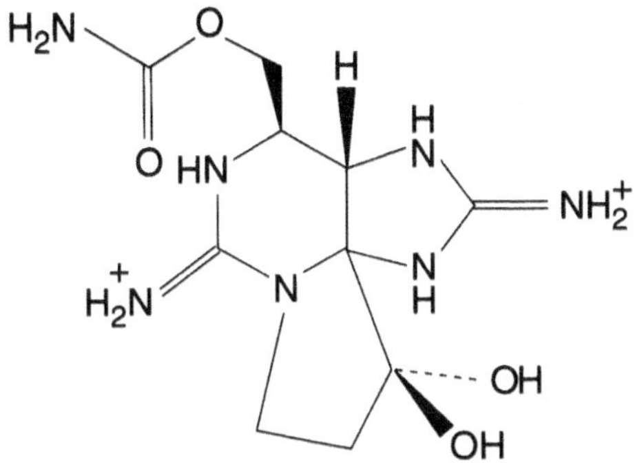

Figure 5–13 Saxitoxin.

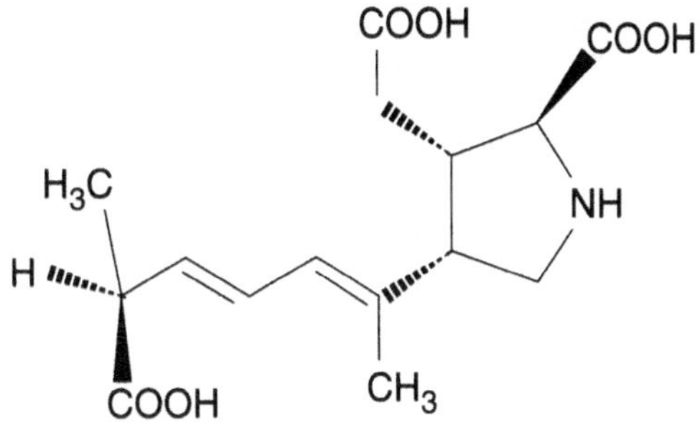

Figure 5–15 Domoic acid.

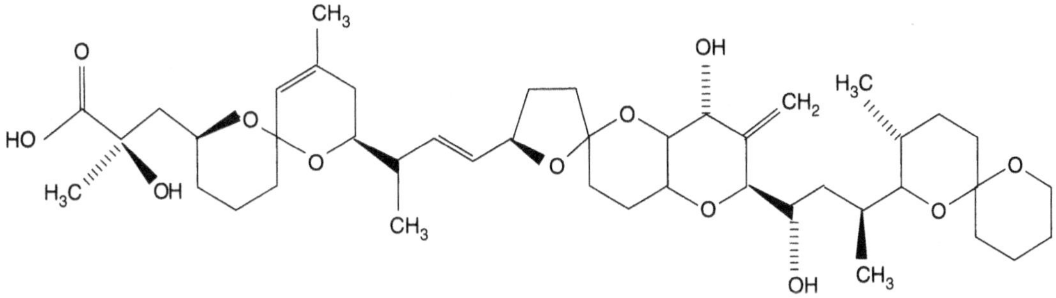

Figure 5–14 Okadaic acid.

species and toxins involved[27] (Figure 5–13, 5–14, and 5–15). These toxins are generally resistant to the temperatures involved in cooking and are likely to be unaffected by the salting and acidification that is associated with fermented fish products. In any case, it seems unlikely that the species of shellfish and fish involved will be used in the production of any fermented foods.

REFERENCES

1. Agata, N., Ohta, M., Mori, M. & Isobe, M. (1995). A novel dodecadepsipeptide, cereulide, is an emetic toxin of *Bacillus cereus*. *FEMS Microbiol Lett* 129, 17–20.
2. Ahokas, J., El-Nezami, H., Kankaanpää, P., Mykkänen, H. & Salminen, S. (1998). A pilot clinical study examining the ability of a mixture of *Lactobacillus* and *Propionibacterium* to remove aflatoxin from the gastrointestinal tract of healthy Egyptian volunteers. *Revue Méd Vét* 149, 568.
3. Aureli, P., Franciosa, G. & Pourshaban, M. (1996). Foodborne botulism in Italy. *Lancet* 348, 1594.
4. Bennett, G. A. & Richard, J. L. (1996, May). Influence of processing on *Fusarium* mycotoxins in contaminated grains. *Food Tech*, 235–238.
5. Bhatnagar, D., Lillehoj, E. B. & Bennett, J. (1991). Biological detoxification of mycotoxins. In *Mycotoxins and Animal Foods*, pp. 815–826. Edited by J. E. Smith & R. S. Henderson. Boca Raton, FL: CRC Press.
6. Bottalico, A. & Logrieco, A. (1998). Toxigenic *Alternaria* species of economic importance. In *Mycotoxins in Agriculture and Food Safety*, pp. 65–108. Edited by K. K. Sinha & D. Bhatnagar. New York: Marcel Dekker.
7. Boudra, H., Le Bars, P. & Le Bars, J. (1995). Thermostability of ochratoxin A in wheat under two moisture conditions. *Appl and Environ Microbiol* 61, 1156–1158.
8. Boutrif, E. & Canet, C. (1998). Mycotoxin prevention and control: FAO programmes. *Revue Méd Vét* 149, 681–694.

9. Breitholtz-Emmanuelsson, A., Olsen, M., Oskarsson, A., Palminger, I. & Hult, K. (1993). Ochratoxin A in cow's milk and in human milk with corresponding human blood samples. *J AOAC Intl* 76, 842–846.
10. Burda, K. (1992). Incidence of patulin in apple, pear, and mixed fruit products marketed in New South Wales. *J Food Prot* 55, 796–798.
11. Carmichael, W. W. (1992). Cyanobacterial secondary metabolites: the cyanotoxins. *J Appl Bacteriol* 72, 445–459.
12. Castelo, M. M., Katta, S. K., Sumner, S. S., Hanna, M. A. & Bullerman, L. B. (1998). Extrusion cooking reduces recoverability of fumonisin B_1 from extruded corn grits. *J Food Sci* 63, 696–698.
13. Chelkowski, J. (1998). Distribution of *Fusarium* species and their mycotoxins in cereal grains. In *Mycotoxins in Agriculture and Food Safety*, pp. 45–64. Edited by K. K. Sinha & D. Bhatnagar. New York: Marcel Dekker.
14. Collins-Thompson, D. L. & Wood, D. S. (1993). Control in dairy products. In *Clostridium Botulinum. Ecology and Control in Foods*, pp. 261–277. Edited by A. H. W. Hauschild & K. L. Dodds. New York: Marcel Dekker.
15. Cruickshank, R. H. & Pitt, J. I. (1990). Isoenzyme patterns in *Aspergillus flavus* and closely related species. In *Modern Concepts in* Penicillium *and* Aspergillus *Classification*, pp. 259–265. Edited by R. A. Samson & J. I. Pitt. New York: Plenum.
16. De Bruijn, J., Frost, D. J., Nugteren, D. H., Gaudemer, A., Lijmbach, G. W. M., Cox, H. C. & Berends, W. (1973). The structure of bongkrekic acid. *Tetrahedron* 29, 1541–1547.
17. De Groene, E. M., Jahn, A., Horbach, G. J. & Fink-Gemmels, J. (1996). Mutagenicity and genotoxicity of the mycotoxin ochratoxin A. *Environ Toxicol and Pharmacol* 1, 21–26.
18. De Nijs, M., Van Egmond, H., Nauta, M., Rombouts, F. M. & Notermans, S. H. W. (1998). Assessment of human exposure to fumonisin B1. *J Food Prot* 61, 879–884.
19. Dickens, F. & Jones, H. E. H. (1961). Carcinogenic activity of a series of reactive lactones and related substances. *Br J Cancer* 15, 85–100.
20. El-Nezami, H., Kankaanpää, P., Salminen, S. & Ahokas, J. (1998). Ability of dairy strains of lactic acid bacteria to bind a common food carcinogen, aflatoxin B_1. *Food and Chem Toxicol* 36, 321–326.
21. El-Nezami, H., Kankaanpää, P., Salminen, S. & Ahokas, J. (1998). Physicochemical alterations enhance the ability of dairy strains of lactic acid bacteria to remove aflatoxin from contaminated media. *J Food Prot* 61, 466–468.
22. Garcia, R. A., Hotchkiss, J. H. & Steinkraus, K. H. (1999). The effect of lipids on bongkrekic (Bongkrek) acid toxin production by *Burkholderia cocovenenans* in coconut media. *Food Add and Contam* 16, 63–69.
23. Harwig, J., Scott, P. M., Kennedy, B. P. C. & Chen, Y.-K. (1973). Disappearance of patulin from apple juice fermented by *Saccharomyces* sp. *J Inst of Technol and Alimentation* 7, 295–312.
24. Hlywka, J. J. & Bullerman, L. B. (1999). Occurrence of fumonisins B_1 and B_2 in beer. *Food Add and Contam* 16, 319–324.
25. Hussein, H. M., Franich, R. A., Baxter, M. & Andrew, I. G. (1989). Naturally occurring *Fusarium* toxins in New Zealand maize. *Food Add and Contam* 6, 49–58.
26. International Agency for Research on Cancer. (1986). Patulin. In *IARC Monographs on the Evaluation of Carcinogenic Risks to Humans. Some Naturally Occurring and Synthetic Food Components, Furocoumarins and Ultra Violet Reductions.* (Lyon, France: IARC) 40, 83–98.
27. International Commission for Microbiological Specifications in Foods. (1996). Seafood toxins of microbiological origin. In *Microorganisms in Foods*, pp. 266–279. London: Blackie Academic and Professional.
28. Jackson, L. S., DeVries, J. W. & Bullerman, L. B., eds. (1996). *Fumonisins in Food.* New York: Plenum Press.
29. Jackson, L. S., Katta, S. K., Fingerhut, D. D., DeVries, J. W. & Bullerman, L. B. (1997). Effects of baking and frying on the fumonisin B_1 content of corn-based foods. *J Agric and Food Chem* 45, 4800–4805.
30. Jørgensen, K., Rasmussen, G. & Thorup, I. (1996). Ochratoxin A in Danish cereals 1986–1992 and daily intake by the Danish population. *Food Add and Contam* 13, 95–104.
31. Krishnamachari, K. A. V. R., Bhat, R. V., Nagarajon, V. & Tilek, T. B. G. (1975). Hepatitis due to aflatoxicosis. *Lancet* i, 1061–1063.
32. Krogh, P. (1987). Ochratoxins in food. In *Mycotoxins in Food*, pp. 97–121. Edited by P. Krogh. London: Academic Press.
33. Krogh, P., Hald, B., Gjertsen, P. & Myken, F. (1974). Fate of ochratoxin A and citrinin during malting and brewing experiments. *Appl Microbiol* 28, 31–34.
34. Kuiper-Goodman, T. & Scott, P. M. (1989). Risk assessment of the mycotoxin ochratoxin A. *Biomed and Environ Sci* 2, 179–248.
35. Le Bars, J. (1979). Cyclopiazonic acid production by *Penicillium camemberti* Thom.,and natural occurrence of this toxin in cheese. *Appl and Environ Microbiol* 36, 1052–1055.
36. Le Bars, J. (1990). Detection and occurrence of cyclopiazonic acid in cheeses. *J Environ Pathol, Toxicol and Oncol* 10, 136–137.
37. Lillehoj, E. B., Ciegler, A. & Hall, H. H. (1967). Aflatoxin B_1 uptake by *Flavobacterium aurantiacum* and resulting toxic effects. *J Bacteriol* 93, 464–471.
38. Lillehoj, E. B., Stubblefield, R. D., Shanon, G. M. & Shotwell, O. L. (1971). Aflatoxin M_1 removal from aqueous solutions by *Flavobacterium aurantiacum.* *Mycopath Mycol Appl* 45, 259–266.

39. Liu, D. -L., Yao, D. -S., Liang, R., Ma, L., Cheng, W. -Q. & Gu, L. -Q. (1998). Detoxification of aflatoxin B_1 by enzymes isolated from *Armillariella tabescens*. *Food and Chem Toxicol* 36, 563–574.

40. Megalla, S. E. & Hafez, A. H. (1982). Detoxification of aflatoxin B_1 by acidogenous yoghurt. *Mycopathologia* 77, 89–91.

41. Megall, S. E. & Mohran, M. A. (1984). Fate of aflatoxin B_1 in fermented dairy products. *Mycopathologia* 88, 27–29.

42. Meredith, F. I., Torres, O. R., De Tejada, S. S., Riley, R. T. & Merrill, A. H. (1999). Fumonisin B_1 and hydrolyzed fumonisin B_1 (AP_1) in tortillas and nixtamalized corn (*Zea mays* L.) from two different geographical locations in Guatemala. *J Food Prot* 62, 1218–1222.

43. Metwally, M. M., El-Sayed, A. M. A., Mehriz, A. M. & Abu Sree, Y. H. (1997). Sterigmatocystin: incidence, fate and production by *A. versicolor* in Ras cheese. *Mycotoxin Res* 13, 61–66.

44. Ministry of Agriculture, Fisheries and Food. (1993). *Mycotoxins: Third Report. The Thirty Sixth Report of the Steering Group on Chemical Aspects of Food Surveillance Sub-Group on Mycotoxins. Food Surveillance Paper* No. 36. London: HMSO.

45. Ministry of Agriculture, Fisheries and Food. (1995). Surveillance of UK cereals for ochratoxin A. *Ministry of Agriculture Fisheries and Food. Food Surveillance Information Sheet No. 48*. London: HMSO.

46. Mirocha, C. J. & Christensen, C. M. (1974). Oestrogenic mycotoxins synthesised by *Fusarium*. In *Mycotoxins*, pp. 129–148. Edited by I. F. H. Purchase. Amsterdam: Elsevier.

47. Moss, M. O. (1996). Mycotoxins. *Mycolog Res* 100, 513–523.

48. Moss, M. O. (1996). Mode of formation of ochratoxin A. *Food Add and Contam* 13 (Suppl.), 5–9.

49. Moss, M. O. (1998). Recent studies of mycotoxins. *J Appl Microbiol* 84 (Suppl.), 62S–76S.

50. Mühlemann, M., Lüthy, J. & Hübner, P. (1997). Mycotoxin contamination of food in Equador. A: aflatoxins. *Mitt Gebiete Lebensm Hyg* 88, 474–496.

51. Nawaz, S., Scudamore, K. A. & Rainbird, S. C. (1997). Mycotoxins in ingredients of animal feeding stuffs: I. determination of *Alternaria* mycotoxins in oilseed rape meal and sunflower seed meal. *Food Add and Contam* 14, 249–262.

52. Northolt, M. D., Van Egmond, H. P., Soentoro, P. S. & Deyll, W. E. (1980). Fungal growth and the presence of sterigmatocystin in hard cheese. *J Association of Official Analytical Chemists* 63, 115–119.

53. Nychas, G. J. E. & Arkoudelos, J. S. (1990). Staphylococci: their role in fermented sausages. *J Appl Bacteriol* 69 (Suppl.), 167S–188S.

54. O'Mahony, M., Mitchell, E., Gilbert, R. J., Hutchinson, D. N., Begg, N. T., Rodhouse, J. C. & Morris, J. E. (1990). An outbreak of foodborne botulism associated with contaminated hazelnut yoghurt. *Epidemiol and Infect* 104, 389–395.

55. Park, J. J., Smalley, E. B. & Chu, F. S. (1996). Natural occurrence of *Fusarium* mycotoxins in field samples from the 1992 Wisconsin corn crop. *Appl and Environ Microbiol* 62, 1642–1648.

56. Piñeiro, M. S., Dawson, R. & Costarrica, M. L. (1996). Monitoring programme for mycotoxin contamination in Uruguayan food and feeds. *Nat Tox* 4, 242–245.

57. Piñeiro, M. S., Miller, J., Silva, G. & Musser, S. (1999). Effect of commercial processing on fumonisin concentrations of maize-based foods. *Mycotoxin Res* 15, 2–12.

58. Pitt, J. I. & Hocking, A. D. (1997). *Fungi and Food Spoilage*. 2nd edn. London: Blackie Academic and Professional.

59. Pittet, A. (1998). Natural occurrence of mycotoxins in foods and feeds—an updated review. *Revue Méd Vét* 149, 479–492.

60. Pourshafie, M. R., Saifie, M., Shafiee, A., Vahdani, P., Aslani, M. & Salemian, J. (1998). An outbreak of foodborne botulism associated with contaminated locally made cheese in Iran. *Scandinavian J Infect Dis* 30, 92–94.

61. Preita, J., Moreno, M. A., Diaz, S., Suarez, G. & Dominguez, L. (1994). Survey of patulin in apple juice and childrens' apple food by the diphasic dialysis membrane procedure. *J Agric and Food Chem* 42, 1701–1703.

62. Rheeder, J. P., Marasas, W. F. O., Thiel, P. G., Sydenham, E. W., Shephard, G. S. & Van Schalkwyk, D. J. (1992). *Fusarium moniliforme* and fumonisins in corn in relation to human esophageal cancer in Transkei. *Phytopathology* 82, 353–357.

63. Riley, R. T. & Norred, W. P. (1999). Mycotoxin prevention and decontamination—a case study on maize. *Food, Nutri and Agric* 23, 25–32.

64. Scott, P. M. (1989). Mycotoxigenic fungal contaminants of cheese and other dairy products. In *Mycotoxins in Dairy Products*, pp. 193–259. Edited by H. P. Van Egmond. London: Elsevier Applied Science.

65. Scott, P. M. & Kanhere, S. R. (1995). Determination of ochratoxin A in beer. *Food Add and Contam* 12, 591–598.

66. Scott, P. M., Kanhere, S. R. & Weber, D. (1993). Analysis of Canadian and imported beers for *Fusarium* mycotoxins by gas chromatography: mass spectrometry. *Food Add and Contam* 10, 381–389.

67. Scott, P. M. & Lawrence, G. A. (1995). Analysis of beer for fumonisins. *J Food Prot* 58, 1379–1382.

68. Scott, P. M. & Lawrence, G. A. (1996). Determination of hydrolysed fumonisin B_1 in alkali processed corn foods. *Food Add and Contam* 13, 823–832.

69. Scott, P. M., Miles, M. F., Toft, P. & Dube, J. G. (1972). Occurrence of patulin in apple juice. *J Agric and Food Chem* 20, 450–451.

70. Scott, P. M., Yeung, J. M., Lawrence, G. A. & Prelusky, D. B. (1997). Evaluation of enzyme linked immunosorbent assay for analysis of beer for fumonisins. *Food Add and Contam* 14, 445–450.

71. Scudamore, K.A. (1999). *Mycotoxins, an Independent Assessment of MAFF-Funded Applied Research and Surveillance 1993–1996*. Report No. PB 4045. London: Ministry of Agriculture Fisheries and Food (MAFF).

72. Skaug, A. S. (1999). Analysis of Norwegian milk and infant formulas for ochratoxin A. *Food Add and Contam* 16, 75–78.

73. Sohn, H. -B., Seo, J. -A. & Lee, Y. -W. (1999). Co-occurrence of *Fusarium* mycotoxins in mouldy and healthy corn from Korea. *Food Add and Contam* 16, 153–158.

74. U.S. Department of Health, Education and Welfare. (1971). Staphylococcal gastroenteritis associated with salami. *MMWR* 20, 253–258.

75. Valenta, H. & Goll, M. (1996). Determination of ochratoxin A in regional samples of cow's milk from Germany. *Food Add and Contam* 13, 669–676.

76. Van Rensburg, S. J. (1984). Subacute toxicity of the mycotoxin cyclopiazonic acid. *Food and Chem Toxicol* 22, 993–998.

77. Wetter, M. T., Trucksess, M. W., Roach, J. A. & Bean, G. A. (1999). Occurrence and distribution of *Fusarium graminearum* and deoxynivalenol in sweet corn ears. *Food Add and Contam* 16, 119–124.

78. Wogan, G. N. (1992). Aflatoxins as risk factors for hepatocellular carcinoma in humans. *Cancer Res* 52 (Suppl.), 2114S–2118S.

79. Yousef, A. E. & Marth, E. H. (1989). Stability and degradation of aflatoxin M_1. In *Mycotoxins in Dairy Products*, pp. 127–161. Edited by H. P. Van Egmond. London: Elsevier Applied Science.

80. Zehren, V. L. & Zehren, V. F. (1968). Examination of large quantities of cheese for staphylococcal enterotoxin A. *J Dairy Sci* 51, 635–644.

81. Zhao, N. X., Qu, C. F., Wang, E. & Chen, W. X. (1995). Phylogenetic evidence for the transfer of *Pseudomonas cocovenenans* to the genus *Burkholderia* as *Burkholderia cocovenenans* comb. nov. *Intl J Sys Bacteriol* 45, 600–603.

82. Zimmerli, B. & Dick, R. (1996). Ochratoxin A in table wine and grape-juice: occurrence and risk assessment. *Food Add and Contam* 13, 655–668.

CHAPTER 6

Toxic Nitrogen Compounds Produced during Processing: Biogenic Amines, Ethyl Carbamides, Nitrosamines

M. Hortensia Silla-Santos

INTRODUCTION

Nitrogen compounds are very important in the nutrition of microorganisms, plants, animals, and humans. On the other hand, nitrogen is also a component of potentially harmful compounds such as amines, amides, and nitrosamines. These compounds can be present in foods as a result of food components and process conditions.

BIOGENIC AMINES

Biogenic amines are basic nitrogenous compounds that are formed mainly by the decarboxylation of amino acids or by amination and transamination of aldehydes and ketones. They are organic bases with low molecular weight and are synthesized by microbial, vegetable, and animal metabolism. Biogenic amines in food and beverages are formed by the enzymes of the raw material or by microbial amino acid decarboxylase activity,[7] but it has been found that some of the aliphatic amines can be formed *in vivo* by the amination of corresponding aldehydes.[66] The chemical structure of biogenic amines (Figure 6–1) can either be aliphatic (e.g., putrescine, cadaverine, spermine, spermidine, agmatine), aromatic (e.g., tyramine, phenylethylamine), or heterocyclic (e.g., histamine, tryptamine). Amines such as polyamines (i.e., putrescine, spermidine, spermine, and cadaverine) are indispensable components of living cells and are important in the regulation of nucleic acid function and protein synthesis, and probably also in the stabilization of membranes.[10]

Formation of Biogenic Amines in Food

In most foods and beverages, amine formation is generated by the decarboxylation of the corresponding amino acid. The precursors of the main biogenic amines involved in food poisoning are shown in Figure 6–1. Prerequisites for biogenic amine formation by microorganisms are the (1) availability of free amino acids, (2) presence of decarboxylase-positive microorganisms, and (3) conditions that allow bacterial growth and decarboxylase synthesis and activity.

Because amines are formed by enzymatic activity of the food or bacterial flora, the inhibition of such activity or the prevention of bacterial growth would be very important in order to minimize the amine content of food. One approach to control the bacterial flora is to use preservatives. Some spices and herbs have been reported to possess antimicrobial activity against food spoilage bacteria. Their use, then, could be considered provided they are compatible with the food. Cloves and cinnamon are reported to be inhibitory to bacterial growth and biogenic amine production by bacteria, but with only a slight effect. Bacteria differ in their sensitivity to spices, and this sensitivity also depends on other factors such as temperature.[138] Manufacturing conditions influence the production of biogenic amines. Thus, tyramine, putrescine, and cadaverine concentrations in tempeh can be low or

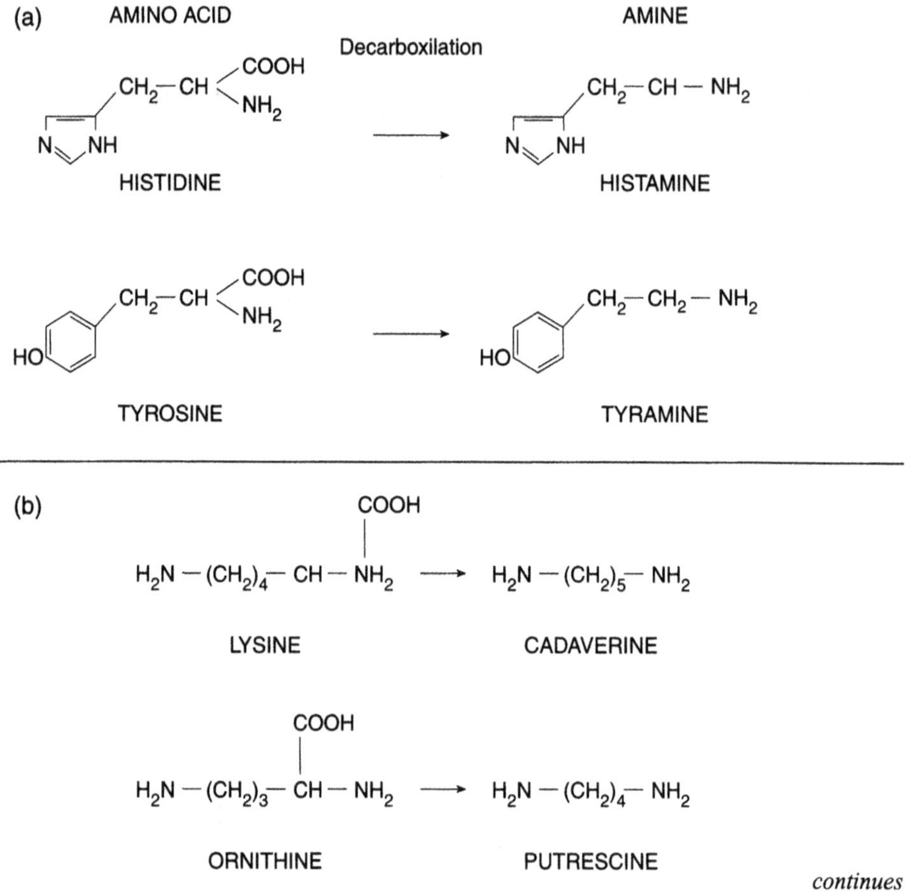

continues

Figure 6–1 Principal biogenic amines in fermented food and their precursors: (a) histamine and tyramine; (b) cadaverine and putrescine; (c) serotonine and tryptamine; (d) spermine and spermidine.

high depending on the manufacturing process (e.g., soaked soybeans, type of fermentative microorganisms, boiling, home cooking by stewing or frying in oil, and storage temperature).[86]

Histamine production in cheese has been related to factors such as the availability of substrate, pH, salt concentration, and temperature (Table 6–1). The proper storage temperature is probably the most important method of prevention.[121] Antibiotics such as penicillin and tetracycline added to scombroid fish repress histamine synthesis,[135] though this is not an approach that is compatible with food use. Amino acid decarboxylase activity is stronger in an acidic environment,[19,77] and bacteria tend to produce these enzymes under acid stress conditions as a part of their defense mechanisms.[19,37,126] Several factors influence the pH decrease in fermented foods, such as starter culture, contaminating flora, temperature/time, and the use of additives, such as glucono-δ-lactone (GDL). The presence of fermentable carbohydrates such as glucose enhances both bacterial growth and their amino acid decarboxylase activity. Histamine formation was inhibited by salting, with the inhibition being proportional to the brine concentration.[113]

There are different opinions as to the influence of time and temperature of storage on the synthesis of biogenic amines. The content of biogenic amines in food seems to increase with both time and temperature.[1] However, it has been reported[103] that histamine levels in raw and cooked ground

Figure 6–1 continued

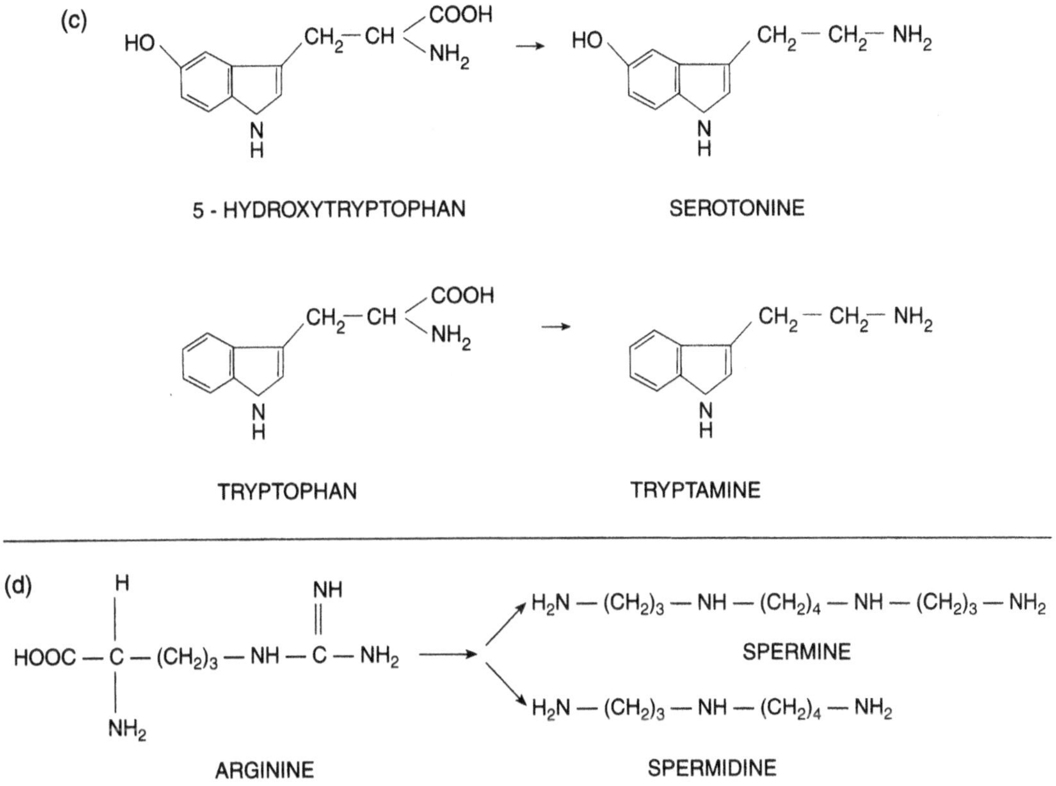

beef were unaffected by storage conditions (4, 7, and 10 °C during 12 days), but when the storage temperature was increased (21 °C), increased amine concentration was observed.[2] Amine concentrations were also unaffected by cooking, with the exception of spermine, which decreased during heat treatment. Wendakoon & Sakaguchi[138] also indicated that histamine is thermally stable during the cooking process.

Oxygen availability also has a significant effect on the biosynthesis of amines. Facultative anaerobic microorganisms produce less biogenic amines in anaerobic conditions as compared with aerobic conditions.[37] Feedback repression of histidine-decarboxylase has also been detected when the amount of histamine increases.[135]

Microorganisms Producing Biogenic Amines

Among the bacteria that are capable of synthesizing biogenic amines, many different species within a group of related genera may possess a specific decarboxylase. Also, large interspecies variations may occur within one genus. Numerous enteric bacteria have been reported to possess histidine decarboxylase activity, but only a few of the bacteria found in foods have been implicated in the formation of toxicologically significant levels of histamine. Enteric bacteria, specifically *Morganella morganii* (*Proteus morganii*), certain strains of *Klebsiella pneumoniae,* and a few strains of *Hafnia alvei*, are prolific histamine producers and are important in the hygiene of fish products.[99,133,138] Likewise, other bacteria have been related to histamine formation in fish, including *P. vulgaris, Escherichia, Clostridium, Salmonella*, and *Shigella*.[47] *Staphylococcus* spp., *Vibrio*, *Pseudomonas,* and *Bacillus* spp. have been identified as histamine-producing bacteria in fermented fish,[99,138,140] as has cadaverine and putrescine

Table 6–1 Factors Affecting Synthesis of Biogenic Amines (BA)

Factors	*Principal Activity*	*References*
Substrate	Free amino acids (increase BA)	68, 83, 118
Microorganisms with necessary enzyme	Amino acid decarboxylase (increases BA)	1, 2, 18, 19, 37, 40, 59
pH	Acid (increases BA)	17, 59, 125, 126, 127
Temperature	Refrigeration (decreases BA)	17, 59, 83, 115
Atmosphere	Aerobiosis (decreased BA in anaerobic and increased in aerobic conditions)	17, 37, 59
Antibiotics	Inhibition of microorganisms (decreases BA)	135
Fermented carbohydrates	Acididification (increases BA)	113
Spices	Inhibition of microorganisms (decreases BA)	138
Oxidase enzymes	MAO, DAO (decrease BA)	66, 67

synthesis capability by *M. morganii*, *Enterobacter cloacae, Citrobacter freundii,* and *Serratia liquefaciens*.[99]

In cheeses, the lactic flora is dominant during the ripening process. *Streptococcus faecalis (Enterococcus faecalis)*[41] has been associated with tyramine in cheese[23] and in other milk products.[35] In miso, tyrosine decarboxylase-producing bacteria have been identified as *E. faecium* and *Lactobacillus bulgaricus,* and histamine decarboxylase has been associated with *Lactobacillus* spp. and *L. sanfrancisco*.[51,52] Fermentation may be important in the formation of biogenic amines through the action of added lactic acid cultures or the natural microflora. Amino acid decarboxylase activity has been shown to depend on the composition of the medium and the growth phase of the microorganisms, with the highest amino acid decarboxylase enzyme activities being detected in the stationary phase.[37]

The presence of high concentrations of free amino acids in meat and meat products that are exposed to microbial degradation enhances the possible formation of large amounts of biogenic amines.[109] Amine-producing lactic bacteria such as *L. brevis, L. buchnerii, L. curvatus, L. carnis, L. divergens* (*Carnobacterium divergens),*[41] and *L. hilgardii* have been isolated from meat and meat products.[73,76,77,110,114,122] In salami, histamine synthesis has been associated with Gram-negative bacteria (*P. fluorescens*, *C. freundii,* and *Acinetobacter calcoaceticus* var. *anitratum*), but also with Gram-positive strains, including micrococci and staphylococci.[127]

The malo-lactic fermentation bacterium *Pediococcus cerevisiae*, which withstands the presence of sulphur dioxide better than *Oenococcus oeni,* can be regarded as the main cause of histamine formation in wine.[12,69] Various histaminogenic abilities of the yeasts have been confirmed in fermentation tests. There is evidence that amino acid decarboxylase enzymes are not well distributed among the yeasts and the most commonly used malo-lactic bacteria. It seems that amino acid decarboxylase is more frequently found in the groups of enterobacteria and lactic acid bacteria (LAB). When LAB proliferated in beer, the formation of biogenic amines could be detected.[26]

Biogenic Amines in Fermented Food

Biogenic amines can be expected in virtually all foods that contain proteins or free amino acids and that are subject to conditions enabling microbial or biochemical activity. The total amount of the different amines formed depends strongly on the nature of the food and the microorganisms present.[18] Biogenic amines are present in a wide range of food products including fish products, meat products, dairy products, wine, beer, vegetables, fruits, nuts, and chocolate[110] (Table 6–2). In nonfermented foods, the presence of biogenic amines above a certain level is considered as indicative of undesired microbial activity.[37] Therefore, the amine level could possibly be used as an indicator of microbial spoilage, although the presence of biogenic amines does not necessarily correlate with the growth of a spoilage flora because not all spoilers are decarboxylase positive. Levels of histamine, putrescine, and cadaverine usually increase during spoilage of fish and meat, whereas levels of spermine and spermidine decrease.[18] Scombroid fish have been associated most commonly with incidents of histamine intoxication (scombrotoxicosis). The formation of histamine in scombroid and other marine fish containing abundant endogenous histidine has been attributed to microbial action rather than to endogenous histidine decarboxylase activity.[37,134] During the preparation of fermented food, one can expect the presence of many kinds of microorganisms, some of them capable of producing biogenic amines. Most products in which LAB grow contain considerable amounts of putrescine, cadaverine, histamine, and tyramine.[18]

Fish Products

The levels of amines can vary extensively in fish products. Trace quantities of putrescine, tyramine, agmatine, and tryptamine have been detected in Ghananian fermented fish.[139] Ornithine and citrulline were detected as decomposition products of arginine in fish sauce, and histamine was also confirmed as a decomposition compound from histidine, but only in trace amounts.[79] In anchovies, a high level of biogenic amines during the manufacturing process can be due to the long processing time, which favors bacterial growth.[99]

Table 6–2 Biogenic Amines Content in Some Foods (ppm)

Food	*Histamine*	*Tyramine*	*Cadaverine*	*Putrescine*	*Reference*
Fish and fish products					
Sailfish	1680.0			145.0	47
Fermented fish paste	640.0	376.0	35.0		121
Anchovies	12.6	21.6	38.3	7.6	134
Cheeses					
Cheddar	1300.0	700.0			121
Swiss	2500.0		490.0	330.0	121
Meat products					
Salchichón	7.3	280.5	11.7	5.5	135
Fuet	2.2	190.7	18.9	71.0	135
Dry fermented sausages	286.0	1500.0		396.0	121
Fermented vegetables					
Sauerkraut	10.0	20.0	25.0	50.0	110
Soy sauce	2740.0	4660.0			121
Beers: top-fermented					
Ale	0.6	5.0	0.9	5.7	55
Kriek	5.6	22.5	6.3	4.5	55
Beers: botton-fermented					
Lager	0.7	4.9	0.8	4.1	55
Pilsner	1.0	5.6	2.0	5.1	55
Nonalcoholic	0.6	6.2	1.0	3.1	55
Wine	3.3			3.3	69

A fishy odor is derived from a variety of components, of which trimethylamine (TMA) is predominant. TMA and its precursor, TMA N-oxide, are used as a nitrogen source by *Aspergillus oryzae*.[63] Possibly, TMA-using molds could be applied to remove TMA from relevant food material.

Cheese and Dairy Products

After fish, cheese is the most commonly implicated food associated with histamine poisoning. The first reported case of histamine poisoning occurred in 1967 in the Netherlands and involved Gouda cheese.[121] Many studies have been undertaken to determine the amine content of cheese products, and a variety of amines, such as histamine, tyramine, cadaverine, putrescine, tryptamine, and phenylethylamine, have been found in different cheeses.[2,13,23,25,64,80,92,104,132]

Meat Products

Biogenic amines can be found in fermented meat products as a consequence of microbial activity related to fermentation, but they may also be found in products made from poor quality raw materials through microbial contamination. Good manufacturing processes for meat products, using amino acid decarboxylase negative starter cultures, results in the formation of very small amounts of biogenic amines.[28,74] Uncooked and ripened meats showed higher amounts of histamine and tyramine than cooked meat products. Maijala *et al.*[74] also detected increased concentrations of histamine and tyramine during sausage fermentation. Amines can also be found after fermentation during storage.[29] Bauer *et al.*[11] reported that the addition of starter culture did not affect their formation. Roig-Sagues & Eerola[100] observed that the influence of starters on biogenic amine formation in minced meat depended on the kind of decarboxylating microorganisms that were present in the raw material. Dry sausages without starter microorganisms showed variable concentrations of biogenic amines. This could be due to variations in the ripening time[22] and differences in the natural microflora responsible for fermentation.[80,100] Polyamines, putrescine, cadaverine, histamine, tyramine, and 2-phenylethyl-amine, have also been detected in fermented sausages by Straub *et al.*[123] Cured meat products such as sausages are frequently implicated in amine poisoning due to histamine and tyramine.[19,73,109,126]

Vegetables

A succession of microorganisms is involved in sauerkraut production. And consequently, biogenic amines, especially putrescine, can be expected in the brine.[18] Other fermented vegetables where amines have been detected are green table olives, cucumbers, and lupin.[42]

Very low levels of biogenic amines were detected in a study of Asian foods. Low levels of tyramine were found in commercial samples of Japanese pickled vegetables *(urume-zuke)* and in *kim chee (kimchi),* a traditional Korean fermented cabbage.[121] In *miso,* tyramine and histamine have been determined by Ibe *et al.*,[48–50] but it seems that these amines are not generated in high quantities in fermented vegetables.

Wine and Beer

Agmatine, cadaverine, ethanolamine, histamine, putrescine, and tyramine are produced in high quantities during alcoholic fermentation.[20] Many kinds of biogenic amines have been detected in both white and red wine: tyramine, histamine, tryptamine, monomethylamine, 2-phenethylamine, putrescine, cadaverine, spermidine, iso- and n-amylamine, pyrrolidine, iso- and n-butylamine, iso- and n-propylamine, and ethylamine.[51,65,94] Putrescine, histamine, methylamine, and tyramine also developed during malo-lactic fermentation and the wine aging process.[12]

In beers, the presence of tyramine[37] and histamine[27,54] has been observed. Biogenic amines are formed by the barley enzymes during malting, even under sterile conditions. The grain microflora and pitching yeast contaminants seemed to be responsible for elevated amine levels, especially histamine, in beer. Malt and hops contribute to the amine content of wort and beer. Spermine and spermidine levels decreased sharply during mashing, whereas the other amines increased, with the exception of putrescine.

Health Effects

Putrescine, spermidine, and spermine are indispensable components of all living cells. Polyamines are very stable compounds that are able to resist heat and survive acidic and alkaline conditions. They can interact nonspecifically with negatively charged structures and, in these nonspecific roles, they can be replaced by metal ions, most often by Mg^{2+} or Ca^{2+}. Because of their structure, they can fulfill a wide range of specific functions in cells, such as the control and inhibition of mRNA translation to proteins and regulation of the fidelity of translation. Polyamines can stimulate ribosome subunit association, stabilize the tRNA structure and reduce the rate of RNA degradation, enhance both RNA and DNA synthesis, help to condense DNA, covalently modify proteins, and regulate the rigidity and stability of cellular membranes.[10]

Certain classes of amines, the catecholamines, indolamines, and histamine, fulfill important metabolic functions in humans, especially in the nervous system and the control of blood pressure. Phenylethylamine and tyramine cause a rise in blood pressure, whereas histamine reduces it. Histamine has a powerful biological function, serving as a primary mediator of the immediate symptoms observed in allergic responses, such as urticarial lesions.[121] Although biogenic amines such as histamine, tyramine, and putrescine are needed for many critical functions in man and animals, the consumption of food containing high amounts of these may have toxic effects. Symptoms that can occur after excessive oral intake include nausea, respiratory distress, hot flush, sweating, heart palpitation, headache, bright red rash, oral burning, and hyper- or hypotension.

The most notorious food-borne intoxications caused by biogenic amines are related to histamine. Food poisoning involving the consumption of food containing abnormally high levels of histamine has been recognized for many years. Numerous outbreaks of histamine poisoning have occurred after eating cheese or fish. Histamine is the most toxic amine detected in food (fish, cheese, wine, meat products), and normally it is associated with food spoilage. The toxic effect depends on histamine intake, presence of other amines, aminooxidase activity, and the intestinal physiology of the individual. The gut can absorb six to ten times more histamine when putrescine, cadaverine, and spermidine are present.[10]

The biogenic monoamines (i.e., serotonin, phenylethylamine, and tyramine) and the diamines (i.e., histamine and tryptamine) are also formed and degraded during normal cellular metabolism, playing a variety of physiological roles, such as the regulation of body temperature, stomach volume, and pH. They can also affect brain activity.[10] Biogenic amines such as putrescine, cadaverine, spermidine, and others have been reported to be radical scavengers. Tyramine has a marked antioxidative activity, which increases with tyramine concentration, and this antioxidative effect may be attributable to its amine and hydroxy groups.[141] Polyamines, spermine, spermidine, and putrescine, inhibit the oxidation of polyunsaturated fatty acids, and this antioxidative effect is correlated with the number of amine groups in the polyamine.[70]

Nitrosable secondary amines (i.e., agmatine, spermine, spermidine) can form nitrosamines by reaction with nitrite and produce carcinogenic compounds.[37] In general, N-nitroso compounds can be formed by the interaction of amino compounds with nitrosating reagents such as nitrite and nitrogen oxides during the storage, preservation, and cooking of foods.[44] The reaction of nitrosating agents with primary amines produces short-lived alkylating species that react with other components in the food matrix to generate products (mainly alcohols) that are devoid of toxic activity at the concentrations produced. The nitrosation of secondary amines leads to the formation of stable N-nitroso compounds; that of tertiary amines produces a range of labile N-nitroso products. Secondary amines such as putrescine and cadaverine can react with nitrite to form heterocyclic carcinogenic nitrosamines, nitrosopyrrolidine and nitrosopiperidine.[46] For this reason, concern over the use of nitrite has promoted many investigations.[43,44,93,108,128]

Fortunately, a fairly efficient detoxification system exists in the intestinal tract of animals, which is capable of metabolizing normal dietary intakes of biogenic amines. Under normal conditions in humans, exogenous amines absorbed from food are detoxified rapidly by the action of amine oxidases or by conjugation; but in allergenic individuals, or if monoamine oxidase inhibitors are applied or when intake levels of amines are too high, the detoxification process is disturbed and biogenic amines accumulate in the body. Amino oxidases are induced in the presence of mono- or diamines.[18] The enzymes monoamine oxidase (MAO) and diamine oxidase (DAO) play an important role in this detoxification process. However, when the activity is suppressed by one or more substances known as potentiators, detoxification is inhibited. This may explain why spoiled fish or aged cheese is more toxic than histamine in aqueous solution.

Alcoholic beverages are good potentiators for sensitivity against biogenic amines. Certain drugs have been implicated as contributing factors in cases of histamine poisoning. Antihistamines, antimalarials, and other medications can inhibit histamine-metabolizing enzymes.[18] Some amines, specifically putrescine and cadaverine, inhibit histamine-detoxifying enzymes. Other amines that may act as potentiators include tyramine (which can inhibit MAO) and tryptamine (which inhibits DAO). Phenylethylamine is a DAO and HMT (histamine N-methyltransferase) inhibitor.[121] Some microorganisms (*Sarcina lutea, Lactococcus lactis, L. lactis* subsp. *lactis*, and *L. lactis* subsp. *diacetilactis*[41]) contain monoamine oxidase or diamine oxidase (*L. lactis* subsp. *cremoris*,[41] *A. niger,* and *Trichosporon* spp[112]). Some types of yeasts found in Roquefort cheese are capable of assimilating cadaverine, putrescine, and histamine.[13]

The levels reported for histamine and its potentiators in food would not be expected to pose any problem if normal amounts were consumed. In susceptible individuals, 3 mg of phenylethylamine causes migraine headaches; 6 mg total tyramine intake was reported to be a dangerous dose for patients receiving monoamine oxidase inhibitors.[110] A level of 1000 mg/Kg (amine/food) is considered dangerous for health. This level is calculated on the basis of food-borne histamine intoxications and the amine concentration in the food. The discrepancy in the toxic histamine level in food might be due to the absence or presence of other synergistic biogenic amines like putrescine and cadaverine. The European Union has proposed that the average content of histamine should preferably be less than 100 ppm, but should not be higher than 200 ppm in fish or fish products belonging to the scombridae and clupedae families.[71] The European Union set a three-class plan for maximum allowable levels of histamine in fresh fish (n = 9 number of units to be analyzed from each lot; c = 2 number of units allowed to contain a histamine level higher than m = 100 ppm; maximum permissible level M = 200 ppm) and fish products (m = 200 and M = 400 ppm).

Conclusions

A reduction in the biogenic amine content of food can be achieved by the following steps.

1. Select raw material without high levels of amines or microbial contamination.
2. Provide systematic control at each processing step with good hygienic practice. This can result in an optimized process with the corresponding inhibition of spoiling microorganisms and low levels of biogenic amines.
3. In fermented food, the selection criteria for bacteria used for starter cultures should include analysis of potential amine production.
4. Taking into account the effect of potentiators (including alcohol and other amines) on the toxicological activity of amines, amine profiles should be determined instead of simply the levels of histamine and tyramine.
5. The original concentration of amines in food can change as a result of storage conditions and these should be controlled.

ETHYL CARBAMIDES

Ethyl carbamate is an ester of cabamidic acid. Whereas carbamidic acid is unstable and dissociates spontaneously into carbon dioxide and ammonia, the esters of carbamidic acid are stable.

$$H_2N\text{-}CO\text{-}OH \rightarrow NH_3 + CO_2$$

carbamidic acid ammonia carbon dioxide

The carbamides can be esterified with different alcohols such as ethanol to form ethyl carbamate.

Formation

The carbamide or carbamidic acid is formed from carboxylic and amino groups. The most common carbamide is urea, the final product of protein metabolism in ureolytic animals.

$$HO\text{-}CO\text{-}OH + NH_3 \rightarrow$$

carbonic acid ammonia

$$H_2N\text{-}CO_2^- NH_4^+ \rightarrow H_2N\text{-}CO\text{-}NH_2$$

ammonium carbamate urea

In fermented beverages, ethyl carbamate is produced during the fermentation process as a product of yeast metabolism. Carbamyl phosphate is produced from adenosine triphosphate (ATP), carbon dioxide, and ammonia by carbamyl-phosphate synthase, a reaction necessary for arginine synthesis, which is enhanced when ammonia levels are high. This compound may react with the ethanol that is formed in fermentation and results in ethyl carbamate, which is a potent carcinogen for humans. Ethyl carbamate can also be formed in fermented food (Figure 6–2) by the reaction of ethanol and urea that is produced naturally from amino acids like citrulline, which is formed via arginine degradation in wine by some LAB during malo-lactic fermentation;[36] or when urea is added as a yeast nutrient.[53]

The overall reaction for these compounds is via an acid-catalyzed ethanolysis. In wines from grapes with low nitrogen content and the addition of sugar and water (to further dilute the significant amino acids), the urea remaining in the wine may

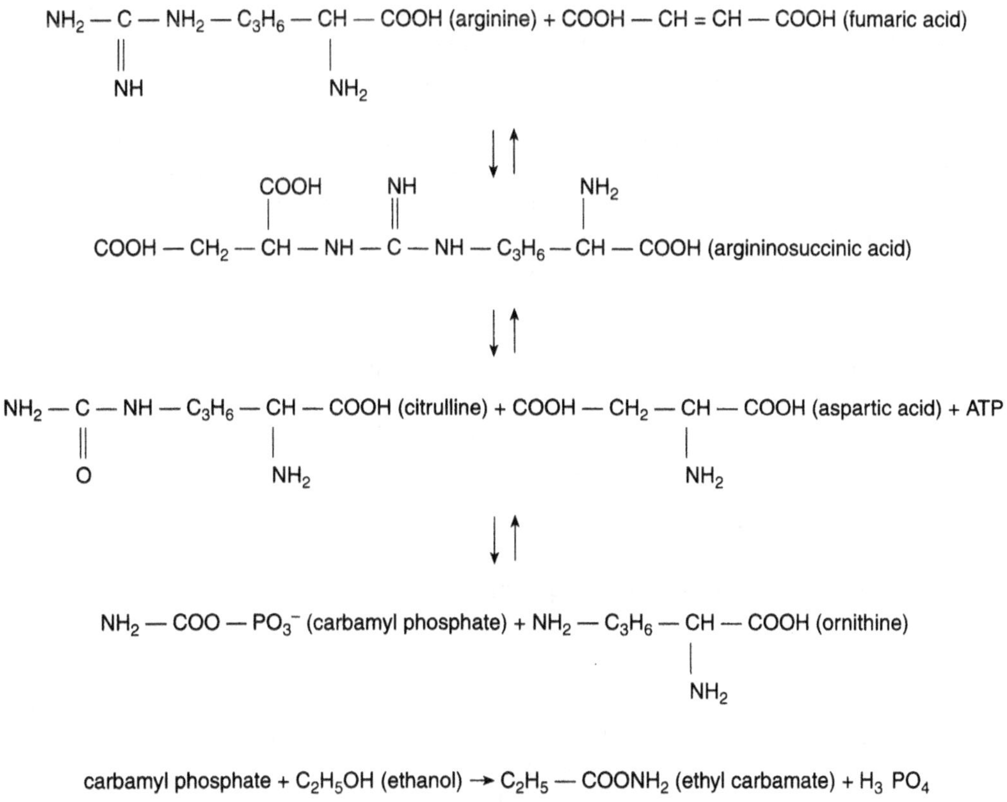

Figure 6–2 Formation of ethyl carbamate via arginine degradation

be negligible.[120] The extent of ethyl carbamate formation is likely to depend on the level of precursors, the yeast strain used, fermentation conditions, and the pH during the process.[85]

Ethyl carbamate was not detected in a range of fermented foods where yeasts were not involved as a major population. Experiments on a model grape juice demonstrated that N-carbamyl-amino acid was produced by yeast during fermentation. Because the conversion of N-carbamyl-amino acid into ethyl carbamate is not favored chemically, N-carbamyl amino acids are not expected to contribute substantially to the formation of ethyl carbamate in wines under normal conditions of wine making and aging.[45,90] Ingledew *et al.*[53] demonstrated that yeast fermentation itself did not lead to the production of ethyl carbamate. However, when subsequent heating steps were applied to fermented beverages in which urea was present or made, ethyl carbamate was formed in significant quantities.

The source of urea is arginine breakdown via arginase activity. Citrulline and other N-carbamyl amino acids can react with ethanol to form ethyl carbamate. Samples of commercially processed juices were fermented under controlled conditions to determine variation in amino acid uptake or excretion during yeast growth and fermentation. There was almost always rapid uptake of all of the amino acids except alanine, proline, citrulline, ornithine, and arginine. Urea excretion and uptake were dependent primarily on yeast strain and on amounts of arginine remaining in the juice during fermentation. Amounts of urea excreted tend to increase at higher temperatures. At very high amino acid levels, arginine is metabolized very slowly, which minimizes urea excretion. Comparison of various yeasts shows that patterns of urea uptake are generally similar.[90]

Studies on factors (Table 6–3) influencing the formation of ethyl carbamate in wines have demonstrated that ethyl carbamate can be formed spontaneously from urea, citrulline, or carbamyl phosphate in acidic ethanolic solutions. At high temperatures, all compounds containing a carbamyl structure have the potential to form ethyl carbamate. Although all wines have the potential to form ethyl carbamate, the likelihood is greatest in wines that have undergone malo-lactic fermentation.[36] Nitrogenous constituents of wine other than urea are not correlated with urea formation, and N-fertilization of the grapes does not affect urea formation in the resulting wines.[119] In Cabernet Sauvignon and Chardonnay wines, ethyl carbamate contents increased over two years of storage. There was no significant effect of wine type (red vs. white) or pH on ethyl carbamate content. Holding temperature had a significant effect on ethyl carbamate formation; increasing the temperature by 10 °C may increase the rate of ethyl carbamate formation up to 3-fold. Thus, knowledge of the urea concentration and wine storage temperature permits a good estimation of the amount of ethyl carbamate that will form over a period representative of wine storage.[120]

Applications

In the past, the American baking industry has used several reducing agents such as ascorbic acid, sodium metabisulphite, and cysteine-N-carbamide as chemical dough developers in-

Table 6–3 Factors Affecting Ethyl Carbamate (EC) Synthesis

Factors	*Causative Effect*	*References*
Substrate	Free urea, carbamyl-P (increase EC)	36, 61, 137
Enzyme activity	Carbamyl-P synthetase, urease (increase EC)	61
pH	Acid (increases EC)	61
Heat process	Toasting (increases EC)	21
Ascorbic acid	Decreases EC	21

stead of cysteine. Cysteine-N-carbamide has been shown to improve sour bread dough noticeably, and the addition of cysteine to the sourdough system significantly reduced mixing requirements and increased bread volume.[130] In chemical dough development, cysteine is used most commonly. It speeds the breadmaking process by expediting disaggregation by splitting disulphide bonds between protein aggregates through reduction and promoting disulphide interchange with less mixing. Carbamide considerably increases the stability of starch phosphate at low temperatures. An addition of a small quantity of carbamide to the starch increases its water absorption power, raises its viscosity, and improves the transparency of its aqueous solution.

In the European Union, the use of cysteine-N-carbamide is forbidden because of its possible role in promoting carcinogenesis. For this reason, when it is used in any food-related material, it must be within the limits permissible by the sanitary authorities. The use of carbamide as a substrate in the growth of yeasts led to improved molasses fermentation. Carbamide addition as a nitrogen source was recommended for molasses with reduced nitrogen content. Molasses-grown yeast reduced leavening time in the baking process and improved the quantity of molasses alcohol with simultaneous production of baker's yeasts of good quality.[102] After molasses fermentation is finished, carbamide can be removed from alcoholic beverages with urease, preferably from LAB; the optimum for this reaction is pH 2–5.[55] Carbamide was the most efficient nitrogenous substance added to manioc (cassava) mash for the nutrition of the yeast cells in the saccharification and fermentation of manioc starch.[84]

Carbamides also find application in packaging material for maintaining more moisture in the product.[6,117] In other cases, carbamide is added in the form of carbamide peroxide to give hydrogen peroxide in processes for protecting milk or milk products containing lactoperoxidase from bacterial spoilage during storage.[15] Carbamide used as a feed supplement or as fertilizer can influence the chemical composition of vegetables and meat.[136] When carbamides are used as insecticides in wheat storage, the control and norms to be followed in the fumigation and storage of wheat to be used as seed or for food are very strict. At certain levels, the presence of carbamide in some foods can be detected by its odor or bitter taste.[5]

Health Effects

Ethyl carbamate (urethane) has mutagenic and carcinogenic properties. Due to the toxicity of carbamide, there are maximum permissible concentrations in some countries set on the basis of animal trials in foods and water. The presence of various levels of ethyl carbamate in distilled spirits and wines, and fermented foods such as bread, soy sauce, miso, and yeast spread has been reported. In most countries, there is no legal limit for ethyl carbamate, but the Food and Agricultural Organization (FAO)/World Health Organization (WHO) suggested a level of 10 ppb for soft drinks; in Canada, the tolerance level ranges from 30 ppb in wines to 400 ppb in distilled spirits.[85]

Conclusions

Ethyl carbamate is a carcinogenic and mutagenic compound that results from the esterification of carbamic acid by ethanol in alcoholic fermentations. The use of proper starter cultures of yeasts in the manufacturing of fermented foods can result in increased levels of safety of these foods. The selection of nontoxigenic starters that antagonize pathogenic microorganisms and have detoxifying ability should be a priority.

NITROSAMINES

Nitrosamines are formed by the reaction of secondary and tertiary amines or amino groups contained in compounds such as dialkylamines, alkylarylamines, piperazine, and pyrrolidine, with nitrite. Two precursors, secondary amines and nitroso acid, are required for the formation of nitrosamines. Certain foods have been found to contain various nitrosamines, many of which are known carcinogens. The formation of nitrosamine has been demonstrated in food-nitrite mixtures. Although amines are present in foods, wine, tobacco products, drugs, and other environmental chemicals such as pesticides, sources of nitrite, the other precursor, vary. Human exposure can be by the ingestion or inhalation of preformed N-nitroso compounds or by endogenous nitrosation. It has been established unequivocally that N-nitroso compounds are also formed in the body from precursors that are present in the normal diet.[44]

Formation

The formation of nitrosamine from nitrite takes place through various steps, as indicated in Figure 6–3.

The rate constant K is independent of pH, but the amine and nitrous acid concentrations change with pH. The first two reactions are favored by acidic conditions. However, if the environment is too acidic, the amine will be protonated and unable to react with the nitrous anhydride (N_2O_3). Nevertheless, nitrosation can also occur under the nonacidic conditions that are found in some foods, and oxides of nitrogen (NO_x) can react directly without the requirement for acid conditions. Some components found in foods catalyze nitrosation reactions. Nitrite in food can be produced by microbial nitrate reduction. Nitrates are naturally present in soil, water, and food, and also in the air at low concentration as a result of industrial pollution. Nitrates in soil can accumulate due to fertilizer treatments. For example, under conditions of excessive fertilization at low temperatures and with low intensity of light, nitrate can accumulate in vegetables.[33]

Nitrosamines can become a part of foods by the use of nitrite as an additive, through processing, especially drying, and through their formation during food preparation, migration from food contact surfaces, or indirect addition along with other additives. Products made from rubber (elastomers) contain volatile nitrosamines that can migrate to food or drinks. The rubber netting used to hold hams and other cured products during processing can also result in the migration of very small amounts of nitrosamines to foods. Fiddler *et al.*[30] confirmed these findings and reported that additional nitrosamines could be formed when rubber products were exposed to solutions that were similar to human saliva. There is a controversy regarding the possible migration of nitrosamines from rubber elastic netting into meat products. Rubber elasticized netting has been in use since the 1960s, and the Keystone Casing Supply Company (Carnegie, Pa) first received confirmation of approval for their Jet Net formulation netting from the U.S. Department of Agriculture (USDA) in 1964. In 1985, a study showed possible migration of nitrosamines from the nipples on baby bottles into saliva or milk.[105] To minimize the potential for nitrosamines formation, the manufacturers of elastic rubber netting reformulated the rubber component to reduce the level of nitrosatable

$$2\,(NO_2^-) \xrightleftharpoons{2H^+} 2\,(HNO_2) \xrightleftharpoons{\text{fast}} N_2O_3 + 2\,H_2O$$

NITRITE — NITROUS ACID — NITROUS ANHYDRIDE — WATER

$$N_2O_3 + R_2NH \xrightarrow{\text{slow}} R_2NN{=}O + HNO_2$$

NITROUS ANHYDRIDE — SECONDARY AMINE — NITROSAMINE — NITROUS ACID

Figure 6–3 Formation of nitrosamine from nitrite at a rate of K (R_2NH) × (HNO_2).

amines.[75] Some paper-based packaging materials contain trace levels of volatile nitrosamines, which can migrate to foods.[44]

Nitroso compounds can be formed inside the body, and the diet influences the nature of nitroso compound formed. Humans are also exposed to nitrosamines through normal physiological processes in the stomach through the reaction between amines and nitrite. The nitrite can be derived from two sources: consumption of foods that contain nitrite and the conversion of nitrate to nitrite in the body. Nitrate can be derived both from normal nitrogen metabolism and from the diet. Dietary nitrate is absorbed quickly after ingestion and is circulated in the blood. Then it is excreted and concentrated in the saliva, where microorganisms in the mouth convert it to nitrite. The nitrite is swallowed, resulting in the formation of nitrosamines in the stomach. The major source of dietary nitrate is vegetables, with cured meats contributing a small amount. Diet may play a role in the types and amounts of nitrous compounds that are formed endogenously. The dietary balance between nitrate and ascorbic acid may be particularly important because ascorbic acid is capable of inhibiting endogenous nitrosamine formation resulting from nitrate exposure.[44]

Nitrosamines in Food

Cured Food

The origin of the use of nitrates as a meat curing agent is lost in antiquity. Humans have been using salt for preserving food for thousands of years, obtaining good products with flavor and color typical of cured foods. Common salt usually contains impurities from other salts, including nitrates, which give the cured color and the cured aroma. The term "curing" came to be used for the first time in northern Germany to describe the treatment of meat and fish with common salt. This has meant that even today, there is some confusion between salting, treating with common salt, and curing, treating with nitrate or nitrite-curing salts. The salt used to preserve meat contains potassium nitrate, which produces the characteristic flavor and color of the cured meats. Later, scientific understanding of the curing process began to evolve, and the role of nitrite was recognized. It was observed that nitrate was reduced to nitrite by bacterial activity, and that nitrite is reduced to nitric oxide to form the typical curing color. These findings led to the use of nitrite directly as a curing ingredient. Nitrite is a highly reactive chemical. In the complex meat system, several reactions can occur, including the reaction of nitric oxide with myoglobin to form the red color of the cured meat.

Concern over nitrosamines stimulated considerable interest in the value of nitrite as an antimicrobial agent. This has focused mainly on the minimum levels necessary for the control of *C. botulinum* and on developing alternatives to replace nitrite wholly or in part. Hauschild[38] suggested that, from a safety point of view, significant reductions in nitrite input could be made for a number of cured meats without the need for compensation. Fermented meats may belong to this group. With respect to the antimicrobial activity of nitrite on other toxigenic or infectious microorganisms in food, the evidence is rather contradictory due in part to the complex interactions occurring during product processing. During the 1970s, as pressure mounted to minimize the allowable level of nitrite for curing meat, there was concern about the risk of botulism. This concern resulted in a large number of collaborative studies conducted jointly by government and industry. The general conclusion was that nitrite is required to maintain safety, but it was also noted that the interaction of other factors such as salt level, pH, and heat treatment is critical. In 1982, the Committee on Nitrite and Alternative Curing Agents in Food[82] issued a report in which the status of the search for alternatives to nitrite was reviewed exhaustively. They listed the following potential alternatives and/or partial substitutes: ascorbate and tocopherol, irradiation, LAB, potassium sorbate, sodium hypophosphite, and fumarate esters. The amount of nitrite added to the bacon can be reduced by the addition of LAB and fermentable sugar, which lower the pH, thereby ensuring that *C. botulinum* will not grow and produce toxin.[38,124]

N-nitrosation is influenced by many factors such as pH, level of precursors, basicity of the nitrosatable amines, and the presence of catalysts and inhibitors. A number of compounds such as sulfur dioxide, bisulfite, α-tocopherol, ascorbic acid, and glutathione are known to inhibit N-nitrosation. Simple phenols and polyphenolic compounds can decrease or increase the rate of N-nitrosation reactions depending on their structure and reaction conditions, especially pH. Under acidic conditions, certain phenolic compounds are known to compete more efficiently for the available nitrite than the amines. The inhibition of nitrosation occurs in most cases because the nitrosating agents are converted to innocuous products such as nitric oxide, nitrous oxide, or nitrogen.[111] Vegetable juices inhibit N-nitrosodimethylamine formation to a certain degree, but lemon juice was found to inhibit its formation to the greatest extent. The inhibition of N-nitrosodimethylamine formation by a solution containing the same amount of ascorbic acid as that of the lemon juice was 0.6%, suggesting that another compound is responsible for the inhibition.[4] Ascorbic acid (vitamin C), tocopherol (vitamin E), and erythorbic acid (isoascorbic acid) are used routinely in meat curing, primarily to speed up the curing process.[91,98,107] Ascorbic and erythorbic acids have the disadvantage of being water soluble, whereas nitrosation occurs primarily in the lipid phase.[44] Tocopherols have the disadvantage that they are lipid soluble and, hence, difficult to disperse in an aqueous cure pump mix. Tocopherol can be coated onto the salt, when salt dissolves in the brine, a fine suspension of tocopherol results.

Temperature influences nitrosamine formation in food that is rich in amines and nitrites.[34] The methods used for heating also have an influence. Cooking by broiling and frying, for example, generates a higher level of nitrosamine than boiling and microwave heating.[89] Cooking can also generate additional amine precursors to contribute to preformed nitrosamines in foods.[31] Smoked products have a higher level of nitrosamines due to the high temperatures and the components of smoke.[96,97] Interest has been focused on cured meats because the potential for nitrosamine formation would be greatest in foods to which nitrite had been added, and because the ingestion of nitrite might lead to the formation of nitrosamines in the stomach. Nitrosamines have been found consistently in bacon only after it has been fried. Fried bacon contains two or three volatile nitrosamines, N-nitrosopyrrolidine and N-nitrosodymethylamine, N-nitrosothiazolidine at lower levels, as well as nonvolatile nitrosamine (e.g., N-nitrothiazolidine carboxylic acid) in lower concentration and less frequently. When bacon is fried, nitrosating agents cannot be extracted from the raw or fried out fat, but the nitrogen oxides formed from the added nitrite react with the unsaturated lipids during the curing process to form compounds that can decompose during frying to form nitrosating agents (e.g., NO, NO_2, N_2O_3).

N-nitrosamine formation occurs primarily in the lipid phase of bacon and can be effectively inhibited by lipophilic free-radical scavengers. This implies a free-radical mechanism that may not be a direct transfer of a nitroso group between the nitrosating agent and the amine.[14] The nitro-nitroso compounds demonstrated the greatest capacity for N-nitrosation. Nitrogen dioxide can also react with unsaturated fatty acids to form uncharacterized nitrosating agents.[78] Dinitrogen trioxide (N_2O_3) easily reacts with methyloleate to form several additional products, some of which are capable of N-nitrosation under conditions similar to frying bacon.[101] Several factors influence nitrosamine formation in fried bacon, including the age and fat content of the bacon when cooked, the amount of nitrite used in the cure, the presence of nitrosation inhibitors, and the temperature of frying. The older the raw belly bacon, the greater protein breakdown is to secondary amines. Bacon with higher fat content tends to produce more nitrosamines when fried. In general, bacon that is fried at lower temperatures contains less nitrosamines, even if it is cooked well done. When bacon is prepared in a microwave oven, it contains less nitrosamine than when it is fried because of the lower temperature. Nitrosamines have not been found routinely in other meats, even when fried. This is because most cured meats, with the exception of bacon, are not fried to complete dryness. The volatile nitrosamines formed in other meat products are carried off with the water vapor during the frying process.[44]

Other factors such as age of food, storage, and processing may affect the amount of amine precursors available. In the Turkish fermented sausage, *sucuk*, nitrite concentration decreased in all raw samples during storage for up to 10 days. Nitrosamines were not detected in either raw or cooked sucuk made with 150 ppm nitrite. Raw samples made with 200 ppm nitrite contained 0.015 ppm N-nitrosodiethylamine and 0.057 ppm N-nitrosodimethylamine; however, no nitrosamines were detected in cooked samples of sucuk made with 200 ppm nitrite. Sucuk made with 300 ppm nitrite was free from nitrosamines before cooking, fried samples contained up to 0.034 ppm N-nitrosodiethylamine, and grilled samples contained up to 0.011 ppm N-nitrosodiethylamine.[24]

Amines have been historically associated with fish and fish products with respect to deterioration of quality. These amines increase the likelihood of nitrosamine formation in the presence of nitrite in curing fish. With fresh fish, cooking increased or caused N-nitrodimethylamine to be formed where none was present originally. This effect was more pronounced with salt-dried seafoods, especially when they were broiled with gas. The concentration of amines present, particularly dimethylamine, a potential precursor of N-nitroso dimethylamine, varies with the fish species and other factors.[31] Other fermented foods in which nitrosamines have been detected are alcoholic beverages such as brandy, vodka, red wine, and vermouth,[3] as well as Bulgarian hard and semihard cheeses.[62]

Noncured Products: Beer

There are many studies on the presence of nitrosamines in beer or in hops, malt, and brewing material.[54,56,116,129,131] The presence of nitrogen oxides (NO_x) in the air in malt kilns may result in the formation of dimethylnitrosamine in the malt, and carryover of nitrosamine into the beer.[34] During the malt drying by the direct-fire kilning process, N-nitrosodimethylamine is produced. Oxides of nitrogen formed from combustion become incorporated into the drying air, where they react with amines in the malt. In response to these findings, the malting industry significantly reduced N-nitrosodimethylamine formation in the process by converting to indirect-fire kilns and/or by introducing sulphur dioxide during kilning.[106] Hordenine and gramine, tertiary amine alkaloids that are biosynthesized from tyrosine in malt, have been demonstrated as amine precursors for N-nitrosodimethylamine[95] (Figure 6–4).

Nitrosamines can also rise to dangerous levels in drinking water,[58,72] snuff of tobacco,[88] and nicotine chewing gum.[88]

Health Effects

Quite large quantities of nitrate are ingested daily with food, particularly from vegetables and, locally, from drinking water. Epidemiological studies have demonstrated a correlation between the concentration of nitrate in potable water and the death rate due to stomach cancer.[33] There is very little nitrate in meat; the same is generally true of meat products. In an overall balance sheet for human nitrate ingestion, meat and meat products play only a subordinate role. As far as nitrite is concerned, however, it is known to have a harmful effect on the human organism. By attaching itself to hemoglobin, the blood pigment, nitrite blocks oxygen supplies in the body. In the 1920s, nitrite was still freely used, added in the wrong dose or inadvertently confused with common salt. As a result of this erroneous application, people died by ingesting nitrite from cured meat products. Because of this, the nitrite regulations were issued at the beginning of the 1930s, banning the use of pure nitrite in cured meat products. Since then, its use has only been permitted in a mixture with common salt. The curing of meat products became a subject of further concern in the 1970s, however, when it became known that nitrosamines are sometimes produced in meat products and can be harmful to human health.

Most nitrosamines are very carcinogenic. The quantity of nitrosamine produced in meat products is variable. But even doses of carcinogenic

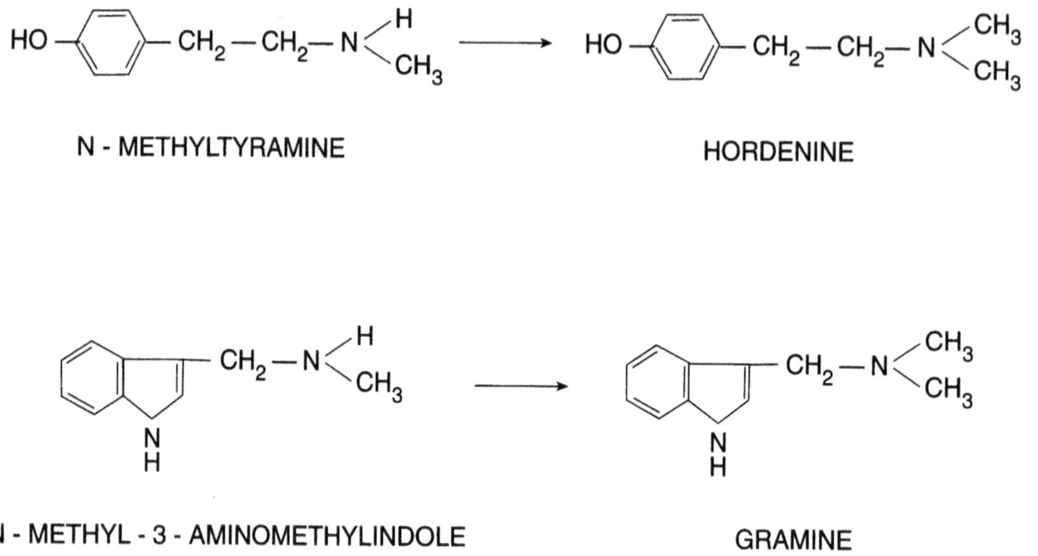

Figure 6–4 Malt alkaloids.

or toxic substances that lie below the threshold may contribute to ill health by their cumulative effects. This should be remembered when curing salt is used. Even residues involving only a slight risk from the toxicological point of view should be regarded as undesirable and avoidable in meat and meat products. It has only been in exceptional cases that tests have been made to determine to what extent a foreign substance is reabsorbed in the human gastrointestinal tract after being bound to muscle protein or animal fat. Substances that have a close binding affinity to macromolecular structures (e.g., ribonucleic acids, proteins) are likely to be reabsorbed only incompletely after ingestion with a full mixed diet. This means that low "secondary" bioavailability is a further safety factor for the consumer.[32]

Epidemiological studies do not produce hard experimental data that prove specific cancer-causative factors. Rather, they provide association hypotheses or "risk factors" for further study. It is well known that nitrosamines are a broad class of compounds that are formed from the nitrosation of substituted amides, urea, carbamates, and guanidine.[9,43] The nitrosamides are direct active carcinogens, meaning that the activation is nonenzymatic, occurring by spontaneous hydrolysis. Bioactivation of nitrosamines, on the other hand, occurs through an initial 2-hydroxylation, catalyzed by cytochrome P450 (Figure 6–5). This distinction between these two classes of N-nitroso compounds correlates with a tendency for nitrosamides to produce tumors at the initial organ exposed (e.g., stomach), whereas nitrosamines can initiate tumors at distal sites. Approximately 300 different N-nitroso compounds have been evaluated for carcinogenicity, and more than 90% have been found positive.[8]

Human exposure to nitroso compounds occurs through three main routes.

1. *Exogenous levels in foods.* These are usually nitrosamines, as the nitrosamides tend to be unstable. The formation of nitrosamines in foods occurs through the reaction of secondary or tertiary amines with a nitrosating agent (commonly, nitrous anhydride), which in turn is formed from sodium nitrite added to a number of foods as a preservative and color enhancer.
2. *Tobacco smoke.* Tobacco smoke contains tobacco-specific nitrosamines, which

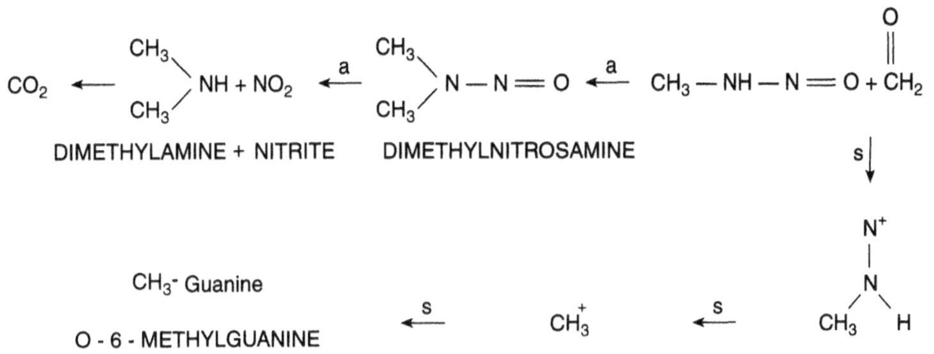

Figure 6–5 Metabolism of food-borne procarcinogens. Note: a = cytochrome P450 2E1; s = spontaneous reaction.

have been implicated in the etiology of cancers associated with smoking.[39] The U.S. National Academy of Sciences[81] estimated that exposure of smokers to volatile nitrosamines through cigarette smoke (17 μg/day) is two orders of magnitude higher than the estimated dietary exposure due to ingestion of food with the highest nitrosamine level (fried bacon, 0.17 μg/day). In fact, foods as sources of human exposure to nitrosamines were ranked lower than either automobile interiors (0.5 μg/day) or cosmetics (0.41 μg/day).

3. *Endogenous formation in the acidic environment of the stomach.* Vegetables contain high levels of nitrate, whereas dietary nitrite comes primarily from cured meats. However, the dietary contribution of nitrite is small compared to the conversion of nitrate to nitrite by microorganisms of the gastrointestinal tract.

In order to control the possible health risks of preservatives, it is of primary importance that the consumption of preservatives in foods must be in accordance with the acceptable daily intake based on the highest intake that does not give rise to detectable adverse effects. The values are often derived from studies with laboratory animals and a safety factor is then applied to the highest level at which no adverse effect is observed. Exposure to food preservatives is unlikely to be constant and will change in the future as a consequence of technological developments in the food industry and changes in dietary habits. There will therefore be a continuing need to update exposure data to monitor the effects of these developments on preservative intakes. The formation of carcinogenic N-nitrosamines from the administration of nitrite and secondary amines in animal studies only occurs at nitrite levels that are far in excess of the dietary intakes that humans are exposed to. In the United Kingdom, the Scientific Committee for Food has recommended that nitrate should not be used as an additive in infant foods in view of the acute effect of infantile methemoglobinemia. Although the nitrate anion is chemically very stable, it is readily reduced microbially to nitrite; approximately 5% of ingested nitrate is so reduced to nitrite. In practice, however, epidemiological investigations have failed to demonstrate an unequivocal link between nitrate exposure and cancer.[75]

Conclusions

Nitrite is one of the most important food additives from both an economic as well as a technical standpoint. The addition of nitrite changes perishable meats and fish into unique cured products such as bacon and ham, which have desirable flavor and appearance characteristics and longer shelf life. It also protects these products against the deadly food-borne bacterium, *C. botulinum*.

The correct use of nitrite in bacon results in the occurrence of very low levels of nitrosamines, which at higher levels have been shown to be carcinogenic in laboratory animals. Beverages such as beer also contain volatile nitosamines. Through modifications in the processing and use of chemical inhibitors, and the addition of lactic starters to meat products, the food industry has substantially reduced the content of volatile nitrosamines in human food.

REFERENCES

1. Ababouch, L., Afilal, M. E., Benabdeljelil, H. & Busta, F. (1991). Quantitative changes in bacteria, amino acids and biogenic amines in sardine (*Sardina pilchardus*) stored at ambient temperature (25–28 degree) and in ice. *Int J Food Sci Technol* 26, 297–306.
2. Abd-Alla, E. A. M., El-Shafei, K., Ibrahim, G. A. & Sharaf, O. M. (1996). Changes in microflora and biogenic amines of some market processed cheeses during storage. *Egypt J Dairy Sci* 24, 217–226.
3. Acet, H. A., Tras, B., Karahan, I. & Bas, A. L. (1993). Determination of nitrosamines in alcoholic beverages. *Doga Turkish J Vet An Sci* 17, 275–279.
4. Achiwa, Y., Hibasami, H., Kada, T. & Komiya, T. (1997). Inhibitory effect of vegetable and fruit juices on formation of N-nitrosodimethylamine. *J Jpn Soc Food Sci Technol (Nippon-Shokuhin-Kagaku-Kogaku-Kaishi)* 44, 50–54.
5. Akachkov,V. P. (1973). Storage of smoked products from cartilaginous fish. *Rybnoe Khozyaistvo* 3, 75–76.
6. Anisinov, I. G., Vyazum, T. M., Soletskii, O. I., Kostin, N. I., Zakharnchenko, A. D., Ostrikova, L. V., Shiler, G. G., Rozdov, I. A., Bankaev, E. S., Usmanov, R. M. & Utkin, K. I. (1990). Coating composition for cheese. URRS Patent SU 1.613.097.
7. Askar, A. & Treptow, H. (1986). Biogene amine in Lebensmitteln: Vorkommen, Bedeutung und Bestimmung. Stuttgart, Germany: Eugen Ulmer GmgH and Company.
8. Bailey, G. S., Scanlan, R. A., Selivonchick, D. P. & Williams, D. E. (1991). Food toxicology. In *Encyclopedia of Human Biology*. Vol. 3, pp. 671–681. Edited by R. Dulbelcco. New York: Academic Press.
9. Bailey, G. S. & Williams, D. E. (1993). Potential mechanisms for food-related carcinogens and anticarcinogens. *Food Technol* 47, 105–118.
10. Bardócz, S. (1995). Polyamines in food and their consequences for food quality and human health. *Trend Food Sci Technol* 6, 341–346.
11. Bauer, F., Seus, I., Paulen, P. & Valis, S. (1994). The formation of biogenic amines in meat products. In *Proc. 40th Int Cong. Meat Sci Technol,* Vol. 25, pp.1–3. Clermont-Ferrand, France.
12. Bauza, T., Blaise, A., Teissedre, P. L., Mestres, J. P., Daumas, F. & Cabanis, J. C. (1995). Changes in biogenic amines content in musts and wines during the winemaking process. *Sciences de Aliments* 15, 559–570.
13. Besançon, X., Smet, C., Chabalier, C., Rinemale, M., Reverbel, J. P., Ratomehenina, R. & Galzy, P. (1992). Study of surface yeast flora of Roquefort cheese. *Int J Food Microbiol* 17, 9–18.
14. Bharucha, K. R., Cross, C. K. & Rubin, L. J. (1985). Ethoxyquin, dihydroethoxyquin, and analogues as antinitrosamine agents for bacon. *J Agric Food Chem* 33, 834–839.
15. Bjorck, K. &Claesson, C. (1980). Method for protecting milk from bacterial spoilage by adding thiocyanate and the hydrogen peroxide in the form of alkali percarbonate, alkali-peroxide or urea hydrogen peroxide. Swedish Patent 412.838.
16. Boskou, G. & Debevere, J. (1998). *In vitro* study of TMAO reduction by *Shewanella putrefaciens* isolated from cod fillets packaged in modified atmosphere. *Int J Food Microbiol* 39, 1–10.
17. Bover-Cid, S., Izquierdo-Pulido, M. & Vidal-Carou, M. C. (1999). Effect of proteolytic starter cultures of *Staphylococcus* spp. on biogenic amine formation during the ripening of dry fermented sausages. *Int J Food Microbiol* 46, 95–104.
18. Brink, B. Ten, Damink, C., Joosten, H. M. L. J. & Huis in't Veld, J .H. J. (1990). Occurrence and formation of biologically active amines in foods. *Int J Food Microbiol* 11, 73–84.
19. Buncic, S., Paunovic, L., Teodorovic, V., Radisic, D., Vojinovic, G., Smiljanic, D. & Baltic, M. (1993). Effects of gluconodeltalactone and *Lactobacillus plantarum* on the production of histamine and tyramine in fermented sausages. *Int J Food Microbiol* 17, 303–309.
20. Buteau, C., Duitschaever, C. L. & Ashton, G. C. (1984). A study of the biogenesis of amines in a Villard noir wine. *Am J Enol Vitic* 35, 228–236.
21. Canas, B. J., Diachenko, G. W. & Nyman, P. J. (1997). Ethyl carbamate levels resulting from azodicarbonamide use in bread. *Food Add and Contam* 14, 89–94.

22. Cantoni, C., Bersani, C., Damenis, C. & Comi, G. (1994). Biogenic amines in typical Italian dry sausages. *Ind Aliment* 33, 1239–1243.

23. Celano, G. V., Cafarchia, C. & Tiecco, G. (1996). Formation of biogenic amines in Pecorino Pugliese cheese. *Ind Aliment* 35, 249–252.

24. Develi, N., Gokturk, S. & Ercan, I. (1997). Effects of cooking methods on N-nitrosamines in sucuk samples containing different nitrite concentrations. *Il-Kontrol-Laboratuvar-Mueduerluegue-Ankara* 36.

25. Diaz-Cinco, M. E., Fraijo, O., Grajeda, P., Lozano-Taylor, J. & Gonzales-de-Mejia, E. (1992). Microbial and chemical analysis of Chihuahua cheese and relationship to histamine and tyramine. *J Food Sci* 57, 355–356, 365.

26. Donhauser, S., Wagner, D. & Geiger, E. (1992). Biogenic amines: importance, occurrence and evaluation. *Brauwelt* 132, 1272–1274, 1276, 1278, 1280.

27. Dumont, E., de Geeter, H. & Huyghebaert, A. (1992). Presence and formation of biogenic amines in local Belgian beers. *Meded Fac Landbouwwet Rijksuniv Gent* 57, 1911–1913.

28. Eerola, S., Maijala, R., Roig-Sagues, A. X., Salminen, M. & Hirvi, T. (1996). Biogenic amines in dry sausages as affected by starter culture and contaminant amine-positive *Lactobacillus*. *J Food Sci* 61, 1243–1246.

29. Eerola, S., Roig-Sagues, A. X., Lilleberg, L. & Aalto, H. (1997). Biogenic amines in dry sausages during shelf-life storage. *Food Res Technol* 205, 351–355.

30. Fiddler, W., Pensabene, J. W., Gates, R. A., Custer, C., Yoffe, A. & Phillpo, T. (1997). N-nitrosodibenzylamine in boneless hams processed in elastic rubber nettings. *J AOAC Int* 80, 353–358.

31. Fiddler, W., Pensabene, J. W., Gates, R. A., Hale, M. & Jahncke, M. (1992). N-nitrosodimethylamine formation in cooked frankfurters containing Alaska pollock *(Theragra chalcogramma)* mince and surimi. *J Food Sci* 57, 569–571, 595.

32. Fink-Gremmels, J. & Leistner, L. (1986). Comparative toxicological evaluation of possible residues in meat. Research and practice in food hygiene. In *27th Meeting of the Food Hygiene Section. Garmisch-Partenkirchen, 9–12 Sept. 1986.*

33. Garcia-Roche, M. O. & Hernandez, A. M. (1989). Sources and presence of, and risk from nitrates and nitrites: situation in Cuba. *Alimentaria* 199, 67–71.

34. Gifhorn, A. & Meyer-Pittroff, R (1997). Measurement of NO_x contamination as a measure for prevention of nitrosodimethylamine formation in kilning of malt. *Brauwelt* 137, 1149–1153.

35. Giraffa, G., Carminati, D. & Neviani, E. (1997). Enterococci isolated from dairy products: a review of risks and potential technological use. *J Food Prot* 60, 732–738.

36. Granchi, L., Paperi, R., Rosellini, D. & Vincencini, M. (1998). Strain variation of arginine catabolism among malolactic *Oenococcus oeni* strain of wine origin. *Italian J Food Sci* 10, 351–357.

37. Halasz, A., Barath, A., Simon-Sarkadi, L. & Holzapfel, W. (1994). Biogenic amines and their production by microorganisms in food. *Trend Food Sci Technol* 5, 42–49.

38. Hauschild, A. H. W. (1982). Assessment of botulism hazards from cured meat products. *Food Technol* 36, 95–97.

39. Hecht, S. S. & Offman, D. (1988). Tobacco-specific nitrosamines: an important group of cancerigens in tobacco and tobacco smoke. *Carcinigenesis* 9, 875–884.

40. Hernandez-Herrero, M. M., Roig-Sagues, A. X., Rodriguez-Jerez, J. J. & Mora-Ventura, M. T. (1999). Halotolerant and halophilic histamine-forming bacteria isolated during the ripening of salted anchovies (*Engraulis encrasicholus*). *J Food Prot* 62, 509–514.

41. Holt, J. G., Krieg, N. R., Sfneath, T. H. A., Stanley, J. T. & Williams, S. T., eds. (1994). *Bergey's Manual of Determinative Bacteriology*, 9th edn. Baltimore: MD: Williams & Wilkins.

42. Hornero-Mendez, D. & Garrido-Fernandez, A. (1997). Rapid high performance liquid chromatography analysis of biogenic amines in fermented vegetable brines. *J Food Prot* 60, 414–419.

43. Hotchkiss, J. H. (1989). Preformed N-nitroso compounds in foods and beverages. *Adv Food Res* 31, 53–115.

44. Hotchkiss, J. H. & Cassens, R. G. (1987). Nitrate, nitrite, and nitroso compounds in foods. *Food Technol* 41, 127–134.

45. Huang , Z. & Ough, C. (1993). Identification of N-Carbamyl amino acids in wine and yeast cells. *Am J Enol Vitic* 44, 1, 49–55.

46. Huis in't Veld, J. H. J., Hose, H., Schaafsma, G. J., Silla, H. & Smith, J. E. (1990). Health aspects of food biotechnology. In *Processing and Quality of Foods.* Vol. 2, *Food Biotechnology: Avenues to Healthy and Nutritious Products*, pp. 2.73–2.97. Edited by P. Zeuthen, J. C. Cheftel, C. Ericksson, T. R. Gormley, P. Link & K. Paulus. London: Elsevier Applied Science.

47. Hwang, D. F., Chang, S. H., Shian, C. Y. & Cheng, C. C. (1995). Biogenic amines in the flesh of sailfish (*Istiophorus platypterus*) responsible for scombroid poisoning. *J Food Sci* 60, 926–928.

48. Ibe, A., Kamimura, H., Tabata, S., Hayano, K., Kimura, Y. & Tomomatsu, Y. (1996). Contents of non-volatile amines in fermented foods. II. *T Annu Rep Tokyo Metrop Res Lab Publ Health* 47, 90–94.

49. Ibe, A., Nishima, T. & Kasai, N. (1992). Bacteriological properties of and amine-production conditions for tyramine and histamine-producing bacterial strains isolated from soybean paste (miso) starting materials. *Jpn J Toxicol Environ Health* 38, 403–409.

50. Ibe, A., Nishima, T. & Kasai, N. (1992). Formation of tyramine and histamine during soybean paste (miso) fermentation. *Jpn J Toxicol Environ Health* 38, 181–187.

51. Ibe, A., Saito, K., Nakazato, M., Kikuchi, Y., Fujinuma, K. & Nishima, T. (1991). Quantitative determination of amines in wine by liquid chromatography. *J Assoc Off Anal Chem* 74, 695–698.

52. Ibe, A., Tamura, Y., Kamimura, H., Tabato, S., Hashimoto, H., Iida, M. & Nishida, T. (1991). Determination and contents of non-volatile amines in soybean paste and soy sauce. *Jpn J Toxicol Environ Health* 37, 379–386.

53. Ingledew, W. M., Magnus, C. A. & Patterson, J. R. (1987). Yeast foods and ethyl carbamate formation in wine. *Am J Enol Vitic* 38, 332–335.

54. Izquierdo-Pulido, M., Hernandez-Jover, T., Mariné-Font, A. & Vidal-Carou, M. (1996). Biogenic amines in European beers. *J Agric Food Chem* 44, 3159–3163.

55. Kamimote, S., Sumino, Y., Yamada, H., Imayasu, S., Ichikawa, E. & Suizu, T. (1989). Quality improvement of alcoholic liquors. U.S. Patent 4.844.911.

56. Kellner, V. (1994). Most significant foreign substances in brewing. *Kvasny-Prumysl* 40, 42–45.

57. Keraby, K. M. K., El-Sonbaty, A. H. & Badawi, R. M. (1999). Effects of heating milk and accelerating ripening of low fat Ras cheese on biogenic amines and free amino acids development. *Food Chem* 64, 67–75.

58. Kevekordes, S., Urban, M., Steffens, W., Janzowski, C., Eisenbrand, G. & Dunkelberg, H. (1996). Testing formation of nitrite and potency of nitrosation in a laboratory construction for heterotrophic biological denitrification of drinking water. *Zentralbl Hyg Umweltmedizin* 198, 462–472.

59. Kim, S. H., An, H. & Price, R. J. (1999). Histamine formation and bacterial spoilage of albacore harvested off the U.S. Norwthest coast. *J Food Sci* 64, 340–343.

60. Kodame, S. & Yotsusuka, F. (1996). Acid urease: reduction of ethyl carbamate formation in sherry under simulated baking conditons. *J Food Sci* 61, 304–307.

61. Koh, E. & Kwon, H. (1996). Determination of fermentation specific carcinogen, ethyl carbamate, in kimchi. *Korean J Food Sci Technol* 28, 421–427.

62. Kozhev, S. & Kozhev, A. (1994). Potassium nitrate residues in hard and semihard cheeses. *Khranitelna-Promishlenost* 43, 18–19.

63. Kunimote, M., Katsuhito, T. & Kaneniwa, M. (1992). Consumption of trimethylamine and trimethylamine N-oxide by *Aspergillus oryzae*. *Nippon Suisan Gakkaishi* 58, 1471–1476.

64. Lavanchy, P. & Sieber, R. (1993). Proteolysis in different hard and semihard cheeses. II. amines. *Schweiz Milchwirtsch Forsch* 22, 65–68.

65. Lehtonen, P., Saarinen, M., Vesanto, M. & Riekkola, M. L. (1992). Determination of wine amines by HPLC using automated precolumn derivation with o-phthalaldehyde and fluorescence detection. *Z Lebensm Unters Forsch* 194, 434–437.

66. Leuschner, R. G. K. & Hammes, W. P. (1998). Degradation of histamine and tyramine by *Brevibacterium linens* during surface ripening of Munster cheese. *J Food Prot* 61, 874–878.

67. Leuschner, R. G. K. & Hammes, W. P. (1998). Tyramine degradation by micrococci during ripening of fermented sausage. *Meat Sci* 49, 289–296.

68. Leuschner, R. G. K., Kurihara, R. & Hammes, W. P. (1998). Effect of enhanced proteolysis on formation of biogenic amines by lactobacilli during Gouda cheese ripening. *Int J Food Microbiol* 44, 15–20.

69. Lonvaud-Funel, A. & Joyeux, A. (1994). Histamine production by wine lactic acid bacteria: isolation of a histamine-producing strain of *Leuconostoc oenos*. *J Appl Bacteriol* 77, 401–407.

70. Lovaas, E. (1991). Antioxidative effects of polyamines. *J Am Oil Chem Soc* 68, 353–358.

71. Luten, J. B., Bouquet, W., Seuren, L. A. J., Burggraaf, M. M., Riekwel-Booy, G., Durand, P., Etienne, M., Gouyou, J. P., Landrein, A., Ritchie, A., Leclerq, M. & Guinet, R (1992). Biogenic amines in fishery products: standardization methods within EC. In *Quality Assurance in the Fish Industry*. Denmark: Ministry of Fisheries.

72. Maanen, J. M. S. van, Welle, I. J., Hageman, G., Dallinga, J. W., Martens, P. L. J. M. & Kleinjans, J. C. S. (1996). Nitrate contamination of drinking water: relationship with HPRT variant frequency in lymphocyte DNA and urinary excretion of N-nitrosamines. *Environ Health Perspect* 104, 522–528.

73. Maijala, R. L., Eerola, S. H., Aho, M. A. & Hirn, J. A. (1993). The effect of GDL-induced pH decrease on the formation of biogenic amines in meat. *J Food Prot* 56, 125–129.

74. Maijala, R., Eerola, S., Lievonen, S., Hill, P. & Hirvi, T. (1995). Formation of biogenic amines during ripening of dry sausages as affected by starter culture and thawing time of raw materials. *J Food Sci* 60, 1187–1190.

75. Massey, R. C. (1997). Estimation of daily intake of food preservatives. *Food Chem* 60, 177–185.

76. Masson, F., Lebert, A., Talon, R. & Montel, M. C. (1997). Effects of physico-chemical factors influencing tyramine production by *Carnobacterium divergens*. *J Appl Microbiol* 83, 36–47.

77. Masson, F. & Montel, M. C. (1995). Biogenic amines in meat products. *Viand Prod Carnés* 16, 3–7.

78. Mirvish, S. S. & Sams, S. S. (1984). N-nitroso compounds: occurrence, biological effects and relevance to human cancer. In *International Agency for Research on Cancer*, pp. 283–289. Edited by I. K. O'Neil, R. C. Von Borstel, C. T. Miller, J. Long & H. Bartsch. Lyon, France: Scientific Publications.

79. Mizutani, T., Kimizuka, A., Ruddle, K. & Ishige, N. (1992). Chemical components of fermented fish products. *J Food Comp Anal* 5, 152–159.

80. Moret, S., Conte, L. S. & Spoto, E. (1996). Biogenic amines in cheese: parameters affecting their analytical determination. *Ind Aliment* 35, 788–792.
81. National Academy Sciences/National Research Council. (1981). The health effects of nitrate, nitrite and N-nitroso compounds. *National Academy Press.*
82. National Academy Sciences/National Research Council. (1982). Alternatives to the current use of nitrite in foods. *National Academy Press.*
83. Neamat-Allah, A. A. (1999). Biogenic amines in Karish and Mish cheese in Egypt. *Egyptian J Dairy Sci* 25, 337–348.
84. Nguen, D. C. & Velikaya, E. (1972). Conditions for saccharification and fermentation of manioc mash. *Ferment Spirtov Promysh Lennost* 1, 17–21.
85. Nout, M. J. R. (1994). Fermented foods and food safety. *Food Res Int* 27, 291–298.
86. Nout, M. J. R., Ruikes, M. M. W., Bouwmeester, H. M. & Beljaars, P. R. (1993). Effect of processing conditions on the formation of biogenic amines and ethyl carbamate in soybean tempe. *J Food Saf* 13, 293–303.
87. Oesterdahl, B. G. & Alriksson, E. (1990). Volatile nitrosamines in microwave-cooked bacon. *Food Add and Contam* 7, 51–54.
88. Offman, D., Djordjevic, M. V. & Brunnemann, K. D. (1991). New brands of oral snuffs. *Food Chem Toxicol* 29, 65–68.
89. Ologhobo, A. D., Adegede, H. I. & Maduagiwu, E. N. (1996). Occurrence of nitrate, nitrite and volatile nitrosamines in certain feedstuffs and animal products. *Nutr Health* 11, 109–114.
90. Ough, C. S., Huang, Z. & Stevens, D. (1991). Amino acid uptake by four commercial yeasts at two different temperatures of growth and fermentation: effects on urea excretion and reabsorption. *Am J Enol Vitic* 42, 26–40.
91. Park, K. Y. (1995). The nutritional evaluation, and antimutagenic and anticancer effects of kimchi. *J Korean Soc Food Nutr* 24, 169–182.
92. Petridis, K. D. & Steinhart, H. (1996). Biogenic amines in hard cheese production. I. factors influencing the biogenic amine content in end-product by way of Swiss cheese. *Dtsch Lebensm Rundsch* 92, 114–120.
93. Pfundstein, B., Tricker, A. R., Theobald, E., Spiegelharder, B. & Preussmann, R. (1991). Main daily intake of primary and secondary amines from foods and beverages in West Germany in 1989–1990. *Food Chem Toxicol* 29, 733–739.
94. Pogorzelski, E. (1992). Studies on the formation of histamine in must and wines from elderberry fruit. *J Sci Agric* 60, 239–244.
95. Poocharoen, B., Barbur, F. J., Libbey, L. M. & Scanlan, R. A. (1992). Precursors of N-nitrosodimethylamine in malted barley. I. determination of ordenine and gramine. *J Agric Food Chem* 40, 2216–2221.
96. Potthast, K. (1991). Nitrite and nitrate—key compounds for a number of reactions during curing of meat and meat products. *Mitt Gebiete Lebensm Hyg* 82, 24–35.
97. Pylypiw, H. M. Jr., Harrington, G. W., Kozeniauskas, R. & Diaz, D. (1992). The formation of N-nitrosodimethylamine and N-nitrodimethylamine from the plant growth regulator Daminozide (Alar). *J Food Prot* 55, 218–219.
98. Reichert, J. E. (1994). Influence of ascorbic acid and sodium ascorbate on quality of meat products. *Aliment Equipos Tecnol* 13, 67–68.
99. Rodriguez-Jerez, J. J., Lopez-Sabater, E. I., Hernandez-Herrero, M. M. & Mora-Ventura, M. T. (1994). Histamine, putrescine and cadaverine formation in Spanish semipreserved anchovies as affected by time/temperature. *J Food Sci* 59, 993–997.
100. Roig-Sagues, A. & Eerola, S. (1997). Biogenic amines in meat inoculated with *Lactobacillus sake* starter strains and an amine-positive lactic acid bacterium. *Food Res Technol* 205, 227–231.
101. Ross, H. D., Henion, J., Babish, J. G. & Hotchkiss, J. H. (1987). Nitrosating agents from the reaction between methyl oleate and dinitrogen trioxide: identification and mutagenicity. *Food Chem* 23, 207–222.
102. Rudnitskii, P. V. (1973). New methods for improving the quality of alcohol and molasses by-products. *Trudy Ukrainskii Nauchno Inst Spirtovoi Likero Promyshlnnosti* 15, 5–9.
103. Sayem-El-Daher, N., Simard, R. E. & Fillion, J. (1984). Changes in the amine content of ground beef during storage and processing. *Lebensm Wiss Technol* 7, 319–323.
104. Schneller, R., Good, P. & Jenny, M. (1997). Influence of pasteurized milk, raw milk and different ripening cultures on biogenic amine concentrations in semi-soft cheeses during ripening. *Food Res Technol* 204, 265–272.
105. Sen, N. P., Kushkawa, S. C., Seaman, S. W. & Clarkson, S. G. (1985). Nitrosamines in baby bottle nipples and pacifiers: occurrence, migration, and effect on infant formulas and fruit juices on *in vitro* formation of nitrosamines under stimulated gastric conditions. *J Agric Food Chem* 33, 428–433.
106. Sen, N. P., Seaman, S. W., Bergeron, C. S. W. & Brousseau, R. (1996). Trends in the levels of N-nitrosodimethylamine in Canadian and imported beers. *J Agric Food Chem* 44, 1498–1501.
107. Shahidi, F. & Pegg, R. B. (1994). Absence of volatile N-nitrosamines in cooked nitrite-free cured muscle foods. *Meat Sci* 37, 327–336.
108. Shahidi, F., Synowiecki, J. & Sen, N. P. (1995). N-nitrosamines in nitrite-cured chicken-seal salami. *J Food Prot* 58, 446–448.
109. Shalaby, A. R. (1993). Survey on biogenic amines in Egyptian foods: sausage. *J Sci Food Agric* 62, 291–293.

110. Shalaby, A. R. (1996). Significance of biogenic amines to food safety and human health. *Food Res Int* 29, 675–690.

111. Shenoy, N. R. & Choughuley, A. S. U. (1989). Effect of certain plant phenolics on nitrosamine formation. *J Agric Food Chem* 37, 721–725.

112. Silla, M. H. (1985). Utilización de microorganismos fermentadores en la conservación de alimentos. *Rev Agroquím Tecnol Aliment* 25, 170–182.

113. Silla-Santos, M. H. (1996). Biogenic amines: their importance in foods. *Int J Food Microbiol* 29, 213–231.

114. Silla-Santos, M. H. (1998). Amino acid decarboxylase capability of microrganisms isolated in Spanish fermented meat products. *Int J Food Microbiol* 39, 227–230.

115. Silva, C. C. G., Ponte, D. J. B. da & Dapkevicius, M. L. A. E. (1998). Storage temperature effect on histamine formation in big eye tuna and skipjack. *J Food Sci* 63, 644–647.

116. Smith, N. A. (1994). Nitrate reduction and N-nitrosation in brewing. *J Inst Brew* 100, 347–355.

117. Soloviev, A. I., Pivovarov, A. I., Knyazev, V. A. & Starushenko, A. K. H. (1987). Storage of sugar beet under carbamide formaldehyde foamed plastics. *Sakkarayc Promyshlennost* 1, 40–41.

118. Soufleros, E., Barrios, M. L. & Bertrand, A. (1999). Correlation between the content of biogenic amines and other wine components. *Am J Enol Vitic* 49, 266–278.

119. Sponholz, W. R., Kuerbel, H. & Dittrich, H. H. (1991). Formation of ethyl carbamate in wine. *Wein Wissensch* 46, 11–17.

120. Stevens, D. F. & Ough, C. S. (1993). Ethyl carbamate formation: reaction of urea and citrulline with ethanol in wine under low to normal temperature conditions. *Am J Enol Vitic* 44, 309–312.

121. Stratton, J. E., Hutkins, R. W. & Taylor, S. L. (1991). Biogenic amines in cheese and other fermented foods: a review. *J Food Prot* 54, 460–470.

122. Straub, B. W., Kicherer, M., Schilcher, S. M. & Hammes, W. P. (1995). The formation of biogenic amines by fermentation organisms. *Z Lebensm Untersuch Forsch* 201, 79–82.

123. Straub, B., Schollenberger, M., Kicherer, M., Luckas, B. & Hammes, W. P. (1993). Extraction and determination of biogenic amines in fermented sausages and other meat products using reversed-phase HPLC. *Lebensm. Untersuch Forsch* 197, 230–232.

124. Surono, I. S. & Hosono, A. (1996). Antimutagenicity of milk cultured with lactic acid bacteria from dadih against mutagenic terasi. *Milchwissensch* 51, 493–497.

125. Teodorovic, V. & Buncic, S. (1999). The effect of pH on tyramine content in fermented sausages. *Fleischwirtsch* 79, 85–88.

126. Teodorovic, V. B., Buncic, S. & Smiljanic, D. (1994). A study of factors influencing histamine production in meat. *Fleistwirtsch* 74, 181–183.

127. Tiecco, G., Tantillo, G., Francioso, E., Paparella, A. & Natale, D. (1986). Qualitative and quantitative determination of some biogenic amines in sausages during ripening. *Ind Aliment* 25, 209–213.

128. Tricker, A. R. & Preussmann, R. (1990). Chemical food contaminants in the initiation of cancer. *Symp Diet Cancer Proc Nutr Soc* 49, 133–144.

129. Trova, C., Gandolfo, G. & Cossa, G. (1993). N-nitrosodimethylamine levels in beer. *Ind Bev* 22, 528–530.

130. Tsen, C. C. (1973). Chemical dough development. *Bakers' Digest* 47, 44–47, 129.

131. Tunon, H. (1994). Hops—not just for adding bitterness to beer. *Kem Tidskrift* 106, 30–32.

132. Vale, S. R. & Gloria, M. B. A. (1997). Determination of biogenic amines in cheese. *J AOAC Int* 80, 1006–1012.

133. Valeiro Novo, C. & Pilas Perez, A. (1994). Problemas sanitarios asociados al consumo de pescados. *Aliment* 250, 55–67.

134. Veciana-Nogues, M. T., Mariné-Font, A. & Vidal-Carou, M. C. (1997). Biogenic amines as hygienic quality indicators of tuna: relationships with microbial counts, ATP-related compounds, volatile amines, and organoleptic changes. *J Agric Food Chem* 45, 2036–2041.

135. Vidal-Carou, M. C., Izquierdo-Pulido, M. L., Martin-Morro, M. C. & Mariné-Font, A. (1990). Histamine and tyramine in meat products: relationship with meat spoilage. *Food Chem* 37, 239–249.

136. Vincze, L. (1979). Factors affecting feed protein utilization by broiler chickens. *Agrar Koezl* 38, 393–398.

137. Watkins, S. J., Hogg, T. A. & Lewis, M. J. (1996). The use of yeast inoculation in fermentation for port production; effect on total potential ethyl carbamate. *Biotechnol Lett* 18, 95–98.

138. Wendakoon, C. N. & Sakaguchi, M. (1993). Combined effect of sodium chloride and clove on growth and biogenic amine formation of *Enterobacter aerogenes* in Mackerel muscle extract. *J Food Prot* 56, 410–413.

139. Yankah, V. V., Ohshima, T. & Koizumi, C. (1993). Effects of processing and storage on some chemical characteristics and lipid composition of a Ghanaian fermented fish product. *J Sci Food Agric* 63, 227–235.

140. Yatsunami, K. & Echigo, T. (1993). Studies on halotolerant and halophilic histamine-forming bacteria. III. Changes in the number of halotolerant histamine-forming bacteria and contents of non-volatile amines in sardine meat with addition of NaCl. *Bull Jpn Soc Sci Fish* 59, 123–127.

141. Yen, G. C. & Kao, H. H. (1993). Antioxidative effect of biogenic amine on the peroxidation of linoleic acid. *Biosci Biotech Biochem* 57, 115–116.

Chapter 7

Microbiological Hazards and Their Control: Bacteria

R. R. Beumer

INTRODUCTION

The inactivation or inhibition of undesirable microorganisms is an essential part of food preservation. Foods can be made safe by treatments such as pasteurization or sterilization (e.g., milk, cooked meat products, and canned foods), or they may possess intrinsic properties that contribute to safety, such as structure, low pH, and/or low water activity (A_w) (e.g., fruit, pickles, and honey). One further way of improving shelf life and safety, through the use of a competitive microbial flora, is widely employed in the production of fermented milks, vegetables, meat, fish, and grains, giving products with changed composition and taste as well as a prolonged shelf life. It is estimated that approximately 30% of our food supply is based on fermented products,[40] though not all fermented foods are universally accepted. The modified sensory properties in some fermentation products are acceptable only in certain regions; elsewhere, they would be considered spoiled or unpalatable.

Fermented foods have generally been considered as less likely to be vehicles for food-borne infection or intoxication than fresh foods due to the competitive activity and metabolites of the functional flora[49] (see Chapter 2). All raw materials used in fermentation processes, such as fruits, vegetables, meat, fish, milk, and eggs, can harbor pathogenic bacteria.[60] Depending on the type of fermentation (e.g., lactic acid, alcoholic, or mold), these pathogens may be (partly) eliminated or their growth inhibited by the antibacterial substances that are produced (e.g., lactic acid, acetic acid, alcohol). To ensure a satisfactory fermentation and to control the growth of food-borne pathogens, the use of specific starters is often recommended.

In this chapter, the safety of fermented products will be discussed in relation to the pathogenic bacteria: *Aeromonas hydrophila*, *Bacillus cereus*, *Campylobacter*, *Clostridium botulinum*, *Escherichia coli* O157:H7, *Listeria monocytogenes*, *Salmonella* spp., *Shigella*, *Staphylococcus aureus*, *Vibrio* spp., and *Yersinia enterocolitica*.

OUTBREAKS RELATED TO FERMENTED PRODUCTS

A great part of food-borne disease is preventable. In addition to good manufacturing practices (GMPs) and the Hazard Analysis Critical Control Point (HACCP) system, domestic hygienic practices have a key role to play in preventing infection.[9,12,52] Most food-borne diseases are attributed to the consumption of, or cross-contamination from, raw or undercooked food products (e.g., vegetables, sprouts, shellfish, and poultry), in many cases with an extended period between preparation and consumption.[14]

The antibacterial factors that are present in fermented foods may affect both the growth and the survival of bacterial pathogens that are present in a raw material. In most fermented foods, the inhibition of growth is more common and can often ensure safety where levels of contamination are low. But with infectious pathogens, particularly those with a small minimum infectious dose, some degree of inactivation may be necessary to provide an acceptable level of safety. So, although fermented foods are generally considered to be safe, process failures and contaminated raw materials have resulted in their being involved in food-borne illness. For example, several outbreaks of illness have been attributed to the consumption of fermented sausages contaminated with *S. aureus* and *Salmonella* spp.,[70] and emerging pathogens, such as *L. monocytogenes* and *E. coli* O157:H7, have been identified as causative organisms in outbreaks involving fermented products such as sausages, cheeses, and yogurt.[8,27,47]

These outbreaks have raised questions regarding the safety of fermented food products and have prompted studies using artificially contaminated raw materials to identify critical control points (e.g., fermentation pH, final heating temperature, heating time) and critical limits (e.g., pH 4.6, 55 °C for 20 minutes) that can be integrated into an HACCP plan to ensure safe products.[23]

BACTERIAL PATHOGENS OF CONCERN IN FERMENTED PRODUCTS

There is extensive literature describing the limits for the growth of bacterial pathogens. Some of the data relevant to fermentation processes are summarized in Table 7–1,[61] and a few key features of these organisms are presented in the following paragraphs.

Table 7–1 Limits for Growth of Some Common Bacterial Pathogens

Organism	*Minimum pH*	*Minimum Temperature*	*Minimum A_w (Max % NaCl)*
Aeromonas hydrophila	< 4.5	> 0.0, < 4.0	–
			(5–6%)
Bacillus cereus	5.0	4.0	0.93
Campylobacter	4.9	30.0	> 0.987
			(1.5%)
Clostridium botulinum			
Group 1: Proteolytic	4.6	10.0–12.0	–
Group 2: Nonproteolytic	5.0	3.3	(10%)
			(5%)
Escherichia coli	4.4	7.0–8.0	0.95
Listeria monocytogenes	4.4	–0.4	0.92
Salmonella	3.8	5.2*	0.94
Shigella	4.9–5.0	6.1–7.9	–
			(3.78–5.18%)
Staphylococcus aureus			
Growth	4.0	7.0	0.83
Toxin production	4.5	10.0	0.87
Vibrio cholerae	5.0	10.0	0.97
			(4.0%)
Vibrio parahemolyticus	4.8	5.0	0.94
			(10%)
Yersinia enterocolitica	4.2	–1.3	–
			(> 5%,< 7%)

*Most serotypes fail to grow below 7 °C.

Aeromonas

Aeromonas are facultatively anaerobic, Gram-negative rods (belonging to the *Vibrionaceae*) that are found in water (including sewage) and in food products that have been in contact with contaminated water (e.g., seafood, vegetables). Some species are well known as fish pathogens; others are human pathogens. *A. hydrophila* seems the most important species; however, its significance in the epidemiology of human illness is still unclear. Symptoms associated with infection include diarrhea, abdominal pain, and headache, and are more severe in children.[58,61]

Aeromonas do not appear to be a serious risk in well-produced fermented foods. The addition of *Aeromonas* to skim milk during lactic fermentation,[57] to yogurt,[6] and to two traditional fermented foods (mahewu and sour porridge) resulted in a sharp decrease in numbers.[65] However, microbiological examination of homemade fermented sorghum porridge[41] and industrially produced fermented sausages (longaniza and chorizo) showed its presence in numbers ranging from 1.0 $\log_{10}$ cfu/g to 4.5 $\log_{10}$ cfu/g. It was concluded that the hygienic status of factories significantly affected the incidence and counts of *Aeromonas*.[24]

Campylobacter

Campylobacter species are micro-aerophilic, Gram-negative curved or spiral rods belonging to the *Campylobacteriaceae*. Members of this genus have been associated with disease in animals for many years. From 1970, the thermophilic campylobacters (growth from 30–45 °C) have been recognized as important food-borne pathogens; they are now the most common cause of gastroenteritis in many developed countries. *C. jejuni, C. coli,* and *C. lari* are responsible for more than 90% of the human illness caused by campylobacters, which are normally characterized by profuse diarrhea and abdominal pain three to five days after consumption. Diarrhea can often be bloody, and more serious sequelae such as reactive arthritis and Guillain-Barré syndrome have also been reported.[1]

Enormous numbers of campylobacters may be found in the gut of chickens (and other birds) and pigs. *C. lari* is often found in seagulls and, as a consequence, in shellfish.[34,61] Raw milk and poorly cooked poultry products are most frequently involved in outbreaks, and cross-contamination from raw poultry to ready-to-eat products may also be an important factor in transmission. Although the infective dose of *Campylobacter* is relatively low (approximately 500 cfu), *Campylobacter* transmission by fermented foods is probably not significant. They do not grow below 30 °C and are sensitive to freezing, drying, low pH, and sodium chloride, and so would not survive well in fermented foods. This has been confirmed by artificial contamination of fermenting products such as salami, weaning foods, and yogurt, which resulted in a rapid decrease in numbers.[39,44,48]

Vibrio spp.

Vibrios are facultatively anaerobic, Gram-negative bacteria belonging to the family of the *Vibrionaceae*, and are commonly found in aquatic environments.

V. cholerae includes the true cholera-producing bacteria ("classic" and "El Tor" biovars) and the so-called "non O1 strains," which lack the somatic "O1" antigen and produce a less severe gastroenteritis. Cholera has been known since ancient times and is still a disease of worldwide significance. The organism is commonly isolated from river waters and coastal marine waters and, as a consequence, from shellfish and other marine animals,[61,71] but its transmission by fermented fish products has not been reported. The use of contaminated water for washing or irrigation may introduce the pathogen onto vegetables. Refrigerated or frozen storage of contaminated food products will not ensure safety because of the good survival of vibrios at low temperatures.[19]

V. parahaemolyticus is a halotolerant pathogen that is commonly found in marine coastal waters in warm water regions. Fish and shellfish are usually associated with outbreaks caused by this pathogen, as are, to a lesser extent, salt-preserved vegetables.[18, 61]

V. vulnificus is very similar to *V. parahaemolyticus* but differs from the other pathogenic, halotolerant vibrios in that diarrhea is not a common symptom of infection. *V. vulnificus* is highly invasive, causing septicemia. Illness is clearly associated with liver diseases such as cirrhosis and hepatitis. The majority of cases of illness have been reported after the consumption of raw shellfish, in particular, raw oysters.[38, 61]

Vibrios are not acid tolerant and lactic acid bacteria (LAB) isolated from fermented fish products inhibited both *V. cholerae* and *V. parahaemolyticus,* indicating that the antibacterial activity of the functional flora is important in ensuring the safety of fermented fish products.[56] This has also been observed in pickled foods and squid shiokara that were artificially contaminated with *V. parahaemolyticus*.[72,73]

E. coli O157:H7, *Salmonella* spp., *Shigella* spp., and *Y. enterocolitica*

All members of the Enterobacteriaceae family are facultatively anaerobic, oxidase-negative, Gram-negative rods that are capable of fermenting glucose. Some are known pathogens, such as *E. coli* (especially *E. coli* O157), *Salmonella, Shigella,* and *Yersinia,* and a great part of the family, including the food-borne pathogens, are components of the fecal flora of animals. As a result, the whole family or parts of it (coliforms and *E. coli*) are used as hygiene indicators for processed foods.

E. coli O157:H7

On the basis of virulence factors and disease patterns, human pathogenic *E. coli* strains are classified in the following principal groups:

- Enteropathogenic *E. coli* (EPEC), acute watery diarrhea; young children particularly susceptible
- Enterotoxigenic *E. coli* (ETEC), acute watery diarrhea (often in travelers)
- Entero-invasive *E. coli* (EIEC), dysentery-like syndrome
- Enterohemorrhagic *E. coli* (EHEC), bloody diarrhea syndrome

Within the EHEC-group, *E. coli* O157:H7 is the organism that is isolated most frequently in cases of hemorrhagic colitis (HC) and hemolytic uremic syndrome (HUS). Because this pathogen is relatively acid resistant, it is less affected by fermentation and can be of concern in fermented foods such as yogurt and fermented sausages.[22,25,43,61] Several food-borne outbreaks of *E. coli* have been reported in recent years, including some involving fermented meats and cheeses.[17,63,64,67,69]

Salmonella

Salmonella is often found on raw meat (poultry) and is widely distributed in the environment and universally recognized as an important cause of food-borne infections. Salmonellas are present in the gut of contaminated animals (and humans) and are shed in the feces. As a result, a great variety of raw materials can be contaminated with this pathogen. Illness is characterized by diarrhea, abdominal pain, and fever generally 6–48 hours after consumption of the organism.[1] Meat, milk, poultry, and eggs are the principle vehicles for *Salmonella* transmission, and fermented foods derived from these materials, such as salami and cheeses, have occasionally been associated with outbreaks of illness. Recently, acid-resistant strains of *Salmonella* have been implicated in food-borne outbreaks,[42,61] which led to many studies on the fate of salmonellas inoculated in fermenting products such as finger millet,[4] Siljo,[21] and fermented sausages.[23,33,47,62]

Shigella

Shigella is closely related to *E. coli* (DNA homology and biochemical pattern). Shigellas are not natural inhabitants of the environment but originate from humans and higher primates in the acute phase of illness. Both person-to-person infections (due to poor personal hygiene) and the consumption of contaminated water or foods washed with contaminated water give rise to distinct symptoms of the disease, such as bloody diarrhea. In some countries, shigellas are epidemic.[61] Artificial contamination of products with low pH values (< 5.0) results in a rapid decrease in numbers of shigellas, suggesting that most acid-fermented foods will not be significant as vehicles for this type of infection.[41,53]

Y. enterocolitica

Y. enterocolitica is a widely distributed psychrotrophic bacterium. Certain serotypes (O:3, O:5, 27; O:8, and O:9) present in raw milk, seafood, and raw pork have been associated with food-borne infections. The tonsil area of pigs is a unique ecological niche with a high incidence of *Y. enterocolitica*. Gastroenteritis is the most common symptom; however, sometimes illness is also characterized by fever and abdominal pain (resembling appendicitis). In some cases, this has led to surgery.[61] There are relatively few outbreaks reported with this pathogen; however, in many of these, milk was the incriminated product. The behavior of *Yersinia* has been studied in fermenting milk[57] and yogurt.[11] During the first few hours of the fermentation, growth was observed, followed by a reduction below the detection level after fermentation and storage for four days.

B. cereus

B. cereus is an aerobic, Gram-positive spore-forming rod (belonging to the *Bacillaceae*) that can cause both food-borne infection (characterized by diarrhea) and intoxication (vomiting after ingestion of the toxin cereulide). As a spore former, this bacterium is ubiquitous and may be found in cereals and their products, such as rice and flour, and spices. Small numbers in foods are not considered significant.[61] The behavior of *B. cereus* was studied in fermenting products such as *mageu*, a sour maize beverage;[15] *kinema*, a fermented soya bean food;[50] tempeh;[51] salads;[13] and fish sausage.[5] In all cases, initial growth was observed, but in products where pH decreased due to the growth of LAB, subsequent inhibition occurred that was correlated to the rate of decrease in pH. In products without a lactic acid fermentation (e.g., tempeh), numbers reached 10^8 cfu/g. However, if the soya beans were soaked, resulting in acidification of the beans to pH less than 4.5, the growth of *B. cereus* was prevented.[51] Interestingly, food-borne illness has also been associated with other *Bacillus* species such as *B. subtilis*, *B. licheniformis,* and *B. pumilis,* and strains of many of these are associated with the production of fermented vegetable protein products such as Japanese *natto*, Nigerian *iru* (*dawadawa*), Indian *kinema,* and Thai *thua-nao,* but, as far as we are aware, there have not been any recorded outbreaks of illness associated with them.

C. botulinum

C. botulinum is an anaerobic, Gram-positive, spore-forming rod (belonging to the *Bacillaceae*). Food-borne botulism is a neuroparalytic syndrome caused by the botulinum neurotoxin that is produced by vegetative cells of *C. botulinum* growing in food. This bacterium is ubiquitous, and spores are widely distributed in the soil and on the shores and bottom deposits of lakes and coastal waters, and in the intestinal tract of fish and animals.[61] The human pathogens are classified into proteolytic and nonproteolytic strains; the former being slightly more acid tolerant and mesophilic, whereas the latter can grow down to 3.3 °C. Most outbreaks of food-borne botulism have been caused by home-processed vegetables, fish, or meat products.[59] Although *C. botulinum* spores are present in most raw agricultural products, the number of outbreaks is relatively small, usually occurring after inadequate processing. Nitrite inhibits *C. botulinum* and is an important safety factor in the production of cured and fermented meats. In Japan, botulism caused by nonproteolytic strains producing type E toxin is often associated with the consumption of *izushi*, which is made by the natural fermentation of raw fish and cooked rice.[32] The largest recorded outbreak of food-borne botulism in the United Kingdom was associated with the consumption of contaminated hazelnut yogurt, but in this case, production of the toxin had occurred in the hazelnut puree that was added to the yogurt rather than in the yogurt itself (see also Chapter 5).[55]

S. aureus

S. aureus is a facultatively anaerobic, Gram-positive coccus belonging to the *Micrococcaceae.* It can produce several heat-stable enterotoxins that, after ingestion of the food, may cause food poisoning with symptoms of nausea and vomiting. The production of staphylococcal enterotoxins occurs mostly in food products that are rich in proteins with a low number of competitive microorganisms (e.g., whipped cream, cooked meat products, smoked fish). *S. aureus* is present on the skin and mucous membranes of warm-blooded animals (including humans). Approximately 50% of the human population are carriers of this pathogen. Contamination of cooked or ready-to-eat food products by a colonized person followed by storage for several hours at ambient temperatures is often implicated in outbreaks of illness.[61] The organism is relatively salt tolerant and will produce toxin down to pH values of 4.5. Fermented sausages[7] and, to a lesser extent, raw milk cheeses[35,36] have been associated with outbreaks of illness, although the organism is generally regarded as a poor competitor and its growth in fermented foods is generally associated with a failure of the normal flora.[54]

L. monocytogenes

L. monocytogenes is a facultatively anaerobic, Gram-positive rod. Species of *Listeria* are ubiquitous and potential food contaminants, but only *L. monocytogenes* is a human and animal pathogen. Approximately 30% of cases of listeriosis are perinatal, and in 20–25% of the cases, infection proves fatal for the fetus or newborn. Most other cases of listeriosis occur in immunocompromised persons, with a death rate varying from 30% to 50%.[61]

Due to its widespread occurrence in the environment, this pathogen can be harbored in low numbers by practically all raw food products, and processed foods can be contaminated from the production environment. *L. monocytogenes* is able to grow at refrigeration temperatures, so a considerable increase in numbers is possible in contaminated products with a long shelf life (e.g., cooked meat products, smoked salmon).

Fermented products such as sausages and soft mold-ripened cheeses have been associated with listeriosis.[26,30] The organism is not particularly acid tolerant, but where mold ripening occurs in a fermented product, the rise in pH can allow surviving *Listeria* to resume growth, causing problems to occur.

PRESENCE OF PATHOGENS IN RAW MATERIALS

Bacteria can enter raw materials or food products at various stages in the food chain. Primary production is a major source of both spoilage and pathogenic organisms. The primary contamination of raw materials includes microorganisms from soil, feces, surface water, and so forth, and will also be influenced by conditions of harvesting, slaughter, storage, and transport. During food production, the extent of secondary contamination will be influenced by factory layout, hygienic design of the equipment, personal hygiene, water, air, (pest) animals, and packaging material.[10]

Many of the bacteria introduced into food are of little concern. They are not well adapted to survival or growth in the product or are easily eliminated in further processing steps. However, a certain proportion of bacteria may cause problems and are a good reason to try to keep the microbial load of raw materials as low as possible, particularly for those products that reach the consumer in a raw state (e.g., meat, poultry, seafood, eggs, and vegetables). Pathogen-free raising of livestock would reduce risk, but this is difficult and expensive to achieve. There is clearly an urgent need for efficient and acceptable methods to decrease the numbers of spoilage microorganisms and pathogens from such products. Ionizing radiation is very effective, but consumer acceptance of this procedure is still a problem. Decontamination by a lactic acid dip or spray decreases the numbers of some pathogens on carcasses, although acid sprays appear to produce little reduction in *E. coli* (including *E. coli* O157) and *Salmonella*.[20,66] Food legislators in many countries also have objections to this method on the grounds that it may mask poor hygiene or lead to the adaptation of microorganisms to low pH.

Table 7–2 presents the relationship between the pathogens discussed in this chapter and raw materials used in fermentation processes.

It has been estimated that approximately 30% of the human food supply is based on fermented products. Raw materials used are vegetables (e.g., cabbage, cucumber, olives), cereals (e.g., rice, maize, wheat), legumes (e.g., soybean), fruit (e.g., grapes), meat (particularly pork), and milk.[16] From Table 7–2, it is clear that these raw materials may be contaminated with usually low numbers of some of the pathogens listed.

Inactivation, survival, and/or growth during and after fermentation depends on intrinsic factors (e.g., pH and A_w), extrinsic factors (e.g., storage temperature and modified air packaging), implicit factors (e.g., antagonism and the production of antibacterial substances), and process factors (e.g., heat treatment), as well as the physiological characteristics of the organisms themselves.

EFFECT OF FERMENTATION PROCESSES ON THE SURVIVAL OF BACTERIAL PATHOGENS

The effect of fermentation processes on the survival of bacterial pathogens is discussed more extensively in Chapter 2. Before the main fermentation process starts, nonmicrobial process factors such as peeling, washing, soaking, and/or grinding may reduce the numbers of pathogens present in food. Removal of the outer layer (peeling) has the greatest effect, whereas washing of the raw materials usually reduces total counts by 1 log cycle. Sometimes, a higher reduction can be obtained by adding (organic) acids or disinfectants; however, this procedure is not allowed in all countries. Moreover, a major disadvantage to adding acids or disinfectants is the possibility of selecting strains adapted to low pH or to disinfectants if concentrations are too low to be detrimental.[37,66,68] In particular, the acid-adapted pathogens have shown better survival in fermented products.[28,42]

Processes such as mixing and grinding have no lethal effect on bacteria. The only result of these steps is a better distribution of cells in the product and the separation of clumps of cells.

The main factors controlling the survival and/or growth of food-borne pathogens are (low) pH,

Table 7–2 Raw Materials and the Presence of Some Food-borne Pathogens

Pathogen	*Meat*	*Milk*	*Vegetables & Fruit*	*Fish & Shellfish*
Aeromonas	++	+	+++	++
Bacillus cereus	–	++	++	–
Campylobacter	+++	+	–	–
Clostridium botulinum	++	+	++	++
Escherichia coli O157:H7	++	++	+	–
Listeria monocytogenes	+	+	+	+
Salmonella	++	+	+	+
Shigella	–	–	+	+
Staphylococcus aureus	++	++	–	–
Vibrio	–	–	+	+++
Yersinia	+	+	–	–

+ = sometimes present (25%), ++ = likely to be present (25–50%), +++ = usually present (> 50%), and – = not likely to be found in these products.

A_w, heat treatment, and cold storage (see Chapter 2). In Table 7–3, the effect of various fermentation processes is shown on pH and A_w values in the final products and the possible production of specific antimicrobial substances.

It is generally accepted that in well-fermented foods, the production of organic acids and a low A_w (due to the addition of salt or drying) may control the growth of pathogens in these products. However, in view of the values of pH and A_w in the end products (Table 7–3), it can be concluded that many food-borne pathogens, especially in products with relatively high final pH values, can survive. They may even grow if the A_w is also high enough. A critical point for the growth of pathogens is the beginning of the fermentation process, when the pH is still high and the competitive effect of the functional flora is low. Some alcoholic fermentations may be a possible exception, where a low substrate pH value may prevent pathogen growth.[2,46]

The abundance of literature on the inhibition of food-borne pathogens by LAB in fermentation processes is confusing. There are variations in the types of products (e.g., a wide variety of sausages, fermented milks, etc.) and considerable differences in protocols for growth and addition of the pathogen, which makes comparison of the results obtained by different researchers very difficult.

During fermentation, antimicrobial products are produced (see Chapter 2). Most of these, including organic acids, hydrogen peroxide, carbon dioxide, diacetyl, and bacteriocins, are produced by LAB. There is considerable evidence on the antimicrobial activity of organic acids such as lactic acid and acetic acid, but effective concentrations of hydrogen peroxide, carbon dioxide, and diacetyl are rarely present at levels sufficient to make a major contribution to antibacterial activity.

The value of bacteriocins is limited by the fact that they are usually only active against Gram-positive cells, although theoretically this could be overcome by the addition of chelating agents such as Ethylenediaminetetraaceticacid, or EDTA (see Chapter 2). In practice, it might be of more advantage to concentrate on other factors such as choice of raw materials and the use of GMP and the HACCP concept.

THE CONTROL OF MICROBIAL HAZARDS

Although many fermented foods are a less likely vehicle for bacterial food-borne illness than fresh foods, hygienic practices during fermentation (especially in traditional household fermentations), the possible use of contaminated raw materials, and the low stability of these

Table 7–3 Effect of Various Fermentation Processes on the Survival and Growth of Food–borne Pathogens

Type of Fermentation Substrates	*pH*	*A_w*	*Antagonism**
Lactic acid fermentation			
Meat, pork, poultry	4.5–6.0	0.85–0.99	B, H, C, D
Cabbage, vegetables	< 4.0–5.5	> 0.96	B, H, C, D
Milk	< 4.0–5.5	> 0.96	B, H, C, D
Alcoholic fermentation			
Grains	4.0–5.5	> 0.96	E
Fruits	< 4.0	> 0.96	E
Mold fermentation			
Soybeans	4.0–5.0†	> 0.96	–

*By production of other substances than organic acids (e.g., bacteriocins (B), hydrogen peroxide (H), ethanol (E), carbon dioxide (C), diacetyl (D), etc.).

†pH value after soaking, just before inoculation with spores

products compared with canned or frozen foods, make control of microbial hazards necessary. The following factors are of importance:

- use of contaminated raw materials
- prevention of contamination (zoning, cleaning, and disinfection)
- poorly controlled fermentations, including ripening (HACCP)
- consumption without prior heating

The Use of Contaminated Raw Materials

The prevention of contamination of raw materials with pathogenic microorganisms should be the goal of everyone involved in the preharvest and postharvest phases of delivering products to the consumer. This is a "mission impossible" because pathogens are normally present in the soil (and therefore on the surface of fruits and vegetables), in surface water (which results in contaminated fish and shellfish), and in the gut of animals (causing contaminated products of animal origin, such as milk and meat).

To minimize the risk of infection or intoxication, pretreatment of the raw materials might be helpful. The simple practice of washing, for example, removes a portion of the pathogens (up to 90%). Another possibility (not for raw meat) is heat treatment (pasteurization) of the raw materials (e.g., milk). If pasteurization of the raw materials is not possible, a postprocess pasteurization might be necessary to eliminate pathogens in the final product. This type of postprocess pasteurization is sometimes practiced with fermented sausages.

Zoning

Raw materials are the primary source of many of the bacteria that are responsible for foodborne infections and intoxications. But, bacteria may also be transferred to food by the production environment and personnel, either directly or by cross-contamination through surfaces, equipment, utensils, and/or hands that have not been adequately cleaned and disinfected. Zoning, that is, dividing the production area into "dry and wet" and/or "high, medium, and low care" areas, is useful in preventing product contamination. The concept of hygiene zones was initially applied to prevent *Salmonella* contamination of very sensitive products such as milk powder. Nowadays, zoning has evolved into a complex set of measures including the layout and design of equipment, air filtration, personnel hygiene routes, and appropriate cleaning and disinfection procedures. However, zoning is only useful when it is applied logically and to the appropriate degree.

Applied in fermentation processes, there should be areas for

- the storage of the raw materials (low hygiene)
- the preparation of the raw materials, that is, washing, cutting, and adding ingredients (medium hygiene)
- the fermentation of the raw materials (medium hygiene)
- the filling of suitable packages (medium hygiene)
- the storage of the final products (medium hygiene)

An example of an area of high hygiene is the room where starter cultures are prepared for the lactic fermentation of milk. Here, a potential danger is the infection of the cultures with bacteriophages. Zoning, the allocation of an area strictly separated from the production process, will help to prevent the introduction of bacteriophages in starter cultures. However, this will be insufficient if it is not accompanied by other preventive measures such as strict cleaning and disinfection procedures and changing shoes when entering the area. Changing shoes cannot be replaced efficiently by covering shoes with plastic covers (that often break) or by using shoe disinfecting systems (improper disinfection due to rapid inactivation of the disinfectant), which can cause potentially contaminated wet areas around the system.

It is not feasible to discuss zoning for all types of fermentation processes here. In practice, zoning should be embedded in HACCP studies and must be acceptable and practicable for all individuals. Zoning should be introduced after adequate training so that personnel know why zoning is important and are motivated to follow its requirements.

Cleaning and Disinfection

Inadequate cleaning and disinfection can lead to reduced food quality, resulting in more rapid food spoilage and greater risk of food-borne diseases. Moreover, in food spills (pathogenic) microorganisms might adapt to extreme conditions such as low pH, giving them the opportunity to survive and/or grow in the final product.

Cleaning and disinfection are two separate but closely related concepts. *Cleaning* is removing dirt and a proportion of the microorganisms present; *disinfection* is treating the surfaces in such a way that the remaining microorganisms are killed or reduced to an acceptable level. Cleaning should be done first; otherwise, the subsequent disinfection will be less effective.

HACCPs

After zoning has been introduced as part of the prevention plan, attention should be paid to the presence and behavior of pathogens in raw materials, the fermentation process, the final product, and the production environment. Microbiological investigation estimating the presence and numbers of pathogens and spoilage organisms in raw materials and food products forms an important part of control programs for ensuring the safety and quality of food products. Until recently, microbial counts were an important though ineffective method for assessing food quality. Nowadays, food safety systems have been introduced based on HACCPs, in which control is exercised throughout the process itself.

An HACCP study will provide information concerning critical control points (CCPs), that is, raw materials, locations, practices, procedures, formulations, or processes where measures can be applied to prevent or minimize hazards.[45]

It is outside the scope of this contribution to identify all CCPs for known fermentation processes. In general, raw materials (contaminated with pathogens), the fermentation process (time to reach final pH, A_w, and final pH), storage and preparation of starter cultures, and heat treatment (of raw materials or of the fermented products) are the most important CCPs.

An HACCP study will not result in control measures that eliminate all safety problems, but it will provide information that can be used to determine how best to control the remaining hazards. It is then up to management to use that information correctly. If control procedures follow clearly established rules, inspectors can have greater confidence in food producers. Governments are also more able to accept the responsibility taken by industry because they can understand why and how controls are made. This has been recognized internationally, and the application of HACCP principles is recommended by the Codex Alimentarius Commission and is mandatory in many countries. In its present format, the application of HACCP principles is best suited to industrial food processing; small-scale producers of fermented foods, especially in nondeveloped countries, will have greater difficulty following this route. Relatively simple hygiene codes, based on HACCPs, could inform food handlers and households about appropriate fermentation techniques[46,52] (see Chapters 3 and 12).

Quantitative Risk Assessment

In many cases, the concept of HACCPs is used in only a qualitative way. By the implementation of quantitative risk analysis (QRA) in existing HACCP systems, a more quantitative approach is possible. A smart, stepwise, interactive identification procedure for food-borne microbial hazards has been developed by van Gerwen & Zwietering.[29] This procedure is based on three levels of detail ranging from rough hazard identification via detailed to comprehensive hazard identification. At first, the most relevant problems are identified before focusing on less important problems.

- The first level (rough hazard identification) selects pathogens that were involved in food-borne outbreaks with the product in the past. These data can be found in the literature, although not all outbreaks are reported, particularly in nondeveloped countries, and in many cases or outbreaks, the causative organism or the food vehicle is not identified.
- The second level (detailed hazard identification) selects pathogens that were reported to be present on the raw materials and ingredients used in the product. The greater part of these data can be found in the literature.
- The last level (comprehensive hazard identification) selects all (human) pathogens as hazardous, including pathogens that unexpectedly recontaminate the product. In this step, it is possible to estimate the effect of unexpected hazards (i.e., emerging pathogens, or pathogens that adapted to or are resistant to intrinsic factors of the product, such as acid-resistant *Salmonella*). Possible future problems can be anticipated in this way.

After this type of hazard identification, knowledge rules should be used to reduce the probably long list to a manageable list of the pathogens that are most likely present in the final product.

The following three types of rules are used.

1. *Selection of pathogens that are present or able to survive in the end product (e.g., removal from the list of vegetative pathogens present in a product that will undergo pasteurization).* Survival after pasteurization due to inappropriate time/temperature combinations and post-process contamination are not included in this rule. This should be ascertained by GMP and HACCP analysis.
2. *Selection of pathogens that are likely to cause problems.* Pathogens that are very rarely transmitted are not likely to cause health problems and can be removed from the list.
3. *Selection of pathogens that are able to grow or produce toxin in the product.* Estimation of capacity to grow is based on the maximum and minimum growth temperature, pH, and A_w. Other growth-determining factors such as the presence of preservatives and natural antimicrobial systems are not taken into account.

The knowledge rules can only be applied appropriately by experienced microbiologists assisted by the use of literature and/or models predicting the growth or inactivation of pathogens.

Simplified Identification Procedure for Food-Borne Microbial Hazards in Fermented Sausage

Food Science and Technology Abstracts (retrospective 1969–1989 and current 1990–1999) were used as a database for the first levels of hazard identification. The results are presented in Table 7–4. With the rough hazard identification, only three pathogens (*E. coli* O157:H7, *Salmonella*, and *S. aureus*) involved in outbreaks were found. The ingredients used in this product are meat (pork, beef), pork backfat, salt, nitrate, and spices (pepper). With a detailed hazard identification for the raw materials, all pathogens listed in Table 7–2 are selected, even some pathogens that are indicated as not likely to be present (e.g., *B. cereus*). This is because the selection for Table 7–4 was based on literature data for the presence of these pathogens (even in low numbers). Table 7–2 only presents information if there is a strong or a weak relationship between the pathogen and the raw material.

Applying type 1 knowledge rules (survival of pathogens in the end product) only results in the removal of *Campylobacter*. This pathogen will not survive fermentation processes due to the low pH. Using type 2 rules removes *B. cereus*, *Shigella*, and *Vibrio*. These pathogens are of little concern in fermented sausages because growth at low pH is not possible (*B. cereus* and *Vibrio* spp.), or because the presence in the raw materials is only accidental (transmission via contaminated persons).

Table 7–4 Results of the Hazard Identification Procedure Applied to Fermented Sausage

Rough Hazard Identification	Detailed Hazard Identification	Knowledge Rules type 1	Knowledge Rules type 2	Knowledge Rules type 3
	Aeromonas	x	x	
	Bacillus cereus	x		
	Campylobacter			
Escherichia coli O157:H7	*Escherichia coli O157:H7*	x	x	x
	Listeria monocytogenes	x	x	x
Salmonella	*Salmonella*	x	x	x
	Shigella	x		
Staphylococcus aureus	*Staphylococcus aureus*	x	x	x*
	Vibrio	x		
	Yersinia	x	x	

*Growth and toxin production of *Staphylococcus aureus* is usually restricted to the start of the fermentation process, when there is a rather slow decrease in pH.

The pathogens likely to cause problems are *Aeromonas*, *E. coli* O157:H7, *L. monocytogenes*, *Salmonella*, *S. aureus*, and *Yersinia*.

To use type 3 rules, it was assumed that the pH of the final product was 5.0 ± 0.1 and the A_w was 0.95 ± 0.01. If all types of knowledge rules are applied, *Aeromonas* and *Yersinia* can be removed because usually high levels are needed for food-borne infection, and growth in fermenting sausage is poor or not possible at all.[61]

The remaining pathogens are those from the rough hazard identification (*E. coli* O157:H7, *Salmonella*, and *S. aureus)* and *L. monocytogenes.*

Under the conditions described, growth and toxin production by these pathogens are usually restricted to the first 6–12 hours of fermentation. This can be followed by a gradual decline or survival of the cells.[25,26,31,33,43,47,56] Because of the presence of competitive microorganisms, growth of *S. aureus* is seldom a problem.[54] The low pH values will affect growth and survival of *L. monocytogenes*.[8,26,27,47] in contrast to the acid-adapted and acid-tolerant strains of *Salmonella* and *E. coli* O157:H7.[2,23,47]

The knowledge obtained from the stepwise identification procedure provides a way to efficiently assess those hazards that need to be controlled during processing.

VALIDATION OF THE SAUSAGE FERMENTATION PROCESS FOR THE CONTROL OF PATHOGENS

As a result of outbreaks of illness caused by *S. aureus* in fermented meat products,[7] the American Meat Institute formulated GMP for fermented dry and semi-dry sausages.[3]

The rapid decrease to low pH values, preferably lower than pH 4.0, that results in a sufficient inactivation of salmonellas and *S. aureus* turned out to be of little effect against *E. coli* O157:H7,[23,47] as was demonstrated by the recent outbreaks of *E. coli* O157:H7 linked to the consumption of fermented sausages. The U.S. Department of Agriculture's Food Safety and Inspection Service (USDA-FSIS) developed guidelines for validating a 5-log reduction of *E. coli* O157:H7 in fermented sausage.[23] To meet the criteria set in the guideline, challenge tests should be performed to investigate the behavior of this pathogen during the fermentation process and a postpasteurization treatment. The information to perform a microbiological storage test or a challenge trial can be gained from the literature or from laboratory test models (e.g., the Food MicroModel). In most cases, the laboratory models provide a more conservative view than the real product situation, as each parameter has usually been studied under otherwise ideal conditions.

Microbiological Challenge Tests

Laboratory simulation of what can happen to a food product during processing, distribution, and subsequent handling is an established technique in the food industry. There are two main reasons why this type of testing is carried out. In general, safety is of prime concern because the food manufacturer must ensure that a product presents a minimum risk to the consumer. The other reason is to establish the shelf life or keeping quality of a product. In this case, quality is of most interest. In both cases, the product is normally stored at specified temperatures for at least the anticipated or prescribed shelf life. Depending on the information available on raw materials, process, organisms of interest, and so forth, the product can be inoculated with (low) numbers of spoilage organisms and/or pathogens before processing or storage. With these laboratory simulation tests, which are mandatory in some markets, and data from computer models (e.g., Food MicroModel), it is possible to demonstrate that everything that can be reasonably expected has been done.

Depending on the background for carrying out a laboratory simulation test (quality or safety), and whether the product is likely to contain the types of microorganisms of concern, these tests should be named differently to avoid confusion. In a *challenge test,* the product is inoculated with either pathogenic or spoilage organisms, depending on whether consumer safety issues or product stability are of most concern. In a *storage test,* sometimes called shelf life or keeping quality test, the product is not inoculated with microorganisms and the aim is to check safety and/or quality. The experimental plan and execution of a storage test may, however, differ, depending on which of the factors is to be studied.

Purpose of Microbiological Challenge and Storage Tests

One of the important tools used to ensure the production of safe food is the identification of CCPs in an HACCP study. CCPs have to be controlled in order to eliminate hazards or minimize their likelihood of occurrence. It is essential that the identification and setting of criteria for CCPs are based on sound knowledge and relevant information so that target levels and tolerance limits can be established. CCPs are under control if the criteria are met. Only if sufficient knowledge concerning the effect of intrinsic, extrinsic, and process factors on the safety of the product is available can the setting of criteria at each CCP be done easily.

However, if information is scarce or unavailable, data have to be obtained by storage testing and/or microbiological challenge testing. A *microbiological challenge test* is an exercise to simulate what can happen to a food product during processing, storage, distribution, and subsequent handling, following inoculation with one or more relevant pathogens.

It is important to realize when a microbiological challenge test should not be performed. If it is evident that the pathogen(s) will grow readily in the product, a challenge test will be a waste of time. Such pathogens must be eliminated by proper heat treatment or by preservation of the product.

Challenge testing will provide information on the types of pathogens capable of growth in the product. With the data obtained, the risks of food poisoning can be assessed and the conditions necessary to prevent food poisoning can be determined. As a consequence, the safety of the product can be determined in terms of intrinsic factors. Therefore, the information obtained is the basis for setting criteria at relevant CCPs.

Challenge testing will provide information about the growth of selected pathogens in a product. However, the likelihood of presence of these pathogens in the product should be obtained from literature or databases.

Planning of a Storage Test and a Challenge Test

The factors below, also relevant for the first part of an HACCP study, must be considered when planning storage tests and challenge tests.

- microbiological status of the raw materials
- product composition and limits for critical hurdles
- process design and parameters
- packaging system
- anticipated storage, distribution, and consumer handling conditions
- identification of the risks of food poisoning from the product
- existing knowledge of the types and properties of organisms relevant to the product

Carrying Out a Microbiological Challenge Test

If the decision is made that a challenge test is required, and after deciding which organisms are relevant for the test, a protocol has to be developed. It is important to ensure that enough samples of the product are available at each time point of the investigation. The number of samples in the test depends on the heterogeneity of the product; all of the technical procedures must be reliable and reproducible.

There are two types of microbiological challenge tests: process challenge tests and product challenge tests. The objective of the *process challenge* test is to determine whether or not a selected pathogen survives a process (e.g., fermentation). The results from such tests are available directly after processing by enumeration of surviving microorganisms. In some cases, such as ripening in fermentation processes, analysis must be done after a few weeks. *Product challenge* tests investigate whether or not the pathogens in artificially contaminated products can grow or under what conditions unacceptable levels will be reached.

The test strains used (preferable a cocktail of strains) should be isolated from identical or similar products. During preparation of the inoculum, stress conditions (that might reduce viability) should be avoided. Because in practice, pathogens entering the product from the environment might be adapted to the intrinsic factors of the product (e.g., poorly cleaned product environment), it might be necessary to train the microorganisms to growth at low pH or low A_w in order to ensure reliable results.

Process Challenge Test for Fermented Products Artificially Contaminated with *E. coli* O157:H7

Among the many different types of fermented sausage, pepperoni is one of the most popular varieties. In both the pepperoni and the Lebanon bologna process, *E. coli* O157:H7 survives the fermentation process.[23,47]

Performing challenge tests as described above might lead to the conclusion that other acid-tolerant pathogens will also survive the fermentation process, which makes a postfermentation heat treatment necessary to produce a safe product. Another possibility, less suitable for raw meat but applicable for items such as raw milk, is pasteurization of the raw materials. Zoning, in combination with good manufacturing processes, should prevent postprocess contamination.

In the future, more research is needed to understand the behavior of (acid-adapted) pathogens in fermentation processes. With this knowledge, hygiene codes should be developed for all types of fermentations in order to ensure safe products, both in industrial and in household technologies.

REFERENCES

1. Adams, M. R. & Moss, M. O. (2000). *Food Microbiology,* 2nd edn. Cambridge, England: Royal Society of Chemistry.
2. Adams, M. R. & Nicolaides L. (1997). Review of the sensitivity of different foodborne pathogens to fermentation. *Food Control* 8, 227–239.
3. American Meat Institute. (1982). *Good Manufacturing Practices: Fermented Dry and Semi-Dry Sausage.* Washington, DC: American Meat Institute.
4. Antony, U., Moses, L. G. & Chandra, T. S. (1998). Inhibition of *Salmonella typhimurium* and *Escherichia coli* by fermented flour of finger millet (*Eleusine coracana*). *World J Microbiol Biotechnol* 14, 883–886.
5. Aryanta, R. W., Fleet, G. H. & Buckle, K. A. (1991).

The occurrence and the growth of microorganisms during the fermentation of fish sausage. *Int J Food Microbiol* 13, 143–156.

6. Aytac, S. A. & Ozbas, Z. Y. (1994). Survey on the growth and survival of *Yersinia enterocolitica* and *Aeromonas hydrophila* in yogurt. *Milchwissenschaft* 49, 322–325.
7. Bacus, J. N. (1986). Fermented meat and poultry products. In *Advances in Meat Research*. Vol. 2, *Meat and Poultry Microbiology*, pp. 123–164. Edited by A. M. Pearson & T. R. Dutson. Westport, CT: AVI Publishing.
8. Beumer, R. R. (1997). Listeria monocytogenes: *detection and behaviour in food and in the environment*. PhD diss., Wageningen University Research Center, The Netherlands.
9. Beumer, R., Bloomfield, S., Exner, M., Fara, G. M. & Scott, E. (1999). The need for a home hygiene policy and guidelines on home hygiene. *Annali di Igiene* 11, 11–26.
10. Beumer, R. R., Vrouwenvelder, T. & Brinkman, E. (1994). Application of HACCP in airline catering. *Food Control* 5, 205–209.
11. Bodnaruk, P. W., Williams, R. C. & Golden, D. A. (1998). Survival of *Yersinia enterocolitica* during fermentation and storage of yogurt. *J. Food Sci* 63, 535–537.
12. de Boer, E. & Hahné, M. (1990). Cross-contamination with *Campylobacter jejuni* and *Salmonella* spp. from raw chicken products during food preparation. *J Food Prot* 53, 1067–1068.
13. Bonestroo, M. H., Kusters, B. J. M., de Wit, J. C. & Rombouts, F. M. (1993). The fate of spoilage and pathogenic bacteria in fermented sauce-based salads. *Food Microbiol* 10, 101–111.
14. Bryan, F. L. (1988). Risk of practices, procedures and processes that lead to outbreaks of foodborne diseases. *J Food Prot* 51, 663–673.
15. Byaruhanga, Y. B., Bester, B. H. & Watson, T. G. (1999). Growth and survival of *Bacillus cereus* in mageu, a sour maize beverage. *World J Microbiol Biotechnol* 15, 329–333.
16. Caplice, E. & Fitzgerald, G. F. (1999). Food fermentations: role of microorganisms in food production and preservation. *Int J Food Microbiol* 50, 131–149.
17. Centres for Disease Control and Prevention. (1993). Update: multi-state outbreak of *Escherichia coli* O157:H7 infections from hamburgers: Western United States, 1992–1993. *MMWR* 42, 258–263.
18. Chan, K. Y., Woo, M. L., Lam, L. Y. & French, G. L. (1989). *Vibrio parahaemolyticus* and other halophilic vibrios associated with seafood in Hong Kong. *J Appl Bacteriol* 66, 57–64.
19. Corrales, M. T., Bainotti, A. E. & Simonetta, A. C. (1994). Survival of *Vibrio cholerae* O1 in common foodstuffs during storage at different temperatures. *Lett Appl Microbiol* 18, 277–280.
20. Cutter, C. N. (1999). Combination spray washes of saponin with water or acetic acid to reduce aerobic and pathogenic bacteria on lean beef surfaces. *J Food Prot* 62, 280–283.
21. Dessie G., Abegaz, K. & Ashenafi , M. (1996). Fate of *Salmonella enteritidis* and *Salmonella typhimurium* during the fermentation of siljo. *East Afr Med J* 73, 432–434.
22. Dineen, S. S., Takeuchi, K., Soudah, J. E. & Boor, K. J. (1998). Persistence of *Escherichia coli* O157:H7 in dairy fermentation systems. *J Food Prot* 61, 1602–1608.
23. Ellajosyula, K. R., Doores, S., Mills, E. W., Wilson, R. A., Anantheswaran, R. C. & Knabel, S. J. (1998). Destruction of *Escherichia coli* O157:H7 and *Salmonella typhimurium* in Lebanon bologna by interaction of fermentation pH, heating temperature, and time. *J Food Prot* 61, 152–157.
24. Encinas, J. P., Gonzales, C. J., Garcia-Lopez, M. L. & Otero, A. (1999). Numbers and species of motile aeromonads during the manufacture of naturally contaminated Spanish fermented sausages (longaniza and chorizo). *J Food Prot* 62, 1045–1049.
25. Faith, N. G., Wierzba, R. K., Ihnot, A. M., Roering, A. M., Lorang, T. D., Kaspar, C. W. & Luchansky, J. B. (1998). Survival of *Escherichia coli* O157:H7 in full- and reduced-fat pepperoni after manufacture of sticks, storage of slices at 4° or 21 °C under air and vacuum, and baking of slices on frozen pizza at 135, 191 and 246 °C. *J Food Prot* 61, 383–389.
26. Farber, J. M, Daley, E., Holley, R. & Usborne, W. R. (1993). Survival of *Listeria monocytogenes* during the production of uncooked German, American and Italian-style fermented sausages. *Food Microbiol* 10, 123–132.
27. Farber, J. M. & Peterkin, P. I. (1991). *Listeria monocytogenes*, a food-borne pathogen. *Microbiol Rev* 55, 476–511.
28. Gahan, C. G. M., O'Driscoll, B. & Hill, C. (1996). Acid adaptation of *Listeria monocytogenes* can enhance survival in acidic foods and during milk fermentation. *Appl Environ Microbiol* 62, 3128–3132.
29. van Gerwen, S. J. C. & Zwietering, M. H. (1998). Growth and inactivation models to be used in quantitative risk assessments. *J Food Prot* 61, 1541–1549.
30. Griffiths, M. W. (1989). *Listeria monocytogenes*: its importance in the dairy industry. *J Sci Food Agric* 47, 133–157.
31. Hinkens, J. C., Faith, N. G., Lorang, T. D., Bailey, P., Büge, D., Kaspar, C. W. & Luchansky, J. B. (1996). Validation of pepperoni processes for control of *Escherichia coli* O157:H7. *J Food Prot* 59, 1260–1266.
32. Hisa, K., Hayasi, K. & Sakaguchi, G. (1999). Proposal of HACCP systems for prevention of foodborne botulism. II. theoretical considerations in izushi production to prevent *Clostridium botulinum* type E botulism based on HACCP. *J Antibact Antifung Ag – Jap* 27, 3–11.

33. Ihnot, A. M., Roering, A. M., Wierzba, R. K., Faith, N. G. & Luchansky, J. B (1998). Behaviour of *Salmonella typhimurium* DT104 during the manufacture and storage of pepperoni. *Int J Food Microbiol,* 40 117–121.

34. Jacobs-Reitsma, W. F., Bolder, N. M. & Mulder, R. W. A. W. (1994). Cecal carriage of *Campylobacter* and *Salmonella* in Dutch broiler flocks at slaughter: a one year study. *Poultry Sci* 73, 1260–1266.

35. Johnson, E. A., Nelson, J. H. & Johnson, M. (1990). Microbiological safety of cheese made from heat-treated milk, part I: executive summary, introduction and history. *J Food Prot* 53, 441–452.

36. Johnson, E. A., Nelson, J. H. & Johnson, M. (1990). Microbiological safety of cheese made from heat-treated milk, part II: microbiology. *J Food Prot* 53, 519–540.

37. Jyhshiun, L., In-Soo-Lin, Frey, J., Slonczewski, J. L. & Foster, J. W. (1995). Comparative analysis of extreme acid survival in *Salmonella typhimurium*, *Shigella flexneri*, and *Escherichia coli*. *J Bacteriol* 177, 4097–4104.

38. Kaspar, C. W. & Tamplin, M. L. (1993). Effects of temperature and salinity on the survival of *Vibrio vulnificus* in seawater and shellfish. *Appl Environ Microbiol* 59, 2425–2429.

39. Kingamkono, R., Sjoegren, E., Svanberg, U. & Kaijser, B. (1996). Inhibition of different strains of enteropathogens in a lactic-fermenting cereal gruel. *World J Microbiol Biotechnol* 11, 299–303.

40. Knorr, D. (1998). Technology aspects related to micro-organisms in functional foods. *Trends Food Sci Technol* 9, 295–306.

41. Kunene, N. F., Hastings, J. W. & von Holy, A. (1999). Bacterial populations associated with a sorghum-based fermented weaning cereal. *Int J Food Microbiol* 49, 75–83.

42. Leyer, G. J. & Johnson, E. A. (1992). Acid adaptation promotes survival of *Salmonella* spp. in cheese. *Appl Environ Microbiol* 58, 2075–2080.

43. Massa, S., Altieri, C., Quaranta, V. & DePace, R. (1997). Survival of *Escherichia coli* O157:H7 in yogurt during preparation and storage at 4 °C. *Lett Appl Microbiol* 24, 347–350.

44. Morioka, Y., Nohara, H., Araki, M., Suzuki, M. & Numata, M. (1996). Studies on the fermentation of soft salami sausage by starter culture. *Animal Sci Technol* 67, 204–210.

45. Mortimore, S. & Wallace, C. (1998). *HACCP: A Practical Approach,* 2nd edn. Gaithersburg, MD: Aspen Publishers.

46. Motarjemi, Y. & Nout, M. J. R. (1996). Food fermentation: a safety and nutritional assessment. *Bull WHO* 74, 553–559.

47. Nissen, H. & Holck, A. (1998). Survival of *Escherichia coli* O157:H7, *Listeria monocytogenes* and *Salmonella kentucky* in Norwegian fermented, dry sausage. *Food Microbiol* 15, 273–279.

48. Northolt, M. D. (1983). Pathogenic micro-organisms in fermented dairy products. *Neth Milk Dairy J* 37, 247–248.

49. Nout, M. J. R. (1994). Fermented foods and food safety. *Food Res Int* 27, 291–298.

50. Nout, M. J. R., Bakshi, D. & Sarkar, P. K. (1998). Microbiological safety of kinema, a fermented soya bean food. *Food Control* 9, 357–362.

51. Nout, M. J. R., Beernink, G. & Bonants-vanLaarhoven, T. M. G. (1987). Growth of *Bacillus cereus* in soya bean tempeh. *Int J Food Microbiol* 4, 293–301.

52. Nout, M. J. R. & Motarjemi, Y. (1997). Assessment of fermentation as a household technology for improving food safety: a joint FAO/WHO workshop. *Food Control* 8, 221–226.

53. Nout, M. J. R., Rombouts, F. M. & Havelaar, A. (1989). Effect of accelerated natural lactic fermentation of infant food ingredients on some pathogenic micro-organisms. *Int J Food Microbiol* 8, 351–361.

54. Nychas, G. J. E. & Arkoudelos, J. S. (1990). Staphylococci: their role in fermented sausages. *J Appl Bacteriol Symp Supp* 19, 167S–188S.

55. O'Mahony, M., Mitchell, E., Gilbert, R. J., Hutchinson, D. N., Begg, N. T., Rodhouse, J. C. & Morris, J. E. (1990). An outbreak of foodborne botulism associated with contaminated hazelnut yoghurt. *Epidemiol Inf* 104, 389–395.

56. Ostergaard, A., Embarak, P. K. B., Wedell-Neergaard, C., Huss, H. H. & Gram, L. (1998). Characterization of anti-listerial lactic acid bacteria isolated from Thai fermented fish products. *Food Microbiol* 15, 223–233.

57. Ozbas, Z. Y. & Aytac, S. A. (1996). Behaviour of *Yersinia enterocolitica* and *Aeromonas hydrophila* in skim milk during fermentation by various lactobacilli. *Z Lebensmittel Forsch* 202, 324–328.

58. Poffe, R. & de Beeck, E. O. (1991). Enumeration of *Aeromonas hydrophila* from domestic wastewater treatment plants and surface water. *J Appl Bacteriol* 71, 366–370.

59. Rhodehamel, E. J., Reddy, N. & Pierson, M. D. (1992). Botulism: the causative agent and its control in food. *Food Control* 3, 125–143.

60. Roberts, D. (1990). Sources of infection: food. *Lancet* 336, 859–861.

61. Roberts, T. A., Baird-Parker, A. C. & Tompkin, R. B., eds. (1996). *Microorganisms in Foods, 5, Microbiological Specifications of Food Pathogens.* London: Blackie Academic and Professional.

62. Sauer, C. J., Majkowski, J., Green, S. & Eckel, R. (1997). Foodborne illness outbreak associated with a semi-dry fermented sausage product. *J Food Prot* 60, 1612–1617.
63. Sharp, J. C. M., Reilly, W. J. & Coia, J. E. (1995). *Escherichia coli* O157 infection in Scotland: an epidemiological overview, 1984–1994. *PHLS Microbiol Digest* 12, 134–140.
64. Shay, B. & Souness, R. (1995). Recent regulations impacting on the small goods industry. *Food Aust* 47, 491–494.
65. Simango, C. & Rukure, G. (1992). Survival of bacterial enteric pathogens in traditional fermented foods. *J Appl Bacteriol* 73, 37–40.
66. Smulders, F. J. M. & Greer, G. G. (1998). Integrating microbial decontamination with organic acids in HACCP programmes for muscle foods: prospects and controversies. *Int J Food Microbiol* 44, 149–169.
67. Swinbanks D. (1996). Outbreak of *E. coli* infection in Japan renews concern. *Nature* 282, 290.
68. Thomas C. & Beirne D. (1999). Microbiological safety in processed vegetables. *Int Food Hyg* 10, 17–23.
69. Verotoxin-producing *Escherichia coli* O157. (1997). *CDR Weekly* 7, 409.
70. Warburton, D. W., Weiss, K. F., Purvis, U. & Hill, R. W. (1987). The microbiological quality of fermented sausage produced under good hygienic practices in Canada. *Food Microbiol* 4, 187–197.
71. West, P. A. (1989). The human pathogenic vibrios: a public health update with environmental perspectives. *Epidemiol Inf* 103, 1–34.
72. Wu, Y. C., Kimura, B. & Fujii, T. (1999). Fate of selected food-borne pathogens during the fermentation of squid shiokara. *J Food Hyg Soc Jpn* 40, 206–210.
73. Yu, S. L. & Chou, C. C. (1987). Survival of some food poisoning bacteria in pickled vegetables stored at different temperatures. *Food Sci China* 14, 296–305.

Chapter 8

Microbiological Hazards and Their Control: Viruses

Michael J. Carter and Martin R. Adams

INTRODUCTION

Viruses differ significantly from bacterial pathogens because they are obligate intracellular parasites and can replicate only within an appropriate living host cell. The source of all viruses is thus a previously infected individual who sheds infectious particles into his or her immediate environment. Transmission to another host can be direct, as in person-to-person spread (e.g., through aerosols created by a sneeze or vomiting), or indirect, involving some other agent as a carrier for the virus. Food-borne transmission is of the indirect type. All viruses that infect via the enteric tract (and are shed in feces) are potentially capable of food-borne transmission. Direct person-to-person spread can also occur if the opportunity arises; direct feces-to-mouth transmission is common in children, and airborne spread, leading to contamination of distant surfaces, has been reported.

Foods are contaminated with viruses as a result of the distribution of fecal- or vomitus-derived viruses through the environment, eventually contaminating the food or water of another potential host. This process is highly variable. It could, for instance, result fairly directly from an infected food handler contaminating food immediately before consumption, or it could be the end result of a prolonged distribution process moving virus from fecal material to river and estuary and, perhaps eventually, to seafood. Generally, these indirect routes require more time than direct spread, and the stability of the virus particle will be key to its successful distribution. Evolution has ensured that viruses of this type are especially fitted for survival in transit, which has a direct influence on their ability to survive in processed food.

Water, either ingested directly or used as an ingredient or washing agent during food processing, is the chief vehicle for disseminating enteric viruses. It is also significant in carrying these viruses to plants in the field (e.g., in irrigation water or sprays), and to the most significant vector of food-borne viral illness, molluscan shellfish. Most food-borne viruses survive well in water. Their survival is assisted by high protein content and high ionic strength, particularly calcium and magnesium ions, which tend to stabilize the particles. These conditions are often found in sewage and sewage-contaminated water. The adsorption of virus to suspended solids such as clays or organic matter can also protect the particles from inactivation.

Viruses contaminating food are there as a result of a dilution and distribution process that may have taken a significant time. Because viruses have an absolute requirement for living cells in which to replicate, they cannot increase in numbers during this distribution phase or within contaminated food during storage. In fact, the amount of contaminating viruses may decrease during storage; this can be assisted by

Acknowledgement—The authors would like to acknowledge Albert Bosch of the University of Barcelona for the supply of unpublished information.

treatment with heat, irradiation, chemicals, or other factors such as the pH changes associated with the fermentation of foodstuffs. Thus, the levels of active viruses in foods must be expected to be generally very low. Consequently, and with very few exceptions (mainly molluscan shellfish), viruses are only rarely detected in foodstuffs before consumption. In fact, it is only molluscan shellfish that are routinely monitored for the presence of viruses; in all other cases, a contaminating virus is usually inferred only retrospectively once consumption has led to illness. Direct contamination of food from an infected food handler could deliver a more concentrated virus loading to a foodstuff, but because this could occur very sporadically, at any stage up to and including serving, testing food for virus contamination from this source is generally not a viable option.

There are problems in assessing the importance of viral food-borne illness. Figures generally suggest that it is relatively low in adults, with the exception of the caliciviruses. However, there is an almost complete lack of data addressing endemic levels of infection. Official figures refer almost entirely to "outbreaks" of food-borne virus illness, that is, to epidemic behavior. Such outbreaks could result from the simultaneous consumption of contaminated food by many people, or equally from the infection of a single person via contaminated food, followed by more direct person-to-person spread within the person's immediate surroundings. It is probably significant that many outbreaks of infection are observed in closed communities where hygiene may not be advanced and where there is increased potential for person-to-person spread, such as in childcare centers or nursing homes. It is this subsequent transmission that renders the original food-borne infection statistically "visible." Food-borne infections such as these would not be recorded if, for instance, they had occurred in a private home and affected one or only a few relatively fit persons. Consequently, the figures available probably represent only the tip of the iceberg as far as the incidence of food-borne virus transmission is concerned. Some measure of the extent of underreporting was gained from the Infectious Intestinal Disease Study conducted in England during the period 1993–1996.[48] This study concluded that underreporting was most severe for diarrhea-causing viruses, and that for every officially reported case of infection with Norwalk-like virus (NLV), there were actually 1,562 cases in the community at large.

FOOD-BORNE VIRUSES

General Features

There are two broad classes of virus that may infect through the gut. The first class uses the enteric tract as the portal of entry to the body, but subsequently spreads elsewhere around the body. Examples of this type of infection would be spread to muscle tissue of the skeleton or heart (as in the Coxsackie B viruses), to the meninges and central nervous system (polio and occasionally other enteroviruses), or to the liver (hepatitis A and E viruses). Viruses in the second class are true inhabitants of the gut. They replicate significantly only in this tissue and induce signs and symptoms of a typical "gastroenteritis." This is mainly diarrhea, but can be accompanied by differing degrees of vomiting. Viruses like this include the caliciviruses, rotaviruses, adenoviruses, and astroviruses. Finally, there are a few agents that are detected routinely in the gut, such as the pico-birna viruses. These are poorly characterized and do not seem to be associated with disease. The general features of these viruses, as well as their size, structure, and genome content, are provided in Table 8–1.

Difficulties in Culture and Diagnosis

The gut is a very specialized habitat in virological terms; all viruses require living cells as hosts and are totally dependent on the processes that their host cells are able to provide (e.g., for protein synthesis and protein processing). There is a multiplicity of different cell types in the gut, each with a different role. This is reflected in a differing complement of enzymes and surface proteins; cells may even vary in these properties

Table 8–1 Food–Borne Viruses

Virus	*Members*	*Features*	*Associated Illness*
Caliciviridae	Norwalk-like and Sapporo-like viruses	34 nm nonenveloped particles with distinctive morphology (surface cups). Single stranded +ve stranded RNA genome.	Norwalk-like: explosive projectile vomiting in older children/young adults. Noncultivable. Sapporo-like: more common in younger children believed to be milder in effect, noncultivable.
Reoviridae	Rotavirus groups A–E; only groups A–C are found in man. Group C is rare and group B is uncommon except in China.	Multi-layered, nonenveloped 70 nm particles, enclosing double-stranded RNA genome. Only type A is cultivable and requires a trypsin supplement.	Diarrhea common in the young, decreasing in frequency with age. Illness returns in the elderly.
Adenoviridae	Adenovirus group F, types 40 and 41	100 nm, nonenveloped icosahedral particles, ds DNA genome. Cultivable in some cells (e.g., Graham 293)	Mild diarrhea, may be prolonged virus shedding. Mainly affects children.
Astroviridae	Types 1–8	28 nm nonenveloped, star motif may be visible. Single-stranded +ve RNA. Appearance can be variable; may be mistaken for calici or parvoviruses. Cultivable in differentiating colonic cells with trypsin supplement.	Mostly infects children, but higher serotypes are rarer and can cause significant disease in adults. Relatively mild, but probably underestimated.
Picornaviridae	(Enterovirus) Poliovirus Coxsackievirus ECHO virus Enterovirus	28 nm nonenveloped particles, single-stranded, +ve sense RNA. Grow relatively easily in a variety of human and primate cell lines.	Mainly asymptomatic, can induce muscle pains (Bornholm Disease), cardiomyopathy, meningitis, and CNS motor paralysis (most common with polio).
	(Hepatovirus) Hepatitis A virus	As above, cultivable with difficulty.	Hepatitis, mild in the young.

continues

Table 8–1 continued

Virus	*Members*	*Features*	*Associated Illness*
Unclassified	Hepatitis E virus	Calicivirus-like particle structure, containing single stranded +ve RNA of unique genomic organization. Noncultivable.	Similar features to HAV but more severe. Often fatal if contracted in late pregnancy.
Parvoviridae	Wollan, Ditchling. Cockle agents	Smooth featureless 25 nm particles. Single-stranded DNA genome. Poorly characterized, noncultivable.	Widespread shellfish-associated outbreaks, largely controlled through cooking.
Coronaviridae	Uncharacterized	Large enveloped virus, fragile, noncultivable. Food-borne transmission unlikely in view of their fragility.	Associated with neonatal necrotising enterocolitis.
Toroviridae	Uncharacterized	Fragile, enveloped virus, may resemble Berne virus in horses. Noncultivable, food-borne transmission unlikely as above.	Unknown

at different stages during their differentiation. Furthermore, all gut cells are bathed in a solution of the various products that are secreted by other specialized gut cells, notably, of course, proteases. These features make the gut a very difficult cellular environment to mimic in culture. For instance, a virus might replicate in one particular type of cell at one stage of its differentiation and could also require extracellular soluble products released from quite different cells entirely. For this reason, viruses that replicate in the gut have been very difficult to culture in the laboratory, and some still cannot be grown. Those that can be cultivated often require that the culture is supplemented with proteases, usually trypsin,[5] or even with duodenal juice.[37] Specialized cells are also sometimes needed (e.g., differentiating colonic carcinoma cells for astroviruses).[49] Viruses that penetrate beyond the gut and invade other tissues are generally (but by no means always) simpler to cultivate than those that reside in the gut.

This difficulty in cultivation has two important effects. First, virus detection can be problematic. Usually, a cause of infectious gut illness can only be established in approximately half of the cases that are investigated. It is likely that a large proportion of those cases of infection that remain unidentified will in fact be found eventually to be caused by viruses, but by viruses that hitherto have been missed because they are not culturable, and because there is no adequate diagnostic test for this particular virus. The only catch-all method is electron microscopy, but this depends on a virus being readily recognizable by its shape and present in large numbers (many workers quote a minimum requirement of 10^6 particles per ml before observation becomes even reasonably likely). Large viruses like the rota- and adenoviruses are distinctive under the microscope, but small, diffuse, or fuzzy viruses may often be overlooked, and their identification makes high demands on the operator. Other tests may be less demanding in use, but require characterized reagents such as antibodies for agglutination or ELISA-based methods and primers for PCR-based detection. Thus, these tests can only find viruses that are already characterized and known; they cannot discover new ones. These shortcomings bias the relative detection frequencies of viruses associated with enteric infection. Although the extent of any such bias is not clear, it is difficult to avoid the conclusion that virus-associated enteric illness is likely to be substantially underestimated, and that this problem will be worse when considering the smaller, less distinct viral agents.

Second, if it is not possible to culture a virus *in vitro*, then that virus can only be investigated *in vivo*. If the only hosts are humans, such experiments become expensive and cumbersome, if not ethically unacceptable. For this reason, many of the fundamental investigations into stability, transmission, and reinfection that are required have not been done or have not been confirmed.

POTENTIALLY FOOD-BORNE VIRUSES

Table 8–1 summarizes the main viruses associated with enteric infection. Several of these are as yet uncharacterized and of necessity will not be considered in this chapter. Any virus that is enterically transmitted is potentially transmissible via food, although person-to-person spread may be more common. This is especially true of viruses that affect children and infants. Most enteric viruses fall into this category. Infection is acquired early in life and becomes less common through late childhood, adolescence, and adult life, increasing in significance once more in the elderly. The significance of food-borne infection can thus depend on the types of food that are usually ingested by children. Chief among infections of adults are the Norwalk-like viruses (NLVs, previously called the small round structured viruses–SRSVs), and the Sapporo-like viruses (SLVs), two related members of a virus family called the *Caliciviridae*. Viruses of the NLV type are the most common viral cause of explosive outbreaks of projectile vomiting in adults. In recent years, reports of viral gastroenteritis, particularly NLV outbreaks, have shown a dramatic increase in England and Wales; 418 cases were identified in 1992 and 2,387 in 1996.[4] In 1994, reports of NLV outnumbered those of *Salmonella* for the first time. This is probably due in part to increased awareness and referral for diagnosis, but it indicates the extent to which such infections may have been missed in the past, and probably continue to be missed today.

Increased surveillance for viral gastroenteritis was introduced in the United Kingdom between 1992–1995, and this has yielded some very interesting results. A preliminary data set was obtained for 2,149 of 2,680 outbreaks during this period. Of these cases, 25% were attributed to *Salmonella*, 33% to NLV, 2% to rotaviruses, and 0.5% each associated with astroviruses and small round (parvovirus) like agents. Twenty percent of cases were of unknown etiology, and approximately half of these were believed due to viruses. The features of all of these viruses associated with food-borne transmission are described briefly in the following sections.

Caliciviruses (NLVs and SLVs)

NLVs and SLVs have different appearances under the electron microscope, and the relationship between them has been debated for some time. SLVs have an obvious and unique structure under the electron microscope; they appear to be covered in cup-like depressions, from which the virus takes its name (calici = a cup). However, the NLVs appear fuzzy. Structural studies using cryo electron microscopy have now revealed considerable underlying similarity in structure, and molecular studies have reinforced these. It is now clear that these two agents are related.[11]

NLVs

NLVs were first identified following an outbreak of enteric illness among adults in the town of Norwalk, Ohio that has now given its name to all viruses of this type. NLVs are associated with sporadic outbreaks of diarrhea and vomiting among young adults and older individuals. The disease is more common in winter than at other times of the year, and had been tentatively named winter vomiting disease in 1940, although its cause was not known at that time. NLV is characterized by an incubation period of up to 48 hours, with illness lasting for 1–2 days. This virus differs from most other agents of viral gastroenteritis in that it mainly attacks adults and it frequently induces a high level of vomiting among infected individuals. In fact, sudden, explosive, projectile vomiting may be the first obvious symptom of infection. For this reason, many cases of NLV are actually identified at work. Thus, the implications if a food handler should be infected are obvious.

In addition, although there are multiple serotypes of NLV (probably more than eight), immunity to each seems to be short lived. Thus, individuals may be protected for only a short period of time (months) following an infection before they become susceptible to the same virus again.[35] Some people appear to have an inherent resistance to infection; these individuals remain symptom free, even in the absence of antibody. However, sensitive persons may require several bouts of infection by the same virus before antibody levels are sufficient for some protection. Even in middle age, only approximately 50% of people are seropositive. Although NLV does cause outbreaks in institutionalized settings where person-to-person spread can be significant, it is also associated with multiple simultaneous exposure events where otherwise healthy adults consume virus-contaminated food. Such outbreaks have minimal person-to-person transmission components, and 7–8% of outbreaks in the United Kingdom are described as mainly or chiefly food-borne in nature. However, this attribution is not particularly robust; the real figure could be significantly higher.

NLV is spread chiefly by water. Significant outbreaks of NLV have been associated with the commercial manufacture of ice from contaminated wells, soft fruits, and fruit juices. However, chief among the sources of NLV outbreaks is the consumption of raw or partially cooked shellfish. Molluscan shellfish are filter feeders, often farmed or growing naturally in estuarine locations. In Europe, most estuaries are contaminated with sewage effluent, and any viruses that are shed in the water from communities living upstream become concentrated within the bodies of the shellfish. Frequent outbreaks of gastroenteritis have been associated with fecal bacteria that was also concentrated in this way, but simply relaying the shellfish in clean water (depuration) is often sufficient to allow them to purge themselves of such bacteria. Unfortunately, this process is inefficient in ensuring the removal of viruses. Current European Union law regulates the sale of shellfish according to the levels of coliform bacteria they may contain, though this is no guarantee of virological cleanliness. Consequently, each year, a number of outbreaks are attributed to the consumption of raw shellfish, particularly oysters. Outbreaks show a slight seasonality, for instance, shellfish seem to clear viruses less efficiently in the winter when sea temperatures are lower, and outbreaks may also increase at such times. Eating habits can also be a contributory factor. For example, in the United States, there are often outbreaks around February 14th and around Thanksgiving because oysters are traditionally eaten there at these times.

SLVs

SLVs cause illness more commonly in children and account for some 3% of hospital admissions for diarrhea in both the United Kingdom and the United States. Most children are seropositive by age 12 and seem to become infected between three months and six years of age. The disease is particularly common in institutionalized settings such as schools and daycare centers. Incubation is between 24–48 hours, and illness is usually mild and short lived, with diarrhea tending to predominate. Any clinical differences reported in the illness induced by SLV and NLV seem to reflect more the age of the patient than an intrinsic property of the virus itself. On those occasions where SLV may cause illness in adults, the signs and symptoms are indistinguishable from those of NLV.

An assessment of the stability of NLV is particularly difficult because the virus can only be cultivated in volunteers. However, despite this limitation, a few valuable studies have addressed this point, and it is worth commenting on the sensitivity of NLV to chlorine used in the treatment of potable water. Reports from the United States show that the virus resists treatment at a peak level of 5 mg/l for 30 minutes. Thus, NLV can survive water chlorination to this level and has even been detected in tap water in the United Kingdom (although it is not known if it was infectious at that point). NLV is inactivated at levels of 10 mg/l, and this is used to decontaminate water supplies if contamination is suspected. The virus is relatively more chlorine resistant than polio and human rotaviruses (rotavirus is inactivated at levels greater than 3.5 mg/l).[26] NLV is acid stable, surviving at pH values as low as 2.7 for three hours at room temperature, and also relatively heat stable (resists 60 °C for 30 minutes).[13] Because the virus cannot be grown, feline calicivirus (FCV) is frequently used as a model.[44] Although it is true that the feline virus will probably mimic the behavior of NLV in terms of where it tends to accumulate, it is a relatively poor model for stability. For example, FCV is considerably more acid labile than NLV. Like all other enterically transmitted viruses, NLV lacks an envelope and thus tends to resist lipid solvents that would otherwise inactivate viruses by stripping their lipid envelopes. For instance, NLV is not inactivated by 20% ether at 4 °C for 18 hours.[13]

Rotaviruses

Rotaviruses are members of the family Reoviridae (Table 8–1) and account for some 3.5 million cases of diarrhea in the United States each year. This equates to 35% of hospital admissions for diarrhea each year. Even in a developed country like the United States, approximately 120 children die each year from this virus, and fatalities are probably far more numerous in less developed countries. The peak age for illness is between six months and two years; by four years of age, most people have been infected.[27] Immunity to rotaviruses is long lasting and secretory immunoglobulin A (IgA) in the gut plays an important role.

Group A rotaviruses exist in nine serotypes, and variations in other proteins increase the antigenic diversity of this group. However, as time progresses, individuals are exposed to many different variants of rotavirus, and immunity gradually accumulates. Thus, frequency of illness decreases with age. However, silent secondary re-infections can occur throughout life (as in parents caring for infants), thereby providing another means for the virus to spread in the community. Illness usually develops after an incubation period of four to seven days and presents as diarrhea and vomiting lasting approximately one week. In the United Kingdom, only some 2% of outbreaks were attributed as food-borne. The viruses are stabilized by calcium ions, which promote integrity of the particle.[42] Ionic detergents can reduce infectivity (e.g., 0.1% sodium dodecyl sulphate, or SDS), but non-ionic detergents can actually increase virus titre by breaking up aggregates and dispersing the particles. Viruses are very resistant to acid conditions and are stable over a wide pH range from 3.0 to 9.0.[14] They can retain infectivity for months at 20 °C[42] and also resist heating to 50 °C. Rotaviruses tend to survive desiccation well and are reduced in titre by less than 10-fold during drying (A. Bosch, personal communication). Once dried, the virus survives well on either porous or nonporous surfaces, being relatively unaffected by temperature or relative humidity. Infectivity of dried viruses reduces in titre by just more than 2 logs over a 60-day period, regardless of substrate, temperature, relative humidity, or the presence of accompanying fecal material. This makes rotavirus one of the most stable of enteric viruses, second only to hepatitis A virus in this respect.[1]

Adenoviruses

There are 47 serotypes of adenoviruses, but only types 40 and 41 cause enteric illness. Adenoviruses account for some 5–20% of U.S. hospital admissions for diarrhea, mainly in children below two years of age.[27] Incubation lasts three to ten days, and illness (watery diarrhea) may last for one week. However, prolonged shedding of the virus has been reported. As children age, exposure to adenovirus infection gradually increases the levels of immunity. Only 20% of children below six months of age have antibody to these viruses, but by age three, this has risen to 50%.[27] In the United Kingdom, adenovirus-related outbreaks are extremely rare; only one case was identified between 1992–1995, and this was not believed to be food-borne.[4] Clearly, this is in stark contrast to the antibody prevalence figures and implies that infection may be mainly in the community and doesn't follow outbreak behavior.

There are few studies that have specifically examined the stability properties of the enteric adenoviruses, but most workers agree that the viruses are certainly stable between pH 5 and 9 and up to 45 °C.[17] Infectivity is rapidly (10 minutes) lost above 56 °C, and this is associated with particle disintegration.[40] Other workers have examined the stability of the enteric adenovirus type 40, when dried on to fomites.[1] Adenovirus is one of the least stable of the enteric viruses under these conditions, and its behavior seems to resemble that of poliovirus rather than rota-, astro-, or hepatitis A viruses, which tend to be more robust. Desiccation itself causes a 100–1000-fold decrease in titre. However, infectivity persists for seven days on nonporous substrates, it is stabilized by the presence of accompanying fecal material, and it is unaffected by levels of relative humidity. Adenoviruses survive less well on porous surfaces, and the accompanying material seems to decrease its stability. Survival is better at 4 °C than at 20 °C on either surface. In liquids, adenovirus stability closely resembles that of the astroviruses, which are small, round RNA containing viruses responsible for enteric infections mainly in the young of animals and humans. Adenoviruses survive relatively well in tap water; log titre reduction values of 3.2 were observed after 60 days at 20 °C. The virus is also disinfected by free chlorine, giving log titre reduction values of 2.5 and 3 after two hours in the presence of 0.5 and 1 mg/l free chlorine, respectively. Again, this behavior mimics that of human astroviruses.[2,3]

Astroviruses

Astroviruses (serotypes 1–8) account for some 5% of hospital admissions in the United States, almost entirely of children.[27] By seven years of age, 50% of children are already seropositive for the most common serotype (type 1), and this reaches 75% by age 10. Illness is generally mild and lasts some two to three days after an incubation period of similar length. This has led many researchers to dismiss these viruses as causative agents of significant disease in humans. However, this is not necessarily true, first because the higher serotypes are less common, thus infection is delayed and occurs among older children and adults. Preexisting antibody to other serotypes may modify the severity of any resulting illness but may not prevent the occurrence of clinical disease. In Japan in 1995, 1,500 older children and their teachers were affected in a widespread food-borne outbreak caused by astrovirus type 4,[33] and symptomatic illness has also been seen in adults in the United Kingdom and France. Second, astrovirus diagnosis can present particular problems that may lead to an underestimation of the number of cases. Diagnosis still largely relies on electron microscopy, but particle morphology alone is not necessarily a good guide. Astroviruses may frequently be mistaken for small round (parvovirus-like) agents and even for NLVs.[23] This is a significant finding, because a recent survey in the United Kingdom identified only nine outbreaks associated with astrovirus (and a further two mixed infections that also contained NLV), of which approximately 20% were thought to be food-borne.[4] A further four outbreaks of illness were associated with small round virus, an agent usually assumed to be parvovirus and routinely dismissed as the causative agent of the outbreak. However, half of these cases were believed to be food-borne, and in view of the potential for misidentification, these agents require further investigation. Should these cases actually represent misidentified astroviruses, then the proportion of astrovirus outbreaks classified as food-borne could be as high as 30–45%. Recently, an ELISA-based detection kit has been produced, which could help to answer these questions if it is widely adopted.

Astroviruses survive desiccation well, dropping by only 10-fold with or without accompanying organic material. If the virus is dried on to a nonporous surface, accompanying fecal matter can boost relative survival by 10–100-fold, with infectivity persisting for up to 65 days and probably much longer at 4 °C. Astrovirus survival is greatly reduced at increased temperatures and can decay completely within 10 days at 20 °C on a nonporous surface. Adenoviruses survive equally well on porous surfaces at 4 °C regardless of the presence or absence of fecal material, infectivity persisting for more than 90 days. This suggests that environmental temperature may play an important role in the seasonality of astrovirus outbreaks because indirect transmission would be assisted by decreased temperature (A. Bosch, personal communication).

Astrovirus survives well in dechlorinated (tap) water[3] and reduces in titre by 100-fold after 60 days at 4 °C. This is increased to a 3.2 log reduction after 60 days at 20 °C, which is a very similar stability to that observed for human rotavirus and adenovirus type 40, where the same reduction was seen at 20 °C. Free chlorine seems more effective at disinfecting astrovirus-contaminated water than water contaminated with hepatitis A or human rotavirus. Astrovirus titres were reduced by 2.5 logs after one hour at a free chlorine concentration of 0.5 mg/l. This increased to 3 logs in the presence of 1 mg/l free chlorine. However, in both cases, residual infectivity was still detected after two hours of treatment when the log titre reduction was 4.17.[3]

Hepatitis Viruses

Hepatitis viruses A and E are classified differently. Hepatitis A virus (HAV) is the only member of a single virus genus (*Hepatovirus*) of the *Picornaviridae.* Hepatitis E virus (HEV), however, has some structural features resembling the caliciviruses, but its unique genomic organization means that it cannot be classified easily within any of the existing virus families.

Both viruses spread predominantly through water and are concentrated by molluscan shellfish. HAV is not uncommon, but HEV is rare in developed countries. It does, however, occur in epidemic form in India and in the former Soviet Union. The worst outbreak of HEV involved 30,000 people in New Delhi in 1955. More limited shellfish-associated outbreaks occur sporadically around the Mediterranean. In the United Kingdom, HEV is limited to returning travelers. Clinical features of both viruses are similar, although HEV tends to be more severe and can be fatal in pregnancy. Convalescence may be prolonged (8–10 weeks), and some 15% of cases of HAV may follow a relapsing course over 12 months or more.

The spread of picornaviruses has been affected profoundly by human activities. In former times, when the quality of water could not be guaranteed, infection occurred early in life through exposure to virus-contaminated water. Under these conditions, infections tended to be mild (often subclinical) and were endemic in the society. However, where water purification and other public health measures have been implemented, the possibilities of infection for all of these viruses are reduced, which has had the effect of increasing the mean interval between contacts with the virus. As a result, infection is delayed, and thus occurs predominantly in older individuals.

This infection delay is well demonstrated in countries where sanitation has improved over recent years.[19,28] In Hong Kong in 1979, 30% of those people under 30 years of age had been infected with HAV; by 1989, this number was 9% and is still falling. Presumably, this will progress eventually to resemble the situation in the developed countries where the vast majority of persons older than 30 have *no* antibody to HAV (i.e., they have never been infected).[18] This is significant for the course of the infection because the virus is more severe if it is contracted in adulthood. People older than 30 years of age account for only some 30% of the cases of HAV, but nearly 80% of the deaths occur in this age range.[19] Shellfish will, of course, concentrate any viruses shed in sewage and contaminating their natural habitat, including picornaviruses, and perhaps more significantly, HAVs and HEVs. Significant outbreaks of illness have been noted in Singapore and have been attributed directly to virus contamination of shellfish;[47] similar outbreaks that are mainly attributable to direct virus contamination of water occur with HEV.[28] Consequently, as water quality improves, it is expected that persons growing older in the absence of HAV infection would develop a taste for shellfish and shellfish products (including fermented foods) and become exposed to the virus later in their lives. They would have no residual immunity from childhood infections (see astroviruses above), which could pose a risk of severe disease later in life.

HAV is more stable to acid than other gut viruses of the picornavirus family. It retained virtually all activity following 120 minutes at pH 1.0 and was still viable (although reduced in titre) after five to eight hours.[41] Other enteroviruses were inactivated virtually completely within two hours at this pH. HAV is also relatively more heat stable and survives heating at 60 °C for 60 minutes,[10,15,21] being only partially inactivated after 12 hours at this temperature at neutral pH.[36,43] In the absence of divalent cations, 50% of particles will disintegrate within 10 minutes at 61 °C, although in the presence of 1 M MgCl, this is not achieved until 81 °C. The equivalent temperatures for poliovirus are 43 °C and 63 °C, respectively, and polio appears far more sensitive to desiccation and storage in the dry state than HAV. This suggests that poliovirus is an inadequate model for the survival of enteric viruses in general, and especially for HAV.[1,46] HAV does, however, become rapidly inactivated (in minutes) at temperatures higher than 98 °C and can also be destroyed by chlorine and hypochlorite (10–15 ppm residual chlorine after 30 minutes; 3–10 ml/l hypochlorite 20 °C for 5–15 minutes). However, HAV is more chlorine resistant than other picornaviruses.[2,38] In general, HAV is the most stable of the enterically transmitted viruses. It survives desiccation with minimal decrease in titre (0.5 log) and, once dried, shows virtually no reduction in titre during a seven-day period regardless of the surface onto which it has been dried, temperature or relative humidity of storage, or the presence of accompanying fecal material. Titres are reduced by only 1–2 logs during the next 60 days after drying.[1] These data suggest that the survival of HAV is likely in fermented foods, although it has not been rigorously assessed, and exposure to the virus from the consumption of shellfish or fermented foods prepared from them remains a possibility.

ASSESSMENT OF VIRUS RISK

Contamination of Foods

Virus levels will not increase during the fermentation of food; therefore, virus content of these materials becomes an issue only where there is significant contamination of the food before it is fermented. Most food-borne viruses are resistant to both acid conditions and mild heat treatment and could survive these aspects of fermentation processes. NLVs represent the greatest risk of gastroenteritis for adults because they are not likely to be immune. Stability data for these viruses are scarce because of the difficulty in assessing their survival, but it seems unlikely that they would be more resistant to physical treatments than HAV. Thus, in assessing the potential for virus survival in food, it would seem that procedures that would inactivate HAV probably give the greatest protection against virus dissemination in the food. The other viruses that could be present in food are largely infections of childhood. With the exception of fermented cereals and dairy curds in some cultures, children may not eat much fermented food and thus the acquisition of infection from this source (even if the viruses survive) is likely to be rare. Adults consuming food that has been contaminated with these viruses would be protected by residual immunity resulting from childhood infection. HAV poses a serious threat to adults with no preexisting immunity. Because this is the virus that is hardest to remove (and thus most likely to survive), HAV contamination of fermented foods (especially of shellfish-derived foods) could pose an increasing threat in areas where sanitation is rapidly improving and seropositivity to HAV is falling in adults.

Improved knowledge of the types of food that are regularly contaminated with viruses should permit an assessment of the relative risks associated with particular foods. In the United Kingdom, rigorous investigation often fails to identify the food vehicle of transmission. However, accepting the significance of NLVs as the most commonly identified cause of illness in adults, a recent survey found that 35 (65%) of the NLV outbreaks could be attributed to a particular foodstuff, the most frequent being oysters.[4] In those cases where a particular food was implicated, virus was detected in the food itself in only two cases, both oysters. This percentage reflects the generally low level of virus contamination in other types of food. Other food vehicles implicated were diverse, including sandwiches, pies, fresh fruit, fruit and vegetable salads, gateaux, fish, lobsters, and prawns. Many of the dishes were not served alone, and the common feature could be fresh fruit or salad vegetables that were often served as a garnish if not a main component of the meals. It is not possible to assess how many of these outbreaks were associated with contamination by food handlers immediately before serving, and how many might have resulted from virus contamination of the food before its preparation. Certainly, food handlers were suspected in 19 cases, but was confirmed in only four. Viral illness has been associated with intrinsic contamination of fruit in the past,[6,31] and the worldwide trade in fruit and vegetables may well increase such a risk in the future. It is generally held, in the United Kingdom at least, that there are two major patterns of transmission of food-borne viruses—those due to intrinsic contamination of shellfish with virus, and those due to other foods probably involving contamination by food handlers. It is not generally believed that virus contamination of vegetables and fruit at source is a major contributor to food-borne viral illness in the United Kingdom.

Food-borne spread of HAV is not common. In the United Kingdom, during the 1980s, seven outbreaks were attributed to shellfish and a similar number to food contamination by handlers. More recently, out of 19,000 cases (mostly involving children), only 0.5% were associated with food, with shellfish again being the major culprit. This risk is far more significant elsewhere where substantial outbreaks have been attributed to shellfish consumption. An outbreak associated with clams in Shanghai may have been the largest recorded outbreak of food-borne illness ever.[20] HEV occurs in the United Kingdom only in one or two cases per year, entirely associated with travelers returning from areas in which the disease is endemic. Waterborne spread would seem to be the major vehicle of transmission, although shellfish-associated cases have been reported.

Virus Survival in Fermented Foods

There are few data that are directly concerned with the survival of viruses in fermented foods. Studies conducted in fermented meats have shown that the viruses responsible for foot and mouth disease (i.e., FMDV), African swine fever, and hog cholera do not survive the production process.[29,34] However, these viruses are of veterinary importance and are known to be more labile than human enteric viruses. FMDV is acid labile, is inactivated below pH 6, and is not like gut viruses of the same family such as polio, whereas the viruses causing hog cholera and African swine fever are both enveloped and therefore more susceptible to the adverse conditions encountered during fermentation.[16] When model enteric viruses such as polio, Coxsackie, and echovirus, which are more relevant to human health, were used, they were found to persist in high titres virtually unaffected by the processing that reduced the pH to 5.0–5.4.[12,22,25] Studies with bacteriophage and simian rotavirus in fermented cereals that are more weakly buffered and consequently have a lower pH have also demonstrated the ability of viruses to survive well in fermented products.[32,50] This relative paucity of information does, however, point to the difficulty of controlling risk from food-borne viruses through fermentation, though further work is clearly needed. Based on all of the information available, however, some general inferences can be drawn concerning the risks posed by viruses and their control.

First, the type of food on which the product is based will be an important factor. Shellfish are an obvious high-risk material because they often live in sewage-contaminated estuarine waters and concentrate available viruses in their tissues. The initial preparation of shellfish involves shucking the shellfish flesh, and even if this is then washed, it is unlikely to remove virus contamination from the shellfish gut. Fermented foods prepared from organisms that consume shellfish (such as octopus or scavenging crabs) may be contaminated indirectly from this source. However, viruses are not believed to be retained with the same efficiency in these creatures, so the initial risk is presumably more short lived.

Vegetable produce may be contaminated either on the surface (e.g., by handling, washing, or spraying with contaminated water)[8] or more deeply within the tissues (e.g., resulting from the uptake of viruses contaminating irrigation waters used in cultivation). This is more likely in the case of produce with a high water content, such as celery, pumpkins, cucumbers, and other soft fruits. Surface contamination can be effectively removed by peeling provided the peelings are not simply allowed to elute any virus into the washing water and thus redistribute the material across freshly peeled surfaces. Similarly, washing itself can be very effective, and efficiency can be increased if small quantities of (preferably ionic) detergent can be used.

Where produce is contaminated with fecal material, viruses may be dried on to the surface as aggregates with other organic material. This tends to promote survival on nonporous surfaces such as a waxy plant cuticle. In these cases, the use of detergent could not only remove the aggregate, but also disperse it, thus increasing the number of potential infectious units present. Under these conditions, thorough rinsing becomes essential. Potable water washing alone can reduce surface contamination by bacteria by 10–100-fold, and the use of hypochlorite is even more effective. Common commercial washing procedures in the developed world use 100 ppm hypochlorite (yielding 30–40 ppm free chlorine) at pH 6.8–7.1 and 4 °C and a contact time of two minutes. Other materials such as soft fruits that could be damaged by this process can be sprayed or immersed for only 10 seconds in 15–20 ppm free chlorine.[7] Tests have found that in general, hypochlorite under these conditions is very effective against viruses in suspension;[24] however, its efficacy against viruses that are adsorbed to a surface or in the presence of other organic materials cannot be assumed as great. Studies suggest that hypochlorite may not be as effective as assumed when it is used to control HAV that was artificially introduced on to strawberries. Hypochlorite treatment was approximately 10-fold less effective against this virus on the fruit than the same virus in suspension, and full control could not be achieved below levels that would render the fruit inedible.[45] Furthermore, this method of cleansing must be limited to an effect on surface contamination only; it could clearly not affect any deeper contamination within the plant tissues.

Fermentation frequently involves salting, particularly fermented shellfish products produced in southeast Asia, where salt levels of up to 30% by weight can be used, producing a saturated brine.[39] Salt (sodium chloride) itself is generally not injurious to virus particles lacking an envelope; other ions are more problematic, with cesium having an adverse and sometimes irreversible effect on virus polymerase function. High salt (especially under acid conditions) can lead to the precipitation of proteins; this is likely to occur and will be aided by a generally high protein concentration in the environment. Precipitation on this scale would certainly cause the precipitation of virus particles and a considerable reduction in their infectivity. Precipitated viruses can redissolve should the salt concentration be reduced at a later stage, but residual clumping and loss of infectivity would mean that the original titre will not be recovered. Other salts have a stabilizing effect on virus particles, particularly divalent cations; magnesium stabilizes polioviruses and calcium stabilizes rotaviruses.

pH reduction in fermentation is typical, and pH could fall to as low as 3.8. However, all enteric viruses are able to infect through the gut, and thus all must be designed for passage through the acid conditions of the stomach. Indeed, stability to acid (pH 4.0) remains one of the routine tests for enteroviruses, differentiating them from the morphologically identical rhinoviruses that infect via the nasal mucosa and are not required to possess acid stability. Acid resistance is thus a common feature of all of the viruses in these groupings, and acid production *per se* is not likely to reduce virus titre significantly unless accompanied by heat. This is supported by the limited number of studies done on virus survival in acid-fermented foods. Nout *et al.*[32] used the bacteriophage (MS-2) as a model for human viruses and found that it survived much better than bacteria in a fermented porridge at pH 3.8; its numbers declining by approximately 0.1 log cycle per hour. Even better survival was found in work using the simian rotavirus SA 11, a good model for the behavior of unculturable human rotavirus. In 24 hours at 30 °C, the virus titre decreased by 0.25 log cycles at pH 4.0 and by 1 log cycle at pH 3.3.[50] Significantly, this work found the effect to be purely pH-related, with no difference when different acidulants were used. The enhanced antimicrobial effect of weak organic acids seen against bacteria clearly does not operate with nonenveloped, nonmetabolically active virus particles.

During fermentation, autolysis of the tissues will lead to protease release. Gut viruses are required to possess some degree of protease resistance because infection through the gut must expose them to these enzymes. In fact, some seem to have evolved to require the addition of proteases such as trypsin during their culture *in vitro*. Trypsin, however, is a specific enzyme recognizing certain sites in the protein, and all viruses are susceptible to prolonged protease treatment. This is especially true of broad-spectrum proteolytic enzymes such as protease K and pronase, and plant enzymes such as bromelain and papain. The enzymes released during autolysis are also broadly reactive against proteins. Regardless of the context of the peptide bond, they too are able to destroy virus particles eventually. Detergents and saponifiers will also assist this action if present. The addition of protease-containing fruit or fruit juice (especially pineapple) would also aid this effect in fermented shellfish products such as *Plaa-mam* and *Khem-mak-nat*.[39] Virus particles would, however, be protected from this process if they are present as aggregates or inclusions (perhaps within shellfish gut tissue) or surrounded by a high local concentration of protein.

Finally, there is heat treatment. The ability of a heat treatment to eliminate risk from a pathogenic organism will depend not only on the intrinsic heat sensitivity of the organism, but also on the initial numbers of the organism present, the heating menstruum, the temperature, and the time of exposure. Although there is a wealth of information on how these factors can affect bacterial survival, the thermal inactivation kinetics of viruses have not been subject to the same kind of scrutiny. A temperature of 65 °C is ineffective against most of the enteric viruses; certainly astroviruses, HAV, and rotaviruses can survive such treatment even if adenoviruses and polioviruses may be substantially reduced in titre. A recent study[9] has shown that less than 0.5 minutes at 85 °C produced a reduction of more than 5 log cycles in the titre of HAV in three different dairy products, whereas heating at conventional pasteurization temperatures of 71–73 °C required 13–18 minutes to achieve the same effect. Some protective effect from fat was seen in cream containing 18% fat, but no significant differences were discernible between skimmed and whole (3.5% fat) milk.

The failure to see outbreaks of viral illness associated with correctly pasteurized milk indicates the very low association of human viruses with this product. Thus, by extension, fermented milk products are likely to be relatively safe with regard to viral infections, particularly when operations such as milking and fermentation are mechanized, thus reducing the risk of contamination from food handlers. FCV could not be cultured from cockles immersed in boiling water for one minute or longer, during which time the average internal temperature reached 78 °C. Immersion for 30 seconds, which achieved an average internal temperature of approximately 63 °C, produced a 2-log reduction in titre.[44] HAV (and polio virus) were completely eliminated from cockles that achieved an average internal temperature of 85–90 °C for one minute. These data were used as the basis for setting heat-treatment regulations for cockles in the United Kingdom (90 °C for 90 seconds).[30] In some products, though, the temperature treatment necessary to eliminate risk may be incompatible with product quality, although there is some evidence that in acid (fermented) foods, the heat resistance of viruses is much reduced.[50]

Viruses clearly pose some special problems. The available data suggest that a fermentation process alone will not ensure safety from viral infection, although the highest risk is associated only with a very limited number of substrates. Further studies are clearly indicated to explore the integrated effect of all aspects of processing on virus survival in order to establish a realistic estimate of risk.

REFERENCES

1. Abad, F. X., Pinto, R. M. & Bosch, A. (1994). Survival of enteric viruses on environmental fomites. *Appl Environ Microbiol* 60, 3704–3710.
2. Abad, F. X., Pinto, R. M. & Bosch, A. (1997). Disinfection of human enteric viruses on fomites. *FEMS Microbiol Lett* 156, 107–111.
3. Abad, F. X., Pinto, R. M., Villena, R., Gajardo, R. & Bosch, A. (1997). Astrovirus survival in drinking water. *Appl Environ Microbiol* 63, 3119–3122.
4. Advisory Committee on Microbiological Safety of Food. (1998). *Report on Foodborne Viral Infections.* London: The Stationery Office.
5. Albert, M. J. & Bishop, R. F. (1984). Cultivation of human rotaviruses in cell culture. *J Med Virol* 13, 377–383.
6. Anon (1997). Hepatitis A associated with consumption of frozen strawberries. *WHO Surveillance Programme for Control of Foodborne Infections and Intoxications in Europe* 54, 5.
7. Beuchat, L. R. (1992). Surface disinfection of raw produce. *Dairy, Food Environ San* 12, 6–9.
8. Beuchat, L. R. (1996). Pathogenic microorganisms associated with fresh produce. *J Food Prot.* 59, 204–216.
9. Bidawid, S., Farber, J. M., Sattar, S. A. & Hayward, S. (2000). Heat inactivation of hepatitis A in dairy foods. *J Food Prot* 63, 522–528.
10. Bradley, D. W., Fields, H. A., McCaustland, K. A., Cook, E. H., Gravelle, C. R. & Maynard, J. F. (1978). Biochemical and biophysical characterisation of light and heavy density hepatitis A virus particles: evidence that HAV is an RNA virus. *J Med Virol* 2, 175–187.
11. Carter, M. J. & Cubitt, W. D. (1995). Norwalk and related viruses. *Curr Opinion Infect Dis* 8, 403–409.
12. Dethmers, A. E., Rock, H., Fazio, T. & Johnston, R. W. (1975). Effect of added sodium nitrate on sensory quality and nitrosamine formation in thuringer sausage. *J Food Sci* 40, 491–495.
13. Dolin, R., Blacklow, N. R., DuPont, H., Bushco, R. F., Wyatt, R. G., Kasel, J. A., Hornick, R. & Chanock, R. M. (1972). Biological properties of the Norwalk agent of acute non-bacterial gastroenteritis. *Proc Soc Exp Biol Med* 140, 578–583.
14. Estes, M. K., Graham, D. Y., Smith, E. M. & Gerba, C. P. (1979). Rotavirus stability and inactivation. *J Gen Virol* 43, 403–409.
15. Feinstone, S. M., Moritsugu, Y., Shih, J. W., Gerin, J. L. & Purcell, R. H. (1978). Characterisation of hepatitis A virus. In *Viral Hepatitis*, pp. 41–48. Edited by G. N. Vyas, S. N. Cohen & R. Schmid. Philadelphia, PA: Franklin Institute Press.
16. Fields, B. N. & Knipe, D. M., eds. (1991). *Fundamental Virology*, 2nd edn. New York: Raven Press.
17. Foy, H. M. (1997). Adenoviruses. In *Viral Infections of Humans Epidemiology and Control,* 4th edn., pp. 119–138. Edited by A. S. Evans & R. A. Kaslow. New York: Plenum Medical Book.
18. Frosner, G. G., Papaevagelou, G., Butler, R., Iwarson, S., Lindholm, A., Courouce-Pauty A., Haas, H. and Deinhardt, F. (1979). Antibody against hepatitis A virus in seven European countries: a comparison of prevalence data in different age groups. *Am J Epidemiol* 110, 63–69.
19. Hadler, S. (1992, June). Hepatitis A: changing epidemiology and the need for a vaccine. *Virus and Life* 14–16.
20. Halliday, M. L., Kang, L. Y., Zhou, T. K., Hu, M. D., Pan, Q. C., Fu, T. Y., Huang, Y. S. & Hu, S. L. (1991). An epidemic of hepatitis A attributable to the ingestion of raw clams in Shanghai, China. *J Infect Dis* 164, 852–859.
21. Havens, W. P. Jr. (1945). Properties of the etiologic agent of infectious hepatitis. *Proc Soc Exp Biol Med* 58, 203–204.
22. Hermann, J. E. & Cliver, D. O. (1973). Enterovirus persistence in sausage and ground beef. *J Milk Food Technol* 36, 426–428.

23. Hermann, J. E., Hudson, R. W., Blacklow, H. R. & Cubitt, W. D. (1987). Marin county agent: an astrovirus. *Lancet* ii, 743.
24. International Committee for Microbial Safety of Food. (1996). *Micro-Organisms in Foods.* Vol. 5, *Microbiological Specifications of Food Pathogens*. London: Chapman & Hall.
25. Kantor, M. A. & Potter, N. N. (1975). Persistence of echovirus and poliovirus in fermented sausage. *J Food Sci* 40, 491–495.
26. Keswick, B. H., Satterwhite, N. R., Cukor, G. C., DuPont, H. L., Secor, S. L., Bitsure, J. A., Gary, G. W. & Hoff, J. C. (1985). Inactivation of Norwalk virus in drinking water by chlorine. *Appl Environ Microbiol* 50, 261–264.
27. LeBaron, C. W., Furutan, N. P., Lew, J. F., Allen, J. R., Gouvea, V., Moe, C. & Monroe, S. S. (1990). Viral agents of gastroenteritis public health importance and outbreak management. *MMWR* 39, 1–24.
28. Margolis, H. S., Alter, M. J. & Hadler, S. (1997). Viral hepatitis. In *Viral Infections of Humans: Epidemiology and Control,* 4th edn., pp. 363–418. Edited by A. S. Evans & R. A. Kaslow. New York: Plenum Medical Book.
29. McKercher, P. D., Hess, W. R. & Hamdy, F. (1978). Residual viruses in pork products. *Appl Environ Microbiol* 35, 142–145.
30. Millard, J., Appleton, H. & Parry, J. V. (1987). Studies on the ehat inactivation of hepatitis A virus with special reference to shellfish. *Epidemiol Infect* 98, 397–414.
31. Niu, M. T., Polish, L. B., Robertson, B. H., Khanna, B. K., Woodruff, B. A., Shapiro, C. N., Miller, M. A., Smith, J. D., Gedrose, J. K. and Alter, M. J. (1992). Multistate outbreak of hepatitis A associated with frozen strawberries. *J Infect Dis* 166, 518–524.
32. Nout, M. J. R., Rombouts, F. M. & Havelaar, A. (1989). Effect of accelerated natural lactic fermentation of infant food ingredients on some pathogenic organisms. *Int J Food Microbiol* 8, 351–361.
33. Oishi, I., Yamazaki, K., Kimoto, T., Minekawa, A., Utagawa, E., Yamazaki, T., Inouye, S., Grohman, G. S., Monroe, S. S., Stine, S. E., Carcamo, C., Ando, T. & Glass, R. I. (1994). A large outbreak of acute gastroenteritis associated with astrovirus among students and teachers in Osaka Japan. *J Infect Dis* 170, 439–443.
34. Panina, G. F., Civardi, A., Massirio, I., Scatozza, F., Baldini, P. & Palmia F. (1989). Survival of foot and mouth disease virus in sausage meat products (Italian salami). *Int J Food Microbiol* 8, 141–148.
35. Parrino, T. A., Schreiber, D. S., Trier, J. S., Kapikian, A. Z. & Blacklow, N. R. (1977). Clinical immunity in acute gastroenteritits caused by the Norwalk agent. *N Engl J Med* 297, 86–89.
36. Parry, J. V. & Mortimer, P. P. (1984). The heat sensitivity of hepatitis A virus determined by a simple tissue culture method. *J Med Virol* 14, 277–283.
37. Parwani, A. V., Saif, L. J. & Kang, S. Y. (1990). Biochemical characterisation of porcine enteric calicivirus: analysis of structural and non-structural viral proteins. *Arch Virol* 112, 41–53.
38. Peterson, D. A., Hurley, T. R. & Wolfe, L. G. (1983). Effect of chlorine treatment on infectivity of hepatitis A virus. *Appl Environ Microbiol* 45, 223–227.
39. Phithakpol, B., Varanyanond, W., Reungmaneepaiton, S. & Wood, H. (1995). *The Traditional Fermented Foods of Thailand.* Bangkok: Institute of Food Research and Product Development.
40. Russel, W. C., Valentine, R. A. & Pereira, H. G. (1967). The effect of heat on the anatomy of adenovirus. *J Gen Virol* 1, 509–522.
41. Scholz, E., Heinricy, U. & Flehmig, B. (1989). Acid stability of hepatitis A virus. *J Gen Virol* 70, 2481–2485.
42. Shirley, J. A. & Beards, G. M. (1981). The influence of divalent cations on the stability of human rotavirus. *Arch Virol* 67, 1–9.
43. Siegl, G., Weitz, M. & Kronauer, G. (1984). Stability of hepatitis A virus. *Intervirology* 22, 218–226.
44. Slomka, M. J. & Appleton, H. (1998). Feline calicivirus as a model system for heat inactivation studies of small round structured viruses in shellfish. *Epidemiol Infect* 121, 401–407.
45. Sobsey, M. (1998). *Surface inactivation of hepatitis A virus on strawberries using chlorine.* Paper presented at the 85th annual meeting of the International Association of Milk, Food and Environmental Sanitarians, October 1998, Nashville TN.
46. Sobsey, M. D., Shields, P. A., Hauchman, F. S., Davis, A. L., Rullman, V. A. & Bosch, A. (1988). Survival and persistence of hepatitis A virus in environmental samples. In *Viral Hepatitis and Liver Disease,* pp. 121–124. Edited by A. J. Zuckerman. New York: Alan R Liss.
47. Tan, Y. W., Wang, J. X., Xu, Z. Y., Guo, Y. F., Qian, W. H. and Xu, J. X. (1991). A serologically confirmed, case-control study, of a large outbreak of hepatitis A in China, associated with consumption of clams. *Epidemiology & Infection* 107, 651–657.
48. Wheeler, J. G., Sethi, D., Cowden, J., Wall, P. G., Rodrigues, L. C., Tompkins, D. S., Hudson, M. J. & Roderick, P. J. (1999). Study of infectious intestinal disease in England: rates in the community, presenting to general practice, and reported to national surveillance. *Br Med J* 318, 1046–1050.
49. Willcocks, M. M. , Laidler, F., Carter, M. J. & Madeley, C. R. (1990). Cultivation of human astrovirus in a continuous cell line. *Arch Virol* 113, 73–82.
50. Wood, G. W. & Adams, M. R. (1992). Effect of acidification, bacterial fermentation and temperature on the survival of rotavirus in a model weaning food. *J Food Prot* 55, 52–55.

CHAPTER 9

Microbiological Hazards and Their Control: Parasites

Mike Taylor

INTRODUCTION

Many types of parasites are food-borne, and humans can become infected following the ingestion of infected or contaminated meat, fish, molluscs, vegetables, or fruit, or products derived from these foods. In most cases, parasitic infections are acquired by eating raw or incompletely cooked food, or food that is partially pickled or smoked or poorly preserved. Most, if not all, infections are preventable if the food is prepared sufficiently to destroy the infective stages of the parasite. However, many infections are commonly associated with cultural and eating habits that have been in practice in populations for generations.

Meat from many species of animals has been a recognized source of many helminth, and some protozoal, infections, in man. In developed countries, the introduction of meat hygiene measures has resulted in reduced incidences of many of the traditionally recognized helminth infections. However, the eating of many traditional raw or lightly cooked meat dishes continues, and may occasionally result in infection in man. Raw, uncooked fish dishes are also commonly eaten in many cultures, and snails, clams, oysters, and a variety of other molluscs are part of the diets of many people worldwide. Although most food-borne helminth infections are reported from Third World countries, increased immigration, tourism, and desire to experience the culinary dishes of other cultures may increase the incidence of parasitic disease in other countries. As with some helminths, protozoan parasites are opportunistic infections that are often acquired as the result of poor hygiene or travel to foreign countries.

Fermented foods that use raw ingredients that have been contaminated or infected with infective or intermediate parasite stages have the potential to cause human infection. Although food-borne parasitic diseases continue to be reported globally, reports of human infection following the ingestion of fermented food are sparse. Cultural dishes such as *som fak* (a fermented Thai minced fish dish) have been reported to result in human infection with the helminth parasite, *Gnathostoma* (see Gnathostomosis). Infection with another protozoan parasite, *Giardia*, has been reported following the ingestion of cheese dip. Fermentation alone may therefore be insufficient to prevent the transmission of many food-borne parasites, and potentially infected material should be avoided wherever possible, or alternatively subjected to freezing or some form of heat treatment.

Many parasites infect humans, but reference is made only to those helminths and protozoa that are recognized as food-borne and a potential source of infection in fermented foods.

NEMATODES

The nematodes are a diverse group of parasitic or free-living unsegmented worms that are usually cylindrical and elongate in shape. With few exceptions, the sexes of nematodes are separate, and the life cycle may be direct or indirect,

---

involving an intermediate host. The general description and classification of the food-borne nematodes discussed in this chapter are summarized in Table 9–1.

Angiostrongylosis

Angiostrongylus cantonensis and *A. costaricensis* are parasites of rodents (predominantly rats) that can cause disease in humans through the ingestion of infected snails. Two other species, *A. malaysiensis* and *A. mackerrasse*, occur but have not been associated with human infection. Most human infections are acquired by eating infected snails, but can occur through eating infected shrimp and crab, or snail- or slug-contaminated raw vegetables.

Public Health Significance

A. cantonensis can cause meningitis and meningoencephalitis with mild to moderate symptoms, often of sudden onset, with intense headaches; vomiting; moderate intermittent fever; and, in approximately 50% of cases, coughing, anorexia, malaise, constipation, and somnolence.[176] In severe cases, coma and death may occur. Accidental ingestion of the slug intermediate host of *A. costaricensis* causes abdominal angiostrongylosis with symptoms similar to appendicitis, including fever, abdominal pain, anorexia, diarrhea, and vomiting.[96] Migratory larvae may cause gastroenteritis, tumor-like masses or abcessation of the intestines, liver enlargement, nervous signs, coma, and, occasionally, death.[108] Infection is normally diagnosed by confirming the presence of parasites or eggs in surgically removed tissues or fluids or by using serological assays.[37,144,169]

Life Cycle

Adult worms are found in the pulmonary arteries of rats. The life cycles of *A. cantonensis* and *A. costaricensis* are similar and are shown in Figure 9–1. The prepatent period (from infection to maturity) is 42–45 days.

Distribution

The distributions of *A. cantonensis* and *A. costaricensis* are summarized in Table 9–2. Most human infections and deaths associated with *A. cantonensis* have been reported from Taiwan, Thailand,[131,176] and some Pacific islands, but human infections have been reported in most countries where the parasite occurs and appears to be spreading.[33] The incidence of angiostrongylosis appears to be spreading, especially in those areas in which snails are an important part of the diet.

Epidemiology and Transmission

Humans become infected with angiostrongylosis by intentionally or accidentally eating infected snails or slugs. Rodents are infected by ingesting infected molluscs or by ingesting infective larvae present in "slime" on plants. In most endemic areas of Asia, both land and aquatic snails (*Achatina* and *Pila*) are eaten regularly. The giant African land snail, *Achatina fulica*, is a particular delicacy in many countries and is a good intermediate host. Slugs and snails (which are used for medicinal purposes in some cultures), land crabs, shrimp, and paratenic hosts such as toads and frogs have also transmitted infection. Sauces prepared from shrimp juices or unwashed contaminated vegetables have also been incriminated. The drinking of untreated water containing larvae released from dead snails has also been suggested as a means of infection. Infection continues to be reported from new areas of the world, in part due to the dissemination of the intermediate snail hosts, but also due to transportation of infected rats on ships. Most terrestrial and aquatic snails are susceptible to infection, and populations can be readily infected from carrier rats.

Prevention and Control

Angiostrongyliosis can be prevented by educating people in endemic areas to avoid eating uncooked molluscs, particularly land snails. Freezing will kill larvae present in snails if they are frozen at –15 °C for 12–24 hours. Paratenic hosts (i.e., shrimp, prawn, crabs) should be cooked before eating, and vegetables should be washed before eating raw. Little or no information is available on the survivability of infective larvae in fermented foods such as *balao-balao* (fermented shrimp).

Table 9–1 Food–Borne Helminths—Classification

Phylum	*Class*	*Order*	*Family*	*Genus*
Nemathelminthes	**Nematoda** (Roundworms) Elongate, cylindrical, unsegmented worms with fluid-filled body cavity. Sexes are separate. Life cycle direct or indirect.	Strongylida	Metastrongyloidae	*Angiostrongylus*
		Ascaridida	Anisakidae	*Anisakis*
		Enophida	Trichinellidae	*Trichinella*
			Capillaridae	*Capillaria*
		Dioctophymida	Dioctophymatidae	*Dioctophyma*
		Spirurida	Thelaziidae	*Gongylonema*
			Gnathostomatidae	*Gnathostoma*
Platyhelminthes	**Cestoda** (Tapeworms) Tape-like segmented body comprising head, neck, and strobila (proglottids). Bothria or suckers for attachment. Hermaphroditic. Indirect life cycles.	Pseudophyllidea	Diphyllobothridae	*Diphyllobothrium*
		Cyclophyllidea	Taeniidae	*Taenia*
	Trematoda (Flukes) Unsegmented leaf or lancet-shaped worms with two muscular suckers for attachment and well-developed oral sucker and pharynx. Hermaphroditic (generally). Indirect life cycles.	Echinostomida	Fasciolidae	*Fasciola*
				Fasciolopsis
			Echinostomatidae	*Echinostoma*
				Hypoderaeum
		Plagiorchiida	Troglotrematidae	*Paragonimus*
				Nanophyetes
		Opisthorchiidae	Heterophyidae	*Heterophyes*
				Metagonimus
			Opisthorciidae	*Opisthorcis (Clonorchis)*

Source: © Crown Copyright.

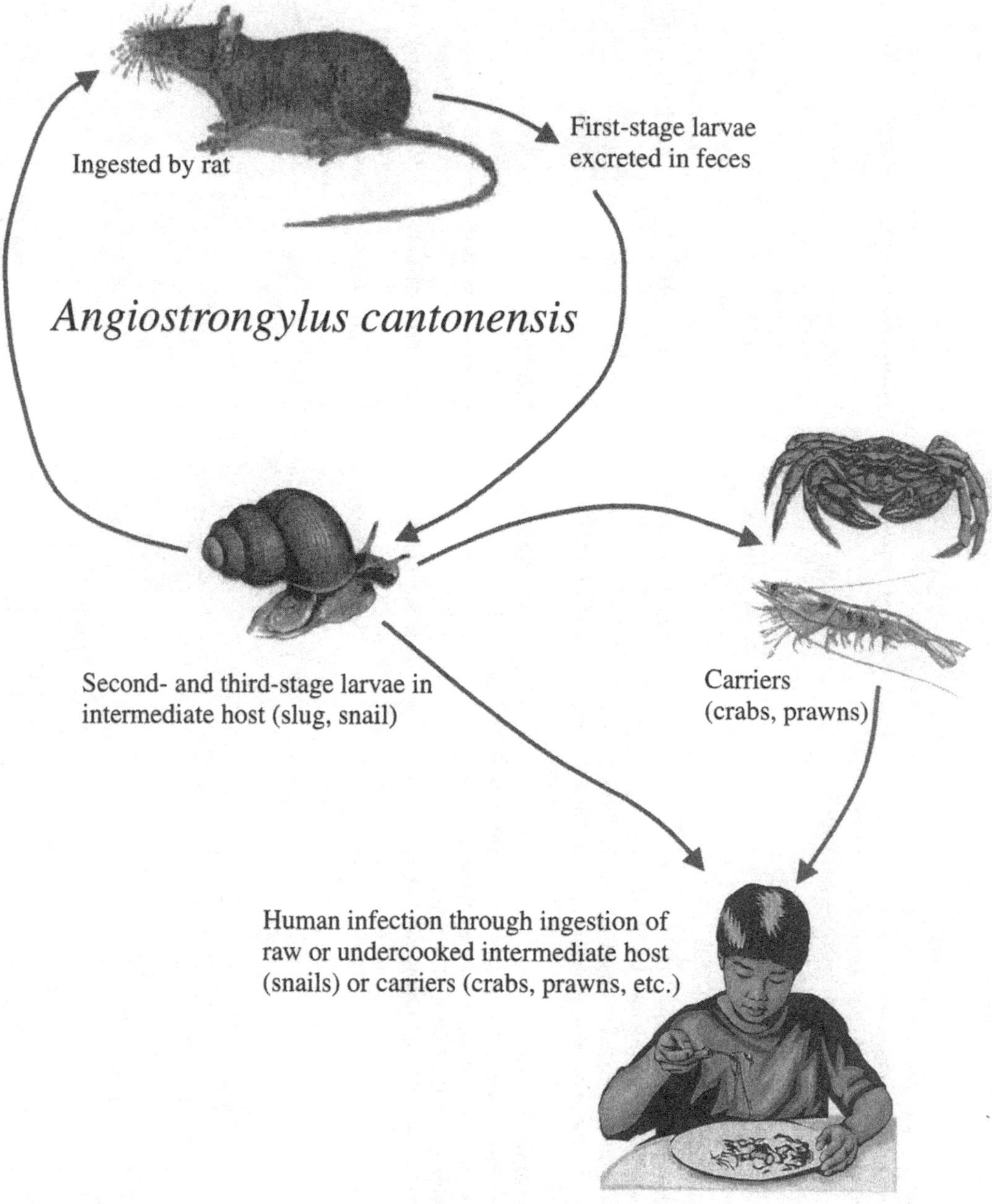

Figure 9–1 Life cycle of *Angiostrongylus cantonensis. Source:* © Crown Copyright.

Anisakiosis

Anisakiosis is a parasitic gastrointestinal disease of man that is caused by the larval stages of nematodes of the family Anisakidae. The adult worms are common parasites of marine mammals (Table 9–2), and the larval stages are found in marine fish and squid. Humans acquire infections by eating the raw or improperly cooked or preserved meat of these animals. There are many

Table 9–2 Food–Borne Helminths (Nematodes, Cestodes)

Parasite Class	*Parasite*	*Distribution*	*Main Hosts (final)*	*Source of Infection to Man*
Nematoda (Roundworms)	*Angiostrongylus*	Asia and Pacific Islands, Australia, India, Africa, Caribbean, parts of United States	Rodents (rats)	Molluscs (snails), shrimps, crabs, amphibians, contaminated vegetables and salads
	Anisakis simplex *Anisakis decipens*	North and South America, Pacific Islands, parts of northern Europe	Pinipeds (whales, dolphins, porpoises) Seals, sealions, walruses	Fish
	Capillaria phillipinensis	Phillipines, Thailand	Man (several species of birds)	Fish
	Dioctophyma renale	Worldwide (North and South America, southern Europe, Asia, Middle East)	Carnivores (mink, ferret, dogs, cats, jackals), man	Fish
	Gnathostoma spinigerum	Thailand, Japan, southeast Asia, China, Mexico (Middle East, Africa, Baltic States, Russia)	Carnivores (dogs and cats), man	Fish , frogs, chickens, ducks, snakes
	Gnathostoma hispidum		Pigs, man	
	Gongylonema pulchrum	United States, former USSR, parts of Europe, Middle East, China, North Africa, New Zealand	Ruminants, pigs, dogs, cats, horses, rodents, primates, man	Salads (insects)
	Trichinella spiralis	Worldwide (except Antartica)	Pigs, rodents, carnivores (mink, fox, badger, bears, walrus, seals), man	Meat
Cestoda (Tapeworms)	*Diphyllobothrium latum*	Northern Hemisphere (northern Europe, Russia, North America, South America, Asia, Africa)	Dog, fox, mink, cat, pig, bear, seals	Fish
	Taenia saginata (Cysticercus bovis)	Central Africa, Asia, South and West Africa, parts of Europe, southeast Asia, Central and South America; also reported in United States, Canada, Australia and Pacific Islands	Man	Meat (beef)
	Taenia solium (Cysticercus cellulosae)	Central and South America, central and east Africa, southeast Asia, southern Europe	Man	Meat (pork)

Source: © Crown Copyright.

species of anisakid nematodes, but those most often associated with human illness are *Anisakis simplex, Pseudoterranova decipiens, Phocanema,* and *Contracaecum* spp.

Public Health Significance

Anisakis causes acute or chronic gastrointestinal disease in man. Migrating larvae cause a foreign-body reaction and eventually necrosis and hemorrhage of the stomach, occasionally creating tunnels and burrows in the stomach mucosa,[139] causing pain, nausea, and vomiting. The acute symptoms subside in a few days, but a vague abdominal pain with intermittent nausea and vomiting persists for weeks, with symptoms resembling those of a peptic ulcer. The condition is often misdiagnosed because of its similarity to other acute gastrointestinal conditions (e.g., gastric ulcers or neoplasm, appendicitis, diverticulitis, Crohn's disease, gallstones, etc.). Diagnosis is normally only made following biopsy and confirmed on histopathology.

Life Cycle

Adult anisakid worms are located in the stomach of marine cetaceans and pinnipeds but do not develop in humans. The life cycle of these worms is shown in Figure 9–2. Humans become infected if an intermediate fish host containing infective third-stage larvae is eaten uncooked.

Distribution

Fish infections are found in most oceans and seas but are highest in areas in which there are high marine mammal populations, such as coastal Japan and Alaska. Many species of fish are naturally infected with anisakiosis, and the prevalence of infection can be very high. In Japanese waters, 123 species of marine fish have been found to harbor the parasite.[118] The global distribution of anisakiosis is summarized in Table 9–2. The highest levels of infection have been reported from Japan and the Netherlands.[88]

Epidemiology and Transmission

The main source of infection from anisakid worms for man is marine fish, many species of which are highly parasitized by anisakid larvae. Humans acquire *A. simplex* by eating raw or poorly salted, pickled, or smoked herring, cod, mackerel, salmon, or squid, and *P. decipiens* from cod, halibut, flatfish, and red snapper. Such traditional preparations as *green herring, lomi lomi salmon, seviche, sushi,* and *sashimi* (i.e., seasoned fish fillets), all of which use raw or uncooked fish, are major sources of infections.[34] In the Netherlands, the occurrence of the disease was due to the habit of consuming raw or lightly salted herring *(green herring)*.[162] Although the habit persists, the incidence of human anisakiosis has been drastically reduced by freezing fish before marketing. In recent years, the highest incidence of the disease has been recorded in Japan, where many fish dishes are eaten raw (sashimi); pickled in vinegar (*sunomono*); or fermented in rice, rice bran, or *koji* (sushi and *zuke*). In the United States, at least two cases have been linked to eating *seviche* (i.e., pieces of raw fish seasoned in lemon juice for several hours), and others to eating Japanese raw fish dishes.

Prevention and Control

The risk of human infection with anisakiosis increases in countries where fish are eaten raw.[172] Anisakiosis is preventable by ensuring that only well-cooked marine fish, octopus, and squid are eaten. Larvae are killed by cooking at 60 °C or above. Freezing fish at –20 °C for 24 hours will kill the larvae, with the exception of some North American species that can survive freezing at that temperature for 52 hours. Cleaning and eviscerating fish immediately after they are caught prevents larvae migrating from the intestine to the muscles. Salt curing, marinating, microwaving, and smoking temperatures are insufficient to kill the parasite.[34] Marination of herring has been a long tradition in parts of northern Europe,[85,106] involving the preservation of herring fillets in salt and acetic acid. The salt/acetic acid marination process produces the typical flavor as a result of denaturing of the fish proteins, lowering of the pH, and addition of sugar and spices.[158] The salt/acid treatment may take up to 42 days to kill *Anisakis* larvae, and at low concentrations may not kill larvae at all.[85] Larvae have also been shown to survive in *izushi*

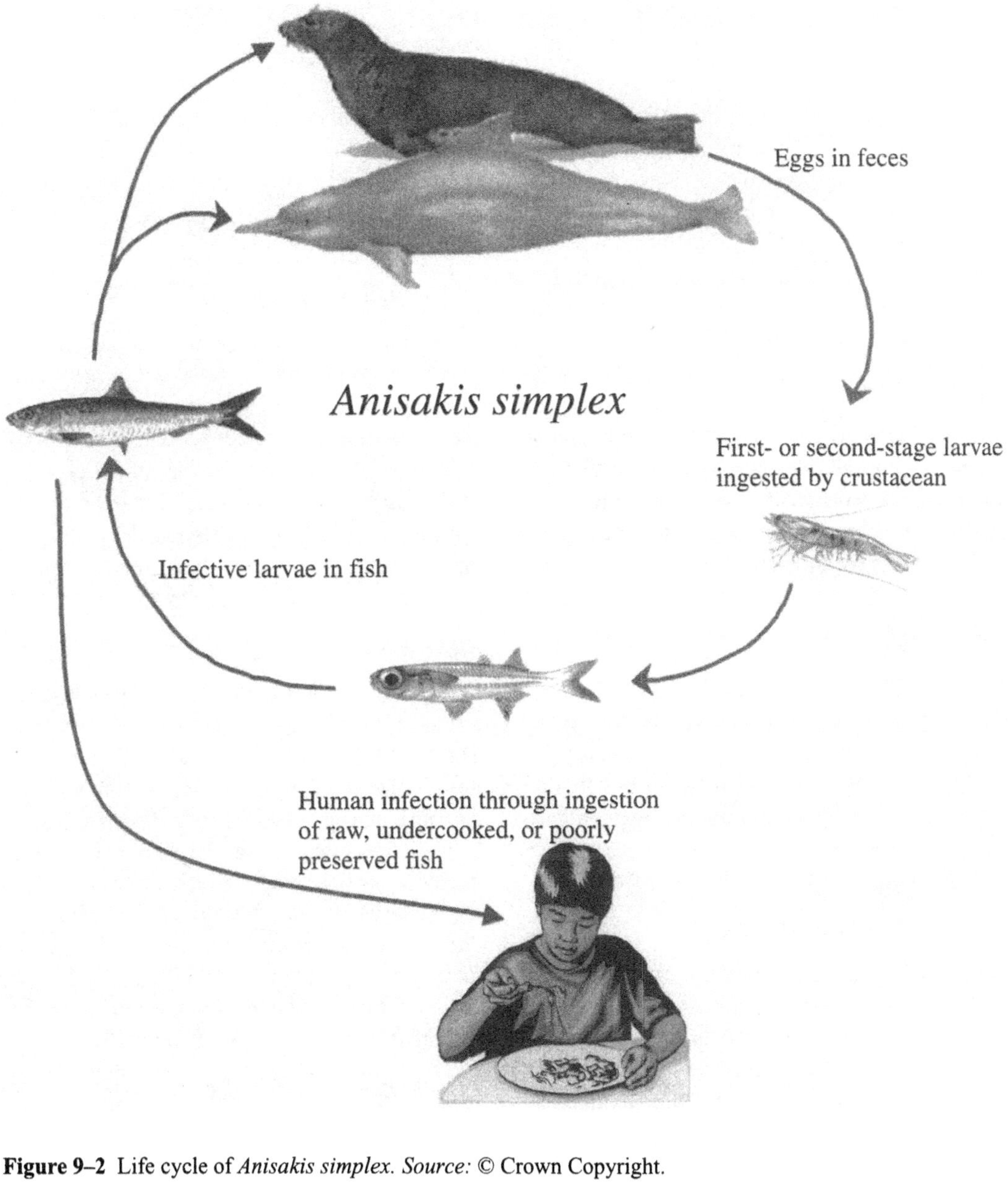

Figure 9–2 Life cycle of *Anisakis simplex. Source:* © Crown Copyright.

(i.e., pickled rice with cod roe, fillets of salmon, and cod) and in sashimi.[118]

Capillariosis

Intestinal capillariosis is caused by a tiny nematode, *Capillaria philippinensis.* The disease was first recognized in the Philippines in the 1960s and has subsequently been reported in other countries.[32,34,93] Man is the main definitive host for the parasite, but several species of birds are now believed to be a natural host and are able to transmit the infection.[35]

Public Health Significance

Infections with *C. philippinensis* can cause diarrhea, anorexia, weight loss, and, if left untreated, death. The parasites cause damage to the mucosa of the small intestine, leading to fluid, protein, and electrolyte loss. Clinical symptoms include abdominal pain and diarrhea. If treatment is not initiated rapidly, patients die because of the irreversible effects of the electrolyte loss, heart failure, or septicemia. In endemic areas, a diagnosis can be made on patients based on clinical signs. In chronic infections, there is weight loss, wasting, and intractable diarrhea. Infection is confirmed by identifying eggs, larvae, and adults microscopically in the feces. Serological tests are not sufficiently specific for routine use.

Life Cycle

C. philippinensis is a small nematode that is found in the small intestines of man. Further development occurs in fish. A number of species of freshwater fish are suspected to serve as intermediate hosts for the parasite. The fish are usually small and are eaten whole in the Philippines and Thailand, leading to human infection.

Distribution

The distribution of intestinal capillariosis is summarized in Table 9–2. Nearly 2,000 cases of capillariosis have been recorded in the Philippines, with more than 100 deaths. In Thailand, there have been only a few hundred cases, with an unknown number of deaths recorded.[36]

Epidemiology and Transmission

Sanitation facilities are poor in many rural areas of southeast Asia and defecation in the fields is a common practice. During the monsoon rains, the feces are broken down and washed away to streams and ponds, resulting in infection of freshwater fish. These and many other foods are eaten uncooked, especially in the endemic areas of the Philippines and in Thailand.

Prevention and Control

Educating both the local population and visitors of the dangers of eating uncooked freshwater fish can help prevent intestinal capillariosis. Improvement of sanitation and the control of indiscriminate disposal of feces would also be beneficial. No information is available on the conditions that kill the parasite in fish and whether they are known to survive in fermented fish products.

Dioctophymosis

Dioctophyma (Dioctophyme) renale is a large kidney-dwelling nematode of carnivores. Infection is acquired by ingesting raw or undercooked fish or frogs. It has only rarely been reported in man.

Public Health Significance

Infection in man can cause renal damage and associated symptoms of renal colic, hematuria, or urinary obstruction. In humans and dogs, *Dioctophyma* usually locates in only one kidney, most often the right one, and in most cases only one parasite is found, causing little or no clinical symptoms. Infection is diagnosed by the presence of the characteristic thick, bipolar eggs in urine (only if female worms are present).

Life Cycle

D. renale is a large, blood-red nematode that is found in the kidneys of carnivores (Table 9–2). The size of the parasite depends on the size of the host species; in dogs, the adult female of the parasite can reach up to 1 m in length. The intermediate host is a free-living, aquatic segmented worm (oligochaete) in which further development to an infective larva occurs. Transmission to a final host occurs following ingestion of the annelid intermediate host, or more usually following ingestion of a fish or a frog, which act as transport hosts. The prepatent period is from $3^1/_2$ to 6 months.

Distribution

D. renale is found on all continents with the possible exception of Africa and Oceania (Table 9–2). The most frequently reported form of infection is canine dioctophymosis, with the highest prevalence of infection in Canadian wild mink *(Mustela vison)*, where 18% of the animals are infected. The disease is very rare in man, with only a few reports worldwide.

Epidemiology and Transmission

In North America, mustelids, especially mink, appear to be the main reservoir. In other areas, it is likely that other species of mustelids or wild canids serve as main definitive hosts. These hosts are infected by ingesting frogs or fish (paratenic hosts) and aquatic oligochaetes (intermediate hosts) that contain the third-stage larvae. Dogs and humans are accidental hosts and are infected by ingesting raw fish and frogs, and almost always harbor only one parasite. The rarity of human infection can be explained by the fact that larvae are located in the mesentery and liver of fish and frogs, which generally are not eaten by man.

Prevention and Control

Infection can be prevented, both in humans and dogs, by avoiding the consumption of raw or undercooked fish and frogs. It is not known what conditions result in larval death.

Gnathostomosis

Gnathostomosis is caused by infection with the larval or immature adult stages of nematodes of the genus *Gnathostoma.* Adult parasites are reported in dogs, cats, and other carnivorous animals worldwide. In Thailand, more than 40 species of vertebrates have been reported to be naturally infected. These include freshwater fish, frogs, snakes, chickens, ducks and other birds, rats, mongooses, and tree shrews.[38] Four species, *G. spinigerum* and *G. hispidum,* and more recently, *G. doloresi* and *G. nipponicum,* have been reported in humans in Japan.[111,152]

Public Health Significance

Man is an abnormal host, with infection resulting in a larval migrans causing red, itchy, and edematous subcutaneous swellings that usually last approximately one week but can recur weeks or months later. More rarely, the parasite may enter the eye, causing subconjunctival edema, exopthalmus, impaired vision or blindness through hemorrhage, and retinal damage. Invasion of the central nervous system (CNS) can produce headaches, neck stiffness, drowsiness, or coma and death. Brain hemorrhage and transitory obstructive hydrocephalus have also been reported.[140] Diagnosis in endemic areas is based on history, symptoms, or serology and can be confirmed following the recovery and identification of parasitic larvae.[114]

Life Cycle

Adult worms of *G. spinigerum* are found in tumor-like masses in the stomach wall of fish-eating mammals. The life cycle of *G. spinigerum* is shown in Figure 9–3. The prepatent period is approximately six months.

Distribution

The distribution of human gnathostomiosis is summarized in Table 9–2.

Epidemiology and Transmission

The main source of human infection of gnathostomosis in Thailand is the snake-headed fish, *Ophicephalus* spp., which is one of the fishes used in *som-fak*, a rice-fermented fish dish with widespread popularity. In Japan, freshwater *Ophicephalus* species are eaten raw as sashimi. The ingestion of raw or inadequately cooked fish is the major source of infection in other areas reporting the disease. Infections in Mexico are attributed to eating raw cycloid fish as ceviche.[99] Human infections are also reported from eating raw or poorly cooked catfish, eels, frogs, chickens, ducks, and snakes.[39] Dogs, cats, and several species of wild mammals are reservoirs of the parasite. These definitive hosts become infected primarily through eating infected fish or other animals that serve as paratenic hosts.

Prevention and Control

Health education programs in endemic areas of Asia are required to control this type of infection. Ensuring that people eat only well-cooked fish, eels, or other intermediate hosts such as snakes, frogs, and poultry can prevent infections. Potentially copepod-infested water should be boiled or treated.

Gongylonemosis

Gongylonema pulchrum is a spiruroid nematode of the Thelaziidae family (Table 9–1). It is found in all domesticated mammals, but is most

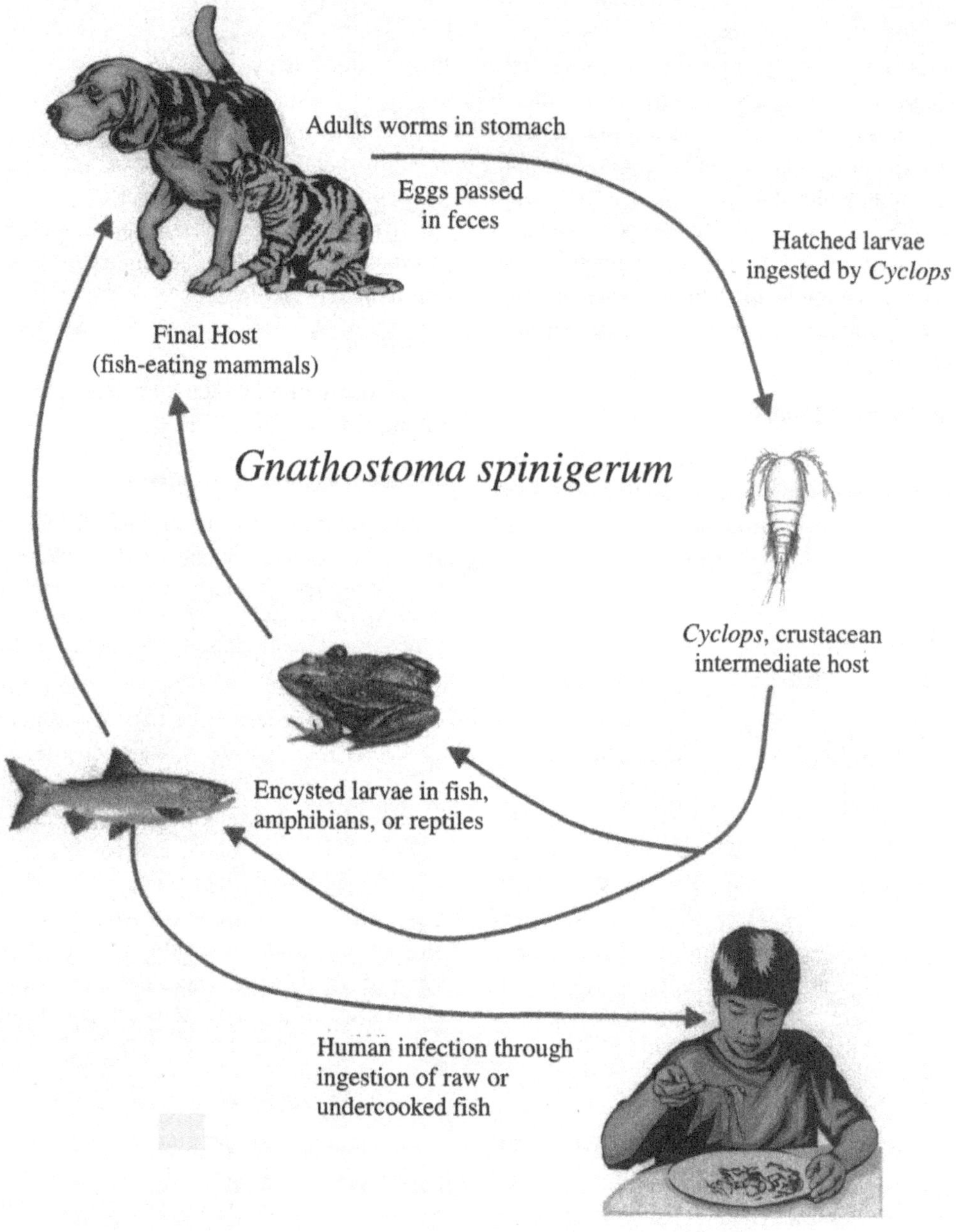

Figure 9–3 Life cycle of *Gnathostoma spinigerum*. *Source:* © Crown Copyright.

prevalent in ruminants (Table 9–2). Cases of human infection are rare.

Public Health Significance

In humans, gongylonemosis parasites have been found in the submucosa of lips, gums, hard palate, soft palate, and tonsils, with pharyngitis and stomatitis reported. Diagnosis is based on history, clinical signs of mouth irritation, and microscopic identification of the parasite. Occasionally, the parasite has been found emerging from the mouth. Infection in animals is usually asymptomatic, but may sometimes cause lesions of the mouth or pharynx.[29] In pigs, the parasite is found in the tongue mucosa and may cause occlusion of the esophagus.[179]

Life Cycle

Adult *Gongylonemosis* worms live in the mouth, esophagus, or rumen of the final hosts. The life cycle of *G. pulchrum* involves coprophagic beetles of the genera *Aphodius, Ontophagus,* and *Blaps* as intermediate hosts. Ruminants become infected by ingestion of the beetles with grass or other infested food.

Distribution

Although *G. pulchrum* is widely distributed geographically, human infection is rare. The distribution of *G. pulchrum* is summarized in Table 9–2. Infection in domestic ruminants varies considerably, with high levels of infection reported in the Ukraine (e.g., 32–94% of adult cattle, 39–95% of sheep, and 0–37% of pigs infected),[29] and Iran (e.g., 49.7% of the cattle).[4] In the United States, the parasite was found in 5.9% of pigs, varying from 0 to 21% according to geographic origin.[179]

Epidemiology and Transmission

Man is an accidental host and is probably infected by ingesting beetles on salads and raw vegetables.

Prevention and Control

Because the *Gongylonemosis* parasite is rare in humans and causes only mild symptoms, special control measures are usually not recommended. Individual protection can be obtained by observing the rules of personal, food, and environmental hygiene.

Trichinellosis (Trichinosis)

Trichinellosis (trichinosis) is a food-borne disease that is caused by infection with parasitic nematodes of the genus, *Trichinella*. Infection results from ingesting meat harboring infective larvae. Until recently, *T. spiralis* was accepted as the sole representative of the genus *Trichinella.* It has now become clear that not all populations of the parasite are the same, and five species have recently been proposed.[24] *T. spiralis* is the most important species, with widespread distribution in domestic pigs.

Public Health Significance

Trichinellosis is a cosmopolitan zoonotic, food-borne, parasitic disease resulting from the ingestion of meat harboring the infective larvae of *Trichinella*. Intestinal trichinellosis is usually mild and therefore not often diagnosed, but may cause diarrhea and abdominal discomfort, sometimes accompanied by nausea and vomiting. Heavy infections can be severe, causing extremely painful muscles, and even fatalities in man. The illness typically begins one to four weeks after ingestion of infective meat, and is characterized by muscle aches and fever. Muscle pain, which may be severe and incapacitating, is an outstanding feature, with the jaw muscles frequently involved. The infection of heart muscle may induce severe, even fatal, myocarditis. Death may also result from encephalitis or pneumonitis. In most parts of the world, the probability of contracting the disease is now low, and its decline, certainly throughout the western world, can be attributed to the introduction of a number of control measures discussed later. Diagnosis may be based on a clinical history of prior consumption of raw or undercooked meat, but this aspect may be missed, particularly in countries where the disease rarely occurs. Clinical signs of muscle aches, fever, and periorbital edema are indicative of the disease. Infection is usually diagnosed by demonstrating larvae in muscle biopsy specimens.

Life Cycle

T. spiralis is a small, nematode parasite that is found in the small intestine of man, pig, rat, and other mammals (Table 9–2). The life cycle of *T. spiralis* is shown in Figure 9–4. Encysted larvae remain infective for months, even years. In some hosts, especially man, they eventually die and become calcified. *Trichinella* is remarkable among parasitic nematodes in having neither a free-living stage between individual hosts nor an intermediate host. The spread of larvae throughout the host musculature permits transmission to

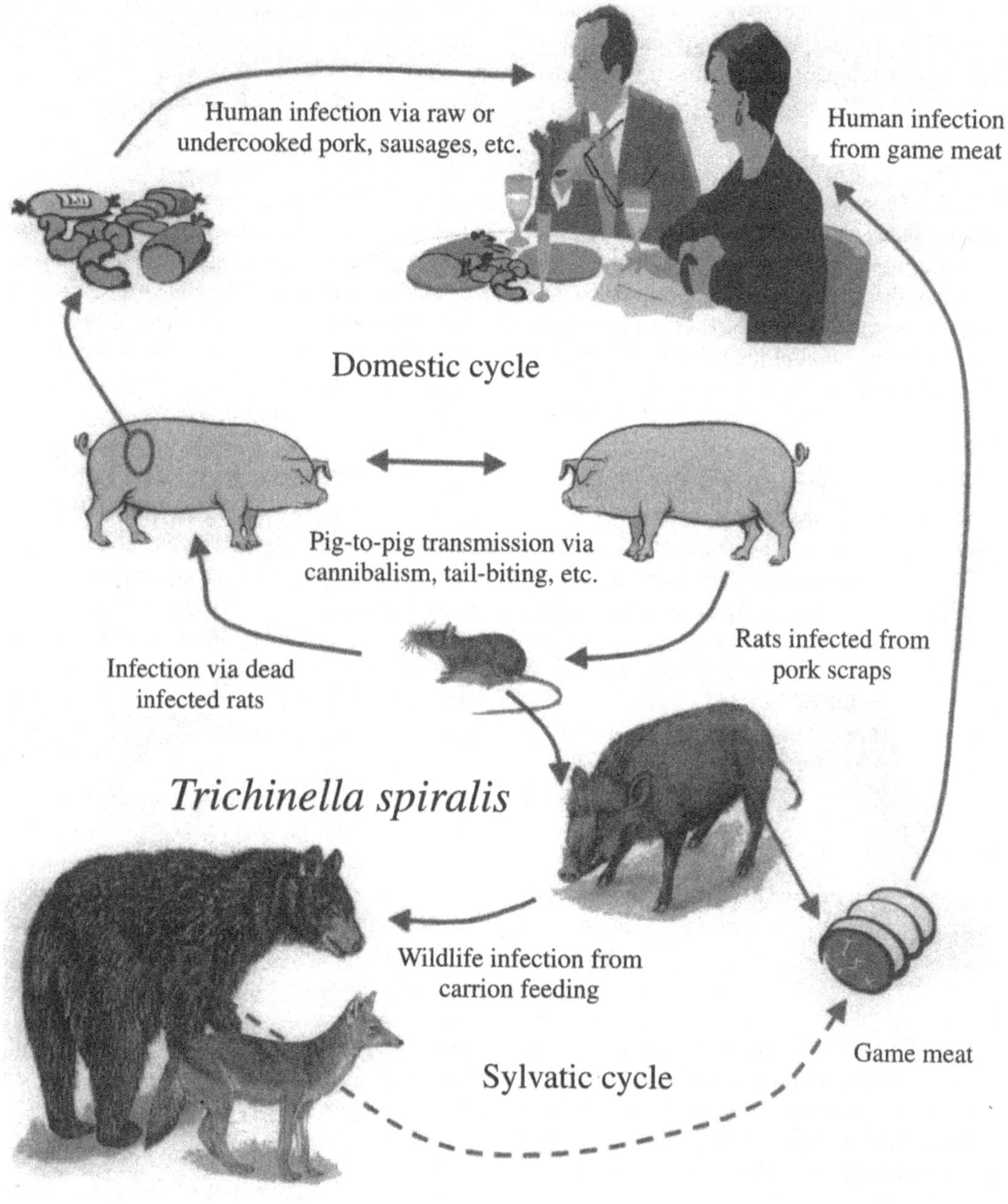

Figure 9–4 Life cycle of *Trichinella spiralis. Source:* © Crown Copyright.

another individual host, but only after the death of the first host.

Distribution

Trichinella has been recorded on every continent except Antarctica (Table 9–2). In many areas, its prevalence in man and pigs has been reduced to a low level, but persists at higher levels in wildlife. Prevalence rates are higher in large carnivorous or scavenging species such as bears, mink, fox, badger, wild boar, and rodents. Information on the global incidence of clinical trichinellosis in man is not readily available, but it is evident that there has been a striking reduction in western Europe and the United States. Several outbreaks in France and Italy have been associated with eating horse meat that was thought to have become infected by intentionally or inadvertently eating fodder that was contaminated with the bodies of dead rats or mice. Outbreaks of varying severity continue to occur in eastern Europe, the former Soviet Union, Asia, and occasionally elsewhere.

Epidemiology and Transmission

Two distinctive epidemiological cycles of *T. spiralis* are recognized, domestic and sylvatic. The domestic cycle involves pigs and rats, through either the feeding of uncooked pork scraps (e.g., in waste feed) to pigs, cannibalism among pigs (e.g., scavenging of dead carcasses and possibly tail biting of live pigs), or infection of rats by disposal of uncooked pork and the subsequent infection of pigs by the ingestion of infected rats. Human infection occurs when undercooked or raw pig meat is eaten.

In the sylvatic cycle, *Trichinella* is transmitted among species of wildlife. Man becomes infected by consuming game meat. The two cycles may be interrelated by the fact that domestic pigs may scavenge on dead wild animals and vice versa. Rats can also be of significance in this respect in that they may become infected either through eating pork scraps or by scavenging wildlife carcasses. Rats may in turn be hunted and eaten by wildlife predators, or when they die, be eaten by wild carrion feeders or pigs. Crossover of *Trichinella* from one transmission cycle to another may occur, with domestic to sylvatic cycle the more likely[109] as wildlife strains may have low infectivity for rats or pigs.

Prevention and Control

Human infection by trichinae is linked to the consumption of undercooked pork or game meat. A number of measures can therefore be taken to control infection. These involve a combination of consumer education and control of infection in pigs. Pig herds can be kept free of infection by ensuring that pork scraps are not fed to pigs. In some countries, laws requiring waste food (swill) to be cooked before being fed to pigs have been introduced to help control a number of important viral pig diseases. Good husbandry and rat-proofing piggeries also help control infection. In many countries, meat inspection procedures have been introduced to identify infection.

Ethnic or cultural practices in which meat is eaten undercooked or raw are a particular risk area and may call for special educational measures. Traditional types of sausage (including fermentated sausage) that receive little cooking have long been associated with human disease. Education draws attention to the need to freeze or cook such foodstuffs. Freezing may be more acceptable because it has less effect on flavor and taste than cooking. To be effective, meat must be frozen throughout and stored for sufficient time to ensure death of encysted larvae. The thermal death point of *T. spiralis* is approximately 57 °C, and cooking pork so that it reaches a temperature of 77 °C gives a margin of safety without destroying the taste of the meat. Slightly lower temperatures may be adequate provided the temperature is achieved uniformly throughout the meat. It can be assumed that pork will be safe if it is cooked until there is no red or pink coloration throughout. In the United States, fermented pork sausages must either originate from *Trichinella*-certified meat or be heated to 58.3 °C at the end of the fermentation process.[89] Although certain factors such as pH may affect survivability of trichinae in fermented sausage, prescribed heat treatment is considered necessary.[31]

CESTODES (TAPEWORMS)

There are a number of food-borne tapeworms that infect humans, of which a few are acquired by eating meat or fish. A general description and classification of food-borne tapeworm species is provided in Table 9–1.

Diphyllobothriosis

Although there are various species of fish tapeworm reported in humans, the most important is *Diphyllobothrium latum*.

Public Health Significance

Diphyllobothriosis is usually asymptomatic, but long-term infection with the tapeworm may produce vitamin B_{12} deficiency leading to anemia. The presence of several worms may cause intestinal obstruction. CNS signs of peripheral and spinal nerve degeneration have been reported. Diagnosis is based on identifying the presence of the characteristic eggs in human feces.

Life Cycle

The fish tapeworms are among the largest worms to infect humans. Adult tapeworms are found in the small intestine, ranging from 2 m to 15 m in length and living as long as 10 years. Many species of fish such as pike, perch, turbot, salmon, and trout serve as second intermediate hosts. The life cycle of *D. latum* is shown in Figure 9–5.

Distribution

The distribution of *D. latum* is widespread in the temperate and subarctic regions of the Northern Hemisphere where freshwater fish are eaten. Distribution is summarized in Table 9–2.

Epidemiology and Transmission

Diphyllobothriosis is transmitted by the ingestion of infected, raw, or improperly cooked freshwater fish. Human cultures with preferences for smoked, pickled, or raw fish (such as sashimi or sushi) are at particular risk.

Endemnicity is maintained where there is poor sanitation, or, due to the practice of lakeside hotels, pumping of raw sewage into freshwater lakes. The parasite may have been spread by the emigration of infected people from endemic areas into lake regions in which susceptible copepods and fish were present. Scandinavian and Russian immigrants are thought to have introduced the parasite into the Great Lakes region of North America and Alaska.

Prevention and Control

Human infection is easily prevented by ensuring that freshwater fish is well cooked or by changing dietary habits through education. If fish is to be eaten raw, smoked, pickled, or fermented, then freezing at –10 °C for 1–3 days should kill pleurocercoid larvae in the fish tissue. Sewage from lakeside hotels and from leisure boats should be treated before release into the lakes.

Taeniosis (Cysticercosis)

Taeniosis is intestinal infection with a preadult or adult stage of a tapeworm of the genus *Taenia*. Only two species, *T. saginata* and *T. solium,* infect human intestines as adult tapeworms, and man is the only known final host. Intermediate stages of these human tapeworms infect the tissues of meat-producing animals. Tissue infection with a metacestode stage of *Taenia* is referred to as cysticercosis. The intermediate hosts of *T. saginata* and *T. solium* are domesticated cattle and pigs, respectively, but several other ruminants, including sheep, goats, and llamas, have been recorded as carriers of *C. bovis,* although the validity of these hosts has been questioned. The reindeer, *Rangifer tarandus*, has been shown to act as an intermediate host in Russia. Cysticerci thought to be *C. cellulosae* have been reported in monkeys, wild pigs, bush babies, camels, rabbits, hares, bears, dogs, foxes, cats, rats, and mice. Four species (i.e., *T. solium, T. saginata, T. multiceps*, and *T. hydatigena*) parasitize human organs as cysticerci; the last three species invading man only exceptionally.[163] Other species, *T. serialis, T. longihamatus, T. crassiceps,* and *T. taeniaeformis,* have occasionally been diagnosed as a cause of cysticercosis in man.[12,14,92]

Public Health Significance

Taeniosis itself causes little if any disability and morbidity. Conversely, human cysticercosis has a definite public health significance, causing disability and death in infected people. Cysticerci may be found in every organ of the body of man, but are most common in the subcutaneous tissue, eye, and brain. In the brain (neurocysticercosis), the tissue reaction that occurs may cause a variety of CNS disorders, some of which may be fatal. In endemic areas, neurocysticercosis may be responsible for 30–60% of cases of epilepsy, and mortality may

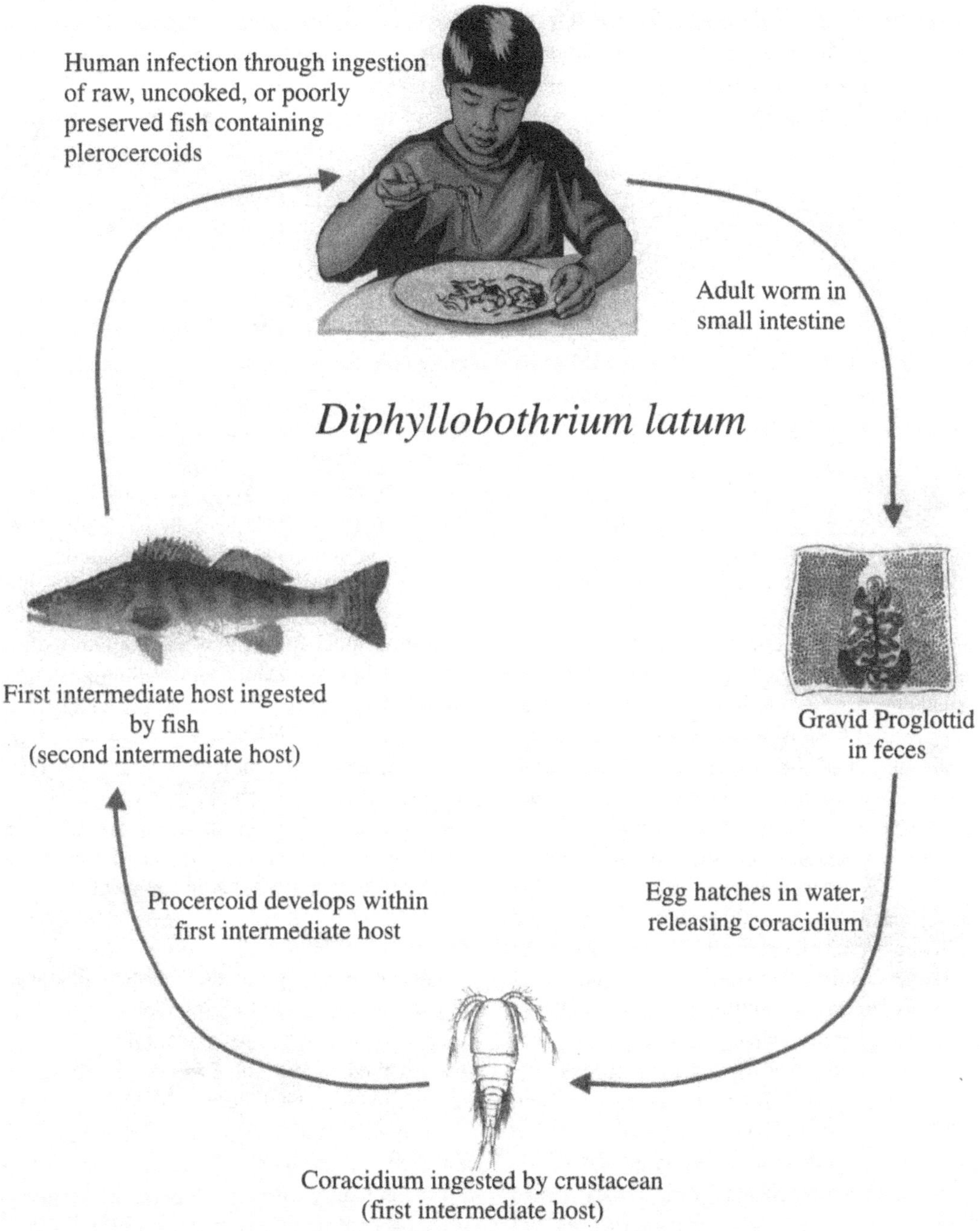

Figure 9–5 Life cycle of *Diphyllobothrium latum. Source:* © Crown Copyright.

range from 1% to 2% of all causes of death. For a long time, treatment of taeniosis was neither safe nor satisfactorily effective; however, the introduction of effective taenicides has seen a dramatic improvement in therapy. Diagnosis of human taeniosis is not always straightforward and is based on clinical symptoms and the identification of tapeworm proglottids or eggs during stool or anal swab examination.[72] Repeated fecal examinations increase the chances of finding eggs, and diagnosis can be improved by taking perianal swabs using sticky cellophane tape. A fecal antigen test has also been reported.[3,41]

Life Cycle

Adult tapeworms of *T. saginata* measure 4–12 m in length and take approximately three months to become fully grown. *T. solium* is shorter, 1.5 m to 8 m, and takes 62–72 days to mature. Both tapeworms are found in the small intestine of man and may live for up to 25 years. The life cycles are similar, involving cattle and pigs respectively (Figure 9–6).

Distribution

T. saginata (*C. bovis*) occurs in many cattle-breeding regions and especially where beef is eaten raw or undercooked. Distribution of *T. solium* (*C. cellulosae*) is usually confined to poor countries because it is mainly related to the low sanitation in pigs' breeding areas. Human migration and increased consumption of pork increases the spread of taeniosis/cysticercosis from the endemic rural areas into urban areas. Bovine cysticercosis has a high prevalence in western and eastern Central Africa and some Asian countries.[60] Moderate infections are seen in other countries[125,173] and are summarized in Table 9–2. The prevalence of bovine cysticercosis in Europe and the United States has decreased throughout the twentieth century, through the introduction of meat inspection procedures in these countries. Fluctuations in prevalence in Europe have occurred as a result of the migration of people following World War II and increased tourism,[124] and in the United States, occasional outbreaks have been attributed to Mexican immigrant workers.[141]

The distribution of pig cysticercosis is also summarized in Table 9–2. In countries that are supposed to be free of *T. solium* infections, human cysticercosis may occur sporadically, mainly due to contact with infected migrants or tourists.[76]

Epidemiology and Transmission

Man becomes infected with taeniosis by eating raw or undercooked meat containing viable cystercerci. The life span of cysticerci varies, probably due to parasite or host strain variations, different infection doses, and host immunological responses. *C. bovis* in cattle, for example, remain viable for nine months to three years. Numbers of cysticerci depend not just on the degree of exposure to infective and viable tapeworm eggs, but also on the level of host immunity that can effectively reduce the number of developing cysticerci.

Adult tapeworms have a high biotic potential. Within infected human populations, the total production of eggs can be enormous. The eggs are sensitive to temperatures higher than 38 °C and to desiccation, but are capable of surviving European winters for 35 days[58] and of surviving in sea or brackish waters for some time. The eggs are also relatively resistant to chemicals. *T. saginata* eggs remain infective for calves after 16 days in sewage, up to 71 days in sludge, and for several months in pastures.[110]

Animal husbandry practices, human eating habits, sanitary education, and willingness to cooperate in control programs are all factors that may further influence the transmission of taeniosis. Poverty, ignorance, and some local customs of eating raw pork or raw meat or viscera may also play an important transmission role. The habit of eating raw meat or semi-raw sausages is strongly rooted in some cultures.

Prevention and Control

Control measures include improved diagnosis and treatment of human taeniosis as well as proper pig and cattle husbandry and the instigation of meat inspection procedures. However, routine meat inspection may be inadequate for light infections and may miss more than half of the infected animals.[165] In cattle, the heart is most frequently infected; in pigs, the tongue is the most commonly infected organ.[125] Poor local sanitation is largely responsible for the ease of transmission in rural endemic areas.

Food transmission can be controlled by avoiding the consumption of raw pork or beef or semi-raw meat products. Freezing infected pork or beef can kill the cysticerci when the internal temperature of meat or a carcass is less than –5

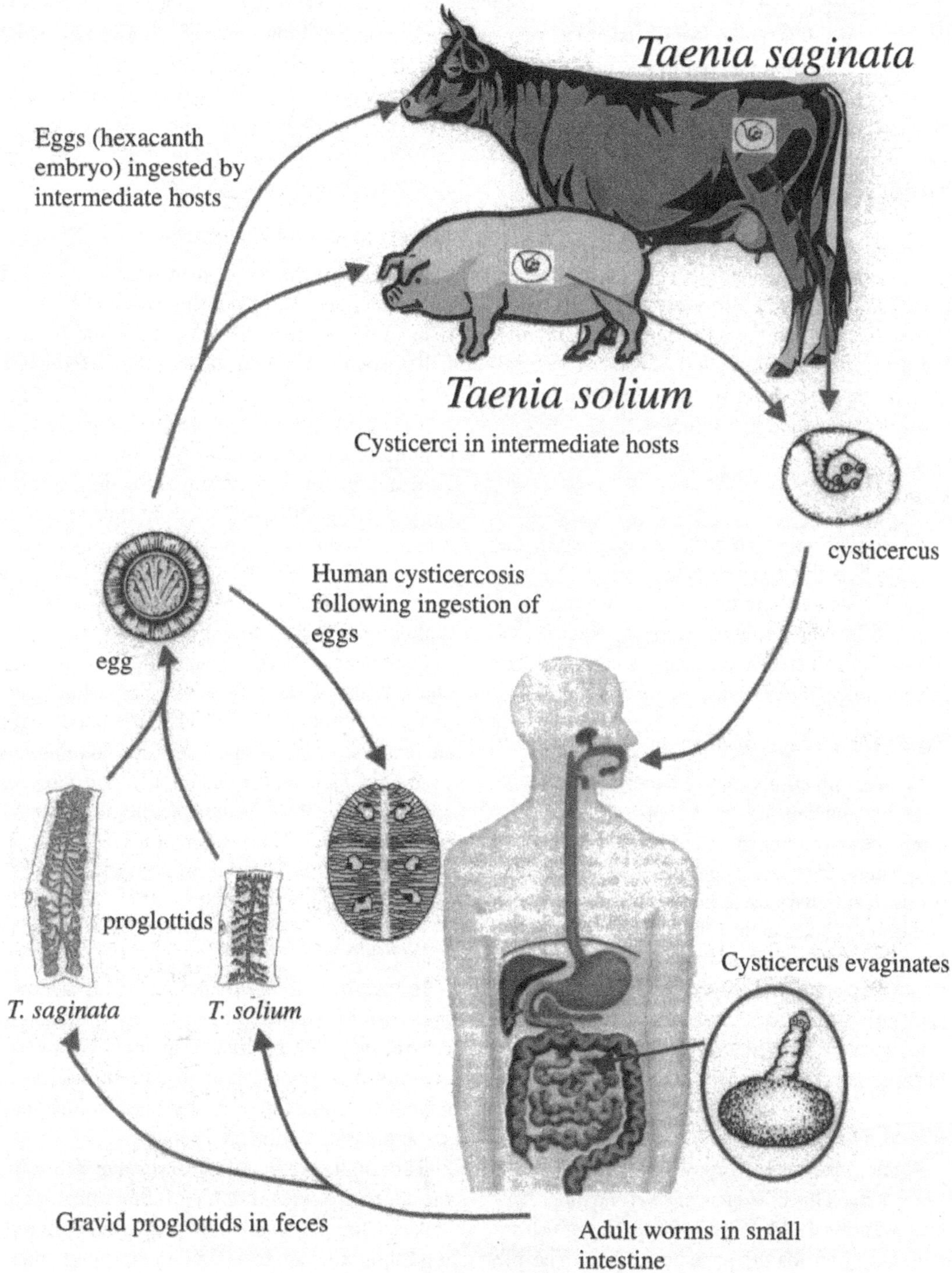

Figure 9–6 Life cycle of *Taenia solium* and *Taenia saginata*. *Source:* © Crown Copyright.

°C for at least four days or around –20 °C for at least 12 hours. It is not known if cysterci can survive meat fermentation conditions, but it would seem sensible to freeze or heat treat flesh ingredients wherever possible.

TREMATODES (FLUKES)

Trematodes or flukes are unsegmented, dorso-ventrally flattened, leaf- or lancet-shaped parasitic worms. The life cycles of the parasitic flukes discussed in the text are indirect, involving a variety of intermediate hosts. A general description of food-borne flukes and their classification is provided in Table 9–1.

Clonorchiosis

Clonorchis sinensis (*Opisthorchis sinensis*), the Chinese liver fluke, is a trematode infection that is found throughout southeast Asia. The parasite is acquired by eating raw or poorly cooked or preserved freshwater fish, particularly carp, and can infect a range of mammals, including man.

Public Health Significance

Human infections may be present for many years without causing overt signs of clinical disease. The development of clinical signs depends on the number of worms present, which can number several thousands in heavy infections. Large numbers of flukes in the bile ducts can cause cholangitis (inflammation of the bile ducts) and gallstones in man,[115] and may possibly predispose to cholangiocarcinoma.[143] The parasite may on occasions enter the pancreatic ducts, causing dilatation, fibrosis, and pancreatic stones.

Life Cycle

Adult flukes, which may live as long as 30 years, inhabit the bile ducts of man and other fish-eating mammals. More than 100 species of fish are reported hosts for the parasite in Asia. The life cycle of adult fluke is shown in Figure 9–7.

Distribution

The parasite distribution of flukes is shown in Table 9–3. Infection rates have decreased in Japan in recent years, but infections remain widespread in China, Taiwan, and Korea. The prevalence and distribution of infection is associated with the presence of susceptible snail and fish intermediate hosts and habits of the indigenous populations. It has been suggested that close to 25 million people are infected in Asia, mostly in southern China and North Korea.

Epidemiology and Transmission

Infection by flukes in man occurs by eating raw, undercooked, or poorly preserved infected fish. In some areas where the parasite occurs, fish is eaten raw in thin slices with rice (*congee*). The eating of raw freshwater fish has become increasingly popular in several Far Eastern countries, and consequently the prevalence of human infection is increasing. Infections outside of Asia are usually imported. In most endemic areas, cats and dogs are infected in high numbers. Levels of infection are maintained by the practice of fertilizing ponds used to raise fish with human excrement.

Diagnosis is based on the identification of the characteristic eggs in fecal samples, which have to be differentiated from other trematode eggs such as *Heterophyes heterophyes, Metagonimus yokogawi, Haplorchis taichui,* and *Opisthorchis* species (Table 9–4). Several serologic tests have been developed, but most are nonspecific. A reported enzyme-linked immunosorbent assay (ELISA) may be of value.[30]

Prevention and Control

In endemic areas, treatment of all infected persons and improved sanitation would help control infection by flukes. In areas where fish are raised in ponds, human and animal feces should be composted or sterilized before they are applied as fertilizer to ponds.

Thorough cooking of all freshwater fish is the most effective means of controlling infection by flukes. Clonorchiosis can be prevented by avoiding eating raw, undercooked or improperly pickled, salted, dried, smoked, or fermented fish in endemic areas. There is little or no information on conditions affecting the survivability of the encysted infective stages in fish flesh. Education projects are of some benefit, but the eating of raw or fermented freshwater fish has been a

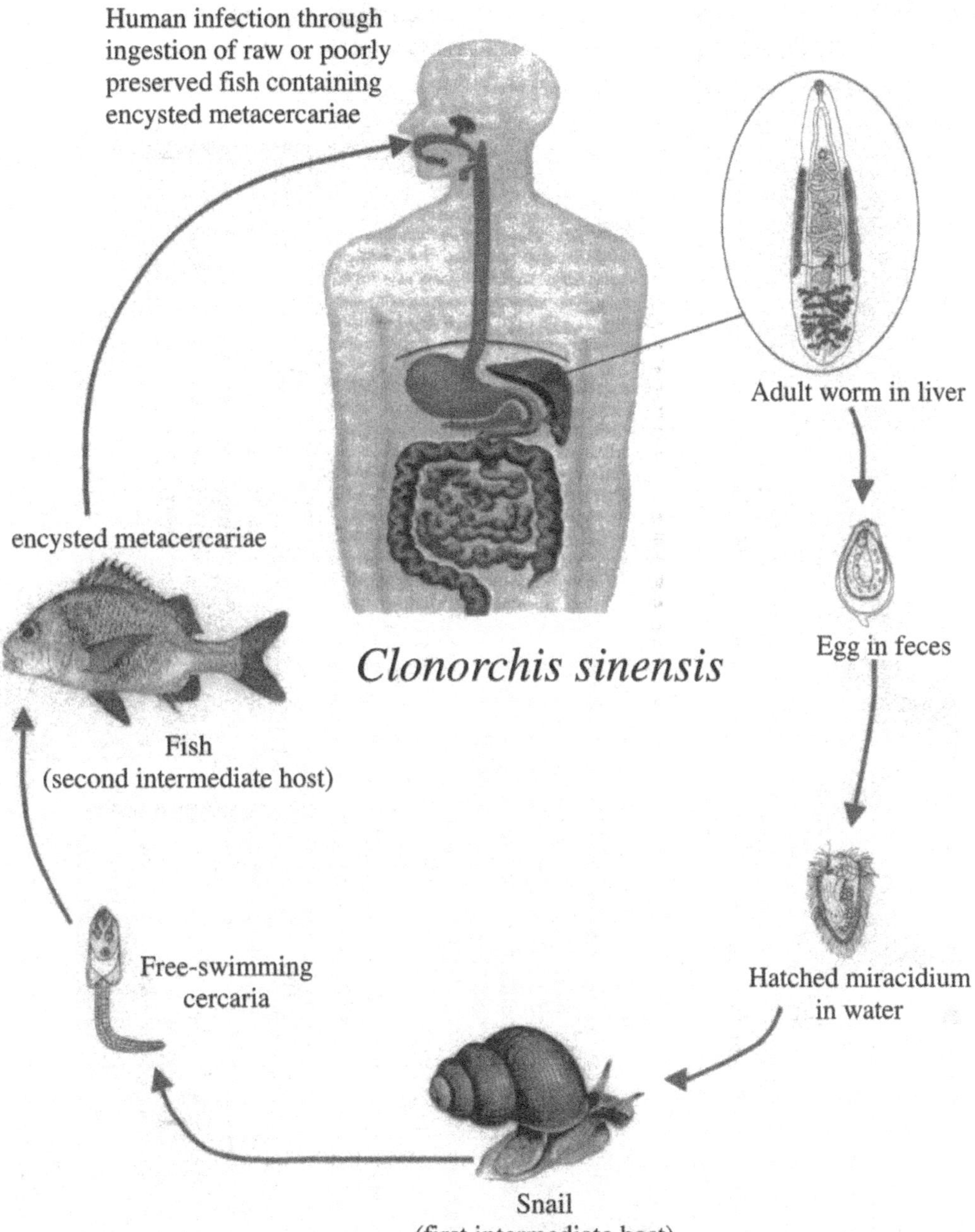

Figure 9–7 Life cycle of *Clonorchis sinensis. Source:* © Crown Copyright.

practice among Asian communities for centuries, and many cultures would be reluctant to change their habits. Freezing fish for a week at −10 °C would be beneficial, but even frozen fish has been incriminated with *Clonorchis* outbreaks in nonendemic areas.

Echinostomosis

Trematodes of the family Echinostomatidae are intestinal parasites of birds and mammals with approximately 16 species reported from humans. Infections occur mainly in southeast

Table 9–3 Food–Borne Helminths (Trematodes)

Parasite Class	*Parasite*	*Distribution*	*Main Hosts (Final)*	*Source of Infection to Man*
Trematoda (flukes)	*Clonorchis (Opisthorcis) sinensis*	China, Korea, Japan, Taiwan, Vietnam	Man, dogs, cats, pigs, rats, mink	Fish
	Echinostoma spp.*	Far East (Phillipines, Thailand, Indonesia, Taiwan, Korea), Africa	Wide variety of birds and mammals, man	Molluscs (snails, clams), fish
	Fasciola hepatica	Worldwide	Cattle, sheep, goats, pigs, horses, camelids, rabbits, wild herbivores, man	Salads (lettuce, watercress), alfalfa juice
	Fasciola gigantica	Africa, parts of Asia, Hawaii	Cattle, sheep, goats, zebras, man	
	Fasciolopsis buski	Asia (China, Taiwan, Thailand, Vietnam, Laos), India, Indonesia	Man, pigs	Aquatic plants (water caltrop, water chestnut)
	Heterophyes heterophyes	Middle East (Egypt, Turkey), Balkans, southern Europe (Spain), and parts of Asia (China, Japan, Korea)	Dogs, cats, fish-eating birds and mammals, man	Fish
	Nanophyetes salmincola	North Amercia, Siberia	Dogs, fish-eating mammals (foxes, racoons, mink)	Fish
	Opisthorcis viverrini	Southeast Asia (Thailand, Laos) Russia, Siberia, Kazakstan, parts of Europe	Dogs, cats, fish-eating mammals, man	Fish
	Opisthorcis felineus	(Poland, Germany) Southeast Asia (Korea, Japan, Phillipines)	Cats and fish-eating mammals	Fish
	Paragonimus westermani	Asia (China, Korea, Japan, Taiwan, Phillipines, Vietnam, India)	Man, dogs, cats, fish-eating mammals (otters, mongooses, tigers, leopards, wolves, mustelids), rats, pigs, monkeys	Crustaceans, crabs, crayfish
	Paragonimus africanus	Central Africa		
	Paragonimus mexicanus	Central and South America (Peru, Ecuador, Colombia, Honduras, El Salvador, Mexico)		

*Approximately 16 species reported in man.
Source: © Crown Copyright.

Asia and involve species of the genera *Echinostoma* and *Hypoderaeum*.

Public Health Significance

Infections with echinostomids are usually of short duration and cause few problems unless large numbers of fluke are present. Inflammatory lesions with shallow ulcers in the mucosa may develop at the site of the fluke attachment. Heavy infections may cause clinical symptoms of abdominal pain, flatulence, diarrhea, anorexia, and edema. Diagnosis is based on finding the eggs in the feces that must be differentiated from those produced by *Fasciola hepatica* or *Fasciolopsis buski* (see Table 9–4).

Life Cycle

These small flukes are found in the small intestine of birds and mammals. The life cycle involves both snails and a range of second intermediate hosts, including other species of molluscs, fish, or tadpoles. Man acquires infections after eating the second intermediate host raw or partially cooked.

Distribution

The distribution of human echinostomid infections is provided in Table 9–3. The highest prevalence rates occur in the Philippines, Thailand, Indonesia, Taiwan, Korea, and Africa.[25]

Epidemiology and Transmission

Infections are acquired by eating raw snails, clams, raw or undercooked frogs and tadpoles, or raw or insufficiently cooked fish. Many people in endemic areas are subclinically infected. Tourists may become infected when visiting endemic areas and sampling local cuisine. An outbreak of gastroenteritis caused by echinostome infection was reported in a group of Americans returning from a tour in Africa.[129]

Prevention and Control

Ensuring that snails, clams, tadpoles, frogs, and fish are well cooked before eating can prevent echinostomid infections. As with other food-borne trematode infections, there is a paucity of information concerning conditions affecting survivability of infective stages in food.

Fasciolosis

F. hepatica and *F. gigantica* (liver flukes) are trematodes that live in the bile ducts of domestic and wild herbivores and occasionally infect man. In animals, fascioliosis occurs predominantly in herbivores, most commonly affecting sheep, followed by cattle.

Public Health Significance

Human infection with *F. hepatica* and *F. gigantica* is often mild or subclinical, with the severity of the infection dependent on parasite numbers and the duration of infection. Serious infections, with a large number of parasites, may produce biliary stasis, liver cirrhosis, and cholecystitis and gallstones, occasionally leading to jaundice.

Diagnosis in humans is based mainly on history and clinical signs and must be differentiated from other hepatic infections. Fecal examination is not appropriate because the flukes rarely develop to maturity.

Life Cycle

Adult flukes are found in the bile ducts of the liver. The life cycles of *F. hepatica* and *F. gigantica* are similar but involve different snail intermediate hosts of the genus *Lymnaea*. The prepatent period is approximately 12 weeks, and adult fluke can live in the bile ducts for several years.

Distribution

F. hepatica is distributed throughout the world; *F. gigantica* occurs in Africa, several Asian countries, and Hawaii. Human *F. hepatica* infections occur sporadically, or in outbreaks, and have been recorded in numerous countries in the Americas, Europe, Africa, and Asia. The most extensive epidemics have occurred in France, with infection due to eating watercress that was contaminated by metacercariae.[98] The distributions of both parasites are summarized in Table 9–3. Fascioliosis is a common disease of

Table 9–4 Worm Eggs in Human Feces

Description	*Size (µm)*	*Parasite*	*Differentiate from*
Small to medium-sized eggs (< 75 µm)			
With weakly protuding polar plugs	35–45 × 21	*Capillaria*	*Trichuris* (protuding polar plugs)
Embryonated			
–spherical with hexacanth embryo	35–40 × 30–35	*Taenia solium* *Taenia saginata*	*Hymenolepis*
–with conspicuous operculum	25–35 × 10–20	*Opisthorcis* *Clonorchis*	*Clonorchis/Heterophytes/Metagonimus* *Opisthorcis/Clonorchis* *Heterophytes/Metagonimus* *Clonorchis/Opisthorcis*
–with inconspicuous operculum	25–30 × 15–17	*Heterophytes* *Metagonimus*	
Large eggs (> 75 µm)			
Unembryonated			
–with operculum	80–120 × 48–60	*Paragonimus*	
	80–120 × 60–90	*Echinostoma*	
	130–145 × 70–90	*Fasciola hepatica* *Fasciolopsis buski*	*Fasciolopsis* *Fasciola*

Source: © Crown Copyright.

sheep, goats, and cattle in many parts of the world. Economic losses due to *Fasciola* infections are the result of liver condemnation, stunted growth, and reduced milk, meat, and wool production.

Epidemiology and Transmission

Man is infected with fascioliosis mainly through consuming salads of watercress *(Nasturtium officinale)* containing metacercariae. In France, where watercress salad is popular, human infection is more frequent than in other European countries. Contaminated lettuce or other plants that are eaten raw can sometimes serve as a source of infection, as can water from irrigation canals or other receptacles. Alfalfa juice has also been implicated in places where it is a customary drink.

Prevention and Control

Human infection can be reduced by avoiding eating watercress of wild or unknown origin and by ensuring that watercress is cultivated under controlled conditions that exclude animals or the possibility of snail infestation. Treating livestock limits levels of contamination, and control of the intermediate snail host can be attempted through the use of molluscides or effective land drainage to reduce available snail habitats.

Fasciolopsiosis

Fasciolopsiosis is a disease that is caused by the largest intestinal fluke, *F. buski*. This parasite infects man through the ingestion of raw, contaminated water plants.

Public Health Significance

Most infections are asymptomatic, but heavy infections may produce diarrhea, abdominal pain, and occasionally intestinal obstruction leading to malnutrition and death.[12,106] Diagnosis is confirmed by identifying the eggs in feces that have to be differentiated from eggs of *Fasciola* spp. (Table 9–4).

Life Cycle

Adult *F. buski* flukes are found in the duodenum. The life cycle is similar to that of *Fasciola,* but intermediate hosts are freshwater snails of the genera *Planorbis* and *Segmentina*. Man (and pigs) becomes infected by eating contaminated aquatic plants.

Distribution

It was estimated that approximately 10 million people were infested with fasciolopsiosis worldwide in the late 1940s.[12] Current estimates on prevalence are not available, but evidence suggests that the endemic range is expanding,[68,71] although it is still confined to southeast Asia. The distribution of *F. buski* is summarized in Table 9–3.

Epidemiology and Transmission

Infection with fasciolopsiosis is transmitted by the ingestion of raw contaminated aquatic plants such as water caltrop, water chestnut, water hyacinth, and water bamboo. Pigs are the reservoir host, but in Muslim countries, humans may be the main reservoir of infection.

Prevention and Control

The disease is easily preventable by avoiding raw or uncooked aquatic plants in endemic areas. The introduction of good sanitation facilities limits contamination of local water courses and ponds.

Heterophyosis

More than 10 species of intestinal flukes of the family Heterophyidae have been reported in man and fish-eating mammals, the most important of which are *Heterophyes heterophyes* and *Metagonimus yokogawai.* These flukes are acquired by eating raw or poorly cooked or preserved freshwater fish.

Public Health Significance

Light infections of intestinal flukes are usually asymptomatic, but heavier infections cause chronic, intermittent, often bloody, diarrhea; upper abdominal pain; anorexia; nausea; vomiting; and weight loss. Eggs or worms entering the CNS may cause clinical symptoms that are similar to those for cerebral hemorrhage or epilepsy.

In the heart, there may be signs of myocarditis or chronic congestive heart failure leading to death.

Diagnosis is based on finding and identifying characteristic eggs in the feces. The eggs of these flukes are very similar, and specific identification is difficult. They also have to be differentiated from those of *C. sinensis* and *Opisthorcis* (Table 9–4). Serological tests are not available.

Life Cycle

These small flukes live in the small intestines of the final hosts. The life cycles are similar to other flukes involving both snail and fish intermediate hosts (e.g., mullet, tilapia, trout, and salmon).

Distribution

The distributions of *H. heterophyes* and *M. yokogawai* are summarized in Table 9–3.

Epidemiology and Transmission

Intestinal fluke infections are acquired by eating raw, undercooked, or poorly preserved freshwater or marine fish and shrimps in endemic areas. Endemnicity is maintained by contamination of water by human or animal feces.

Prevention and Control

The prevention of human infection relies on avoiding eating raw or improperly cooked, salted, or pickled infected freshwater fish in endemic areas. Again, there is little or no information concerning conditions affecting survivability of infective stages in fish or crustaceans. Education and good sanitation would be helpful for control. Raw fish should not be fed to dogs or cats.

Nanophyetiosis

The intestinal trematode, *Nanophyetes salmincola* (*Troglotrema salmincola*), is found in dogs and fish-eating mammals of eastern Siberia and North America. It transmits a rickettsia, *Neorickettsia helminthoeca*, the cause of "salmon poisoning" in these animals. Infection with this parasite has been reported in humans in North America.[51]

Public Health Significance

Mild fluke infections in man are asymptomatic. Heavy infections (> 500 fluke) cause frequent bowel movements or diarrhea, abdominal discomfort, nausea, vomiting, weight loss, and fatigue. Human cases of salmon poisoning due to the rickettsia, *N. helminthoeca,* have not been reported. Diagnosis is based on the identification of eggs in the feces or recovery of the adult worms following treatment.

Life Cycle

The small adult flukes live in the intestinal wall of the host. The life cycle involves both snail and fish intermediate hosts. Infection in man is acquired by eating uncooked infected fish (salmon and trout) containing encysted metacercariae.

Distribution

The parasite has been reported from the Pacific coast of North America and eastern Siberia. Its prevalence is unknown.

Epidemiology and Transmission

Animals and man acquire the infection by eating raw or uncooked fish, although there is one report of the infection being acquired by handling salmon.[73]

Prevention and Control

In endemic areas, the main preventative measure is education in an effort to change the habit of eating raw or undercooked fish. There is no information on survivability of metacercariae in fish-based fermented foods. Dogs and cats should not be fed raw fish and should be kept away from salmon rivers and streams.

Opisthorchiosis

Two species of *Opisthorchis* infect humans and cause a disease similar to clonorchiosis: *O. viverrini* is found in southeast Asia, and *O. felineus* (the cat liver fluke) is found in parts of Europe.

Public Health Significance

Infection with *Opisthorchis* flukes has a significant public health importance because their presence often causes serious disease and is linked with cholangiocarcinoma and hepatic carcinoma. Most infections are asymptomatic, depending on the level and duration of infection. Adult fluke in the bile ducts, gallbladder, and occasionally the pancreatic duct cause thickening of the ducts, predisposing to cholangiocarcinoma and hepatocellular carcinoma.[74] Diagnosis is made by identifying eggs in the feces, although eggs cannot be differentiated from those of *C. sinensis* and other fluke species present in the intestines. An antibody ELISA[59] and a molecular DNA method[146] have been reported for diagnosis.

Life Cycle

Opisthorchis species are similar morphologically and in life cycle to *Clonorchis sinensis.* Adult fluke can live as long as 20 years.

Distribution

The distributions of *O. viverrini* and *O. felineus* are shown in Table 9–3. It is estimated that more than 7 million people are infected with *O. viverrini* and 4 million with *O. felineus.* Prevalence rates in highly endemic areas are greater than 90% for *O. viverrini* and 85% for *O. felineus.* In some parts of Europe, more than 85% of cats are infected with *O. felineus*.

Epidemiology and Transmission

Man and other definitive hosts become infected with *Opisthorchis* flukes by eating raw or undercooked fish. In Europe, metacercariae are common in freshwater fish such as orfe, bream, tench, and barbel. In parts of Asia, fish are eaten raw, especially in the rainy season when the fish are most abundant. In Thailand, fish are raised in ponds and the ponds are fertilized with human and animal feces. In areas endemic for *O. felineus,* the freshwater fish is eaten raw, dried, salted, or fermented. Cats and dogs in southeast Asia are important reservoirs of *O. viverrini*.

Prevention and Control

The main preventative measure is through education of the population to avoid eating raw or poorly preserved fish. Food habits are traditional and are often difficult to break. Salting fish in 5–15% saline for three to ten days destroys metacercariae, as does freezing at −10 °C for five days. It is not known if fermentation conditions kill metacercariae. Changes in sanitation may limit contamination of fish ponds. Snail control and eradication of reservoir hosts are not feasible. Mass anthelmintic treatment with praziquantel targeted at heavily infected individuals as part of community-based control programs may be of benefit, although re-infection after treatment is common.[160]

Paragonimosis

Approximately 40 species of *Paragonimus* have been described from around the world, of which only 8 species are considered to be important. *P. westermani* is widespread in the Far East and is responsible for most cases of pulmonary paragonimosis. Other species that infect man include *P. heterotremas*, *P. pulmonalis*, *P. skryjabini*, *P. miyazakii* in Asia, *P. uterobilateralis and P. africanus* in Africa, and *P. mexicanus* in Latin America. *P. kellicotti* is a North American species reported from animals and only rarely in humans.

Public Health Significance

Pulmonary infections with *Paragonimus* are usually asymptomatic except for occasional hemoptysis. The pathology associated with infections of *Paragonimus* varies depending on the location and numbers of flukes present. Adult flukes in the lung parenchyma may cause general malaise, coughing, and dyspnoea due to inflammation and cyst formation around the fluke. Extrapulmonary infections may produce a cutaneous larval migrans and abscess formation in the skin and viscera. Brain and spinal cord involvement may lead to headaches, seizures, paraplegia, and occasional deaths. Examination of sputum or feces will usually reveal the presence of the typical brown, thick-shelled, operculate eggs (Table 9–4).

Life Cycle

Adult flukes are usually located in pairs in granulomatous capsules in the lungs. The life cycle, which involves snail and crustacean intermediate hosts, is shown in Figure 9–8. Some infections are reported to persist for 10 or even 20 years.

Distribution

The distributions of *P. westermani*, *P. africanus,* and *P. mexicanus* are summarized in Table 9–3. Data on prevalence rates are limited, but it has been estimated that 1–1.5 million people are infected with *P. westermani* in Korea.[106] A study in Cameroon showed a prevalence of 5.6% for *P. africanus*.[90]

Epidemiology and Transmission

The source of infection for man and other definitive hosts is infected freshwater crustaceans. In man, this usually occurs following the ingestion of infected raw or poorly cooked, pickled, or salted crabs or crayfish. In some areas of China, crabs are soaked in rice wine for a period of time before eating (i.e., *drunken crabs).* In other parts of China, crayfish curd, raw crab sauce, or jam are eaten. In Thailand, raw freshwater shrimp salad and crab sauce are eaten. Koreans eat uncooked crabs immersed in soy sauce. Crabs are eaten roasted or raw in the Philippines, and crab juice is used in preparing some meat dishes. Infections are also possible by eating animal meat containing immature flukes. Human infection has been reported in Japan following the ingestion of uncooked slices of wild boar meat.[106] Contaminated water and cooking utensils are also a source of infection.

Domestic and wild animals act as a reservoir of infection for *Paragonimus* spp. In endemic areas, human infection rates are sufficient to maintain the cycle of infection through contamination of water courses.

Prevention and Control

As with other food-borne infections, paragonimosis can be prevented through education to prevent the consumption of raw or undercooked crabs and crayfish. The cooking of crustaceans at 55 °C for five minutes will kill metacercariae. Fermentation processes may be insufficient to kill metacercariae, although direct evidence is lacking. Improved sanitation helps reduce the infection of intermediate hosts.

PROTOZOA

Protozoa are a diverse group of microorganisms with simple or complex structures contained within a single cell and equally simple or intricate life cycles that vary considerably depending on the species. General characteristics of the protozoa described in the text are provided in Table 9–5.

Cryptosporidiosis

Protozoa of the coccidial genus, *Cryptosporidium,* are small, intracellular parasites that occur throughout the animal kingdom and have been reported in many species of mammals, birds, reptiles, and fish.[57,156] There are at least nine recognized species of *Cryptosporidium,* of which the most important is *C. parvum,* which infects a wide range of mammals including man.[56] *C. parvum* has been reported in 79 species of mammals and is highly prevalent in ruminants, particularly in young calves, and appears to be age related.[57,153] It is common in young livestock, especially cattle and sheep, although pigs, goats, horses, and deer can be infected. Dogs, cats, and other pets are occasionally infected, but they do not appear to be an important source of infection to other hosts.

Public Health Significance

Infections with *C. parvum* are a significant cause of diarrhea, abdominal pain, nausea, vomiting, and fever in humans.[159] Diarrhea is more profuse in the immunocompromised, such as those with AIDS; in these patients, infections are

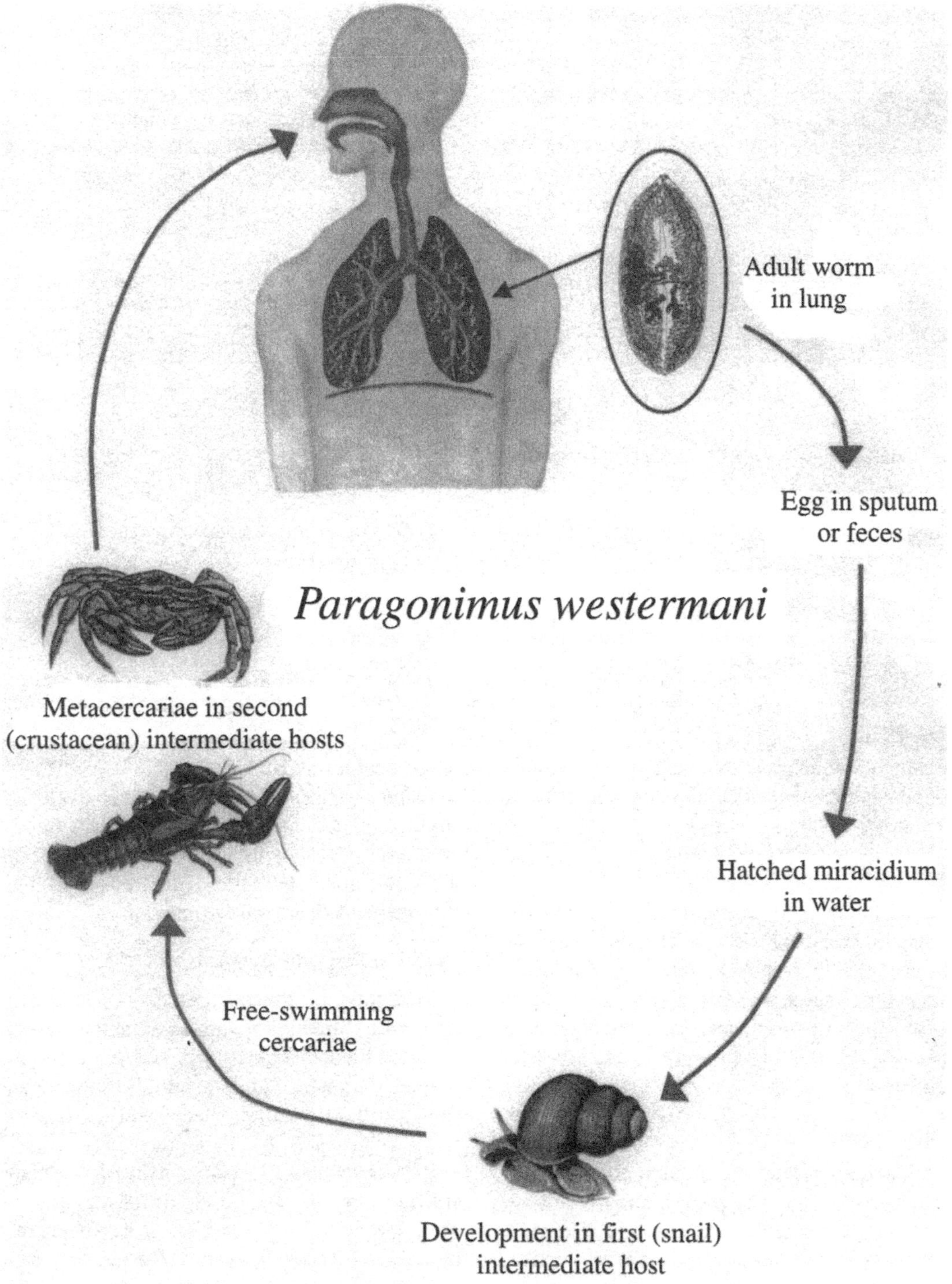

Figure 9–8 Life cycle of *Paragonimus westermani. Source:* © Crown Copyright.

Table 9–5 Food-Borne Protozoa—Classification

Phylum	Genera*	General Characteristics
Apicomplexa	*Cryptosporidium* *Cyclospora* *Sarcocystis* *Toxoplasma*	Apical complex (visible with electron microscope) generally consisting of polar ring, rhoptries, microneme, conoid; all species parasitic
Sarcomastigophora		
Subphylum **Mastigophora**	*Giardia*	One or more flagella usually present in trophozoites; asexual reproduction longitudinal
Subphylum **Sarcodina**	*Entamoeba*	Pseudopoda or locomotive protoplasmic flow without discrete pseudopodia; asexual reproduction by fission
Subphylum **Blastocysta**†	*Blastocystis*	Pleomorphic, brightly refractile cysts with a central vacuole; amoeboid form may occur
Ciliophora	*Balantidium*	Cilated organisms with micronucleus and macronucleus

*Genera of parasitic protozoa referred to in text only

†Jiang and He (1993) suggest that *Blastocystis* should be classified in its own subphylum within the phylum sarcomastigophora[81]

Source: © Crown Copyright.

of long duration, often followed by death. In the immunocompetent, infections are self limiting and last 7–10 days.

Life Cycle

The life cycle of infection with *C. parvum* is typically coccidian, with all development stages occurring within the same host. The infective stage is the oocyst, which is passed into the environment by infective hosts in the feces. For more detailed descriptions of the life cycle, see Taylor.[154]

Distribution

Human infection with *C. parvum* has worldwide distribution and has become a common cause of gastrointestinal infection (Table 9–6). In more countries of Europe and North America, the incidence of clinical infection is generally between 1–3%; elsewhere, rates of between 4.9% in Asia to 10.4% in Africa have been suggested.[159] The higher prevalence in less developed countries may relate to lack of clean water, sanitation, or general crowding. Seroprevalence studies suggest a higher rate of exposure to cryptosporidiosis, with levels between 25% and 35% in Europe and North America, suggesting that many infections are asymptomatic.[159]

Epidemiology and Transmission

Transmission from host to host is via the oocyst stage, usually through fecal contamination of water, food, or animal-to-person or person-to-person contact. Many mammals have been found to be naturally infected with *C. parvum* and have the potential, therefore, to act as reservoir hosts. Person-to-person transmission is considered to be the main route of transmission,[147] although zoonotic transmission has been reported from calves and lambs, particularly following educational visits to farms.[148] Recent research suggests that two distinct strains of *C. parvum* (type 1, human, and type 2, animal) exist, with differing epidemiological cycles.[7,102]

Table 9–6 Food–Borne Protozoa

Parasite Class	*Parasite*	*Distribution*	*Main Hosts (Final)*	*Source of Infection to Man*
Apicomplexa (Coccidia)	*Cryptosporidium parvum*	Worldwide	Man and mammals (cattle, sheep, pigs, goats, horses, dogs, cats)	Animal products (meat, milk, offal) unpasteurized cider, direct fecal contamination, water
	Cyclospora cayetanensis	Worldwide? Americas, Europe, India, parts of Africa and Southeast Asia	Man, primates (?insecti-vorous mammals, snakes)	Fruit (raspberries), water, salad
	Sarcocystis bovihominis	Worldwide but low prevalence (Europe—France, Germany, Poland, Czech Republic, Slovakia)	Man	Meat (beef)
	Sarcocystis suihominis	Eastern Europe	Man	Meat (pork)
	Toxoplasma gondii	Worldwide	Cat, wild felids	Meat (lamb, mutton, pork), water, direct fecal transmission (cat feces)
Mastigophora (Flagellates)	*Giardia intestinalis*	Worldwide	Man, cattle, sheep, goats, horses, dogs, cats	Contaminated salad, fruit, processed food, water, direct fecal contamination
Sarcodina (Amoebae)	*Entamoeba histolytica*	Worldwide	Pigs, cattle, dogs, cats, man	Contaminated food and water, direct fecal contamination
Blastocysta	*Blastocystis hominis*	Worldwide?	Man, pigs, monkeys	Contaminated food and water, direct fecal contamination
Ciliophora (Ciliates)	*Balantidium coli*	Worldwide	Pigs, dog, rat, man	Contaminated food and water, direct fecal contamination

Source: © Crown Copyright.

Food-borne transmission of cryptosporidiosis has been reported following the consumption of certain foods, notably raw sausage, offal, chicken salad, and milk.[15,26–28,65,67,83,112,157,175] An outbreak of cryptosporidiosis has been reported among people drinking freshly pressed, unpasteurized cider, possibly resulting from fecal contamination of fallen apples by animals.[107,121] Outbreaks of cryptosporidiosis associated with contaminated drinking water have been reported from a number of countries, including large outbreaks in the United States and the United Kingdom.[17]

Prevention and Control

Contaminated water is a major source of infection and appears relatively common. It can result in large-scale outbreaks of disease. Improvement in water catchment may reduce the potential health risks associated with the presence of oocysts in water supplies, but would probably not eliminate the problem. The safety of bottled water varies from supplier to supplier. Heating water to temperatures greater than 72.4 °C for one minute, or greater than 64.2 °C for two minutes of a five-minute heating cycle inactivates *C. parvum* oocysts.[54] Where ooysts of *Cryptosporidium* are detected in domestic water supplies, boil notices may be issued with the recommendation to boil water before drinking.[17] Concentration of oocysts from water is possible by filtering large volumes of water using cartridge or membrane filtration, or smaller volumes using a calcium carbonate flocculation technique.[64]

The potential risks from food-borne cryptosporidiosis come from the ingestion of fresh, raw, or uncooked foods. Effective means of control include the avoidance of contamination of vegetables and fruits in the field, strict hygiene measures involving handling of food, and the susceptibility of the organism to freezing and cooking. Evidence suggests that untreated milk, undercooked sausage meat, and offal have been implicated in outbreaks, but heat treatments such as those used for the pasteurization of foods are likely to destroy oocysts.[9] Oocyst infectivity is also destroyed by freezing or by freeze drying. Dried or frozen foods are therefore unlikely to be sources of food-borne cryptosporidiosis. The main risks would appear to be from fresh or refrigerated food. The effects of fermentation processes and other nonheat processes on oocyst survivability are not known.[42] Concentration techniques have been described for the detection of oocysts in food.[91]

Cyclosporiosis

Cyclospora are coccidian parasites reported from a number of animals that have assumed importance more recently in man, with worldwide reports of the presence of "cyanobacterium-like bodies" in the feces of immunocompetent and immunocompromised patients with diarrhea.[117] These organisms are now recognized as *Cyclospora cayetanensis* and are a cause of protracted diarrhea in humans.[174] It is thought that initial cases of infection in man were either reported as unidentified coccidian parasites[6] probably misidentified as *Isospora*[136] or *Cryptosporidium*[150] or described as cyanobacterium-like organisms.[95] Parasites of this genus have been described in insectivorous mammals,[50,61,62] nonhuman primates, and snakes.[126]

Public Health Significance

Infections with *C. cayetanensis* in man can cause protracted diarrhea with nausea, vomiting, anorexia, bloating, abdominal cramps, fatigue, and malaise.[6,149] Among immunocompromised hosts, *Cyclospora* infection has occurred predominantly in HIV-infected individuals. Clinical illness in these patients is prolonged and severe,[77] and is associated with a high rate of recurrence.[123,145]

Life Cycle

The life cycle of *C. cayetanensis* in animals is typically coccidian and is likely to be similar in humans.[154] Oocysts may require from two weeks to four to six months to become infective, depending on temperature.[150]

Distribution

Although *Cyclospora* appears to be widely distributed throughout the world, the prevalence of infection in man and animals worldwide is unknown. Distribution is summarized in Table 9–6. In Canada and the United States, approximately 1 in 1,000 human fecal samples from diarrheic patients were positive,[18,116] although the prevalence in the general population was considered to be generally lower. Prevalence rates of 11% have been reported among allochthonous adults and children in Nepal during the rainy season.[77] Similar levels of infection have been reported in HIV-positive adults in Haiti.[123]

Epidemiology and Transmission

Waterborne outbreaks of cyclosporiosis have been suspected,[150] and drinking untreated water during foreign travel has been incriminated as a factor in the epidemiology of the disease.[77] More recently, there have been several reports of food-borne transmission associated with eating fruit in North America. A large outbreak of cyclosporiosis in the United States was associated with the consumption of imported raspberries.[75] *Meselun* (a mixture of various types of baby leaves of lettuce) was suspected of being the vehicle for one or possibly two outbreaks in the United States in 1997.[161] The consumption of the herb, basil, was suspected as as a source of infection at a luncheon in Virginia.[120] Commercial freezing may kill the oocysts but has yet to be proven.[79] Although there have been no reports of cyclosporidiosis transmission in fermented foods, the distinct possibility exists, and little if any studies have been conducted on the survivability of oocysts under fermentation conditions.

Prevention and Control

Precautions similar to those recommended for cryptosporidiosis should be taken wherever possible. In Third World countries with poor sanitation, either water should be boiled or only bottled water should be consumed. As most reports of cyclosporiosis occur following the consumption of fresh fruit and salads, these should be washed thoroughly and food handlers should ensure good personal hygiene. The detection of *Cyclospora* on food, particularly fruit, is difficult, even though methods for detecting protozoan contamination exist.

Sarcocystosis

Sarcocystis is one of the most prevalent parasites of livestock infecting mammals, including man, birds, and lower vertebrates.[48] Approximately 130 different species have been reported. The parasites derive their name from the intramuscular cyst stage (sarcocyst) of the life cycle, which are found in virtually all skeletal muscles of the body, including the tongue, esophagus, diaphragm, cardiac muscle, and brain of the intermediate host. *Sarcocystis* is found widely throughout the animal kingdom. In domestic animals, for example, there are three species of *Sarcocystis* in cattle (one species of which, *S. hominis* [*S. bovihominis*], is infective for man); four species in sheep (*S. tenella* and *S. arieticanis* [*S. ovicanis*] infect dogs and *S. gigantea* and *S. medusiformis* [*S. ovifelis*] infect cats); and three species in pigs, one of which, *S. suihominis*, is infective for man.[154] Food-borne transmission occurs through the ingestion of raw or undercooked meat containing mature sarcocysts.

Public Health Significance

The ingestion of raw beef containing *S. hominis* can produce clinical signs of nausea, stomachache, and diarrhea within three to six hours.[5,137] The ingestion of infected pork containing *S. suihominis* is much more dramatic, producing clinical signs of bloat, nausea, loss of appetite, stomachache, vomiting, diarrhea, difficulty breathing, and rapid pulse within 6–42 hours. *Sarcocystis* may be responsible for several idiopathic diseases in man, including cardiac diseases such as cardiomyopathy[8,70] and rheumatic diseases.[70] It has also been suggested that *Sarcocystis* may be associated with muscle aches and fatigue as part of the chronic fatigue syndrome (CFS).[128]

Life Cycle

The life cycle of *Sarcocystis* involves an intermediate herbivorous host and a final carnivorous host.[154] The final host becomes infected by ingesting cysts contained in the muscle or nervous tissue of an intermediate host.

Distribution

There are two known species of *Sarcocystis* that infect man— *S. hominis* (*S. bovihominis*) and *S. suihominis* (Table 9–6). There is evidence to suggest that oocysts of these species were previously referred to as "*Isospora hominis*."[1] Humans also serve as the accidental intermediate host for several unidentified species of *Sarcocystis*.[11,48] Levels of infection in man may reflect cooking habits and preferences for undercooked meat.

Examination of various tissues suggests that most cattle worldwide are infected with *Sarcocystis*,[48] with *S. cruzi* (*S. bovicanis*), which is infective to dogs, being the most prevalent. Infection of cattle with *S. (bovi)hominis* appears to be fairly low in many countries, with the possible exception of Germany, where up to 63% of cattle have been shown to be infected.[142] Almost all adult sheep can be infected with *Sarcocystis,* but none are infective for man.[48] The overall prevalence of *Sarcocystis* in pigs is low, between 3–36% worldwide.[48] *S. suihominis* has been reported in European countries, with the highest prevalence in Germany.[16]

Epidemiology and Transmission

In most cases, humans acquire gastrointestinal sarcocystosis by ingesting raw or undercooked meat from cattle or pigs harboring mature *S. (bovi)hominis* or *S. suihominis* cysts. Macroscopic cysts of *Sarcocystis* spp. are often removed at meat inspection, although these cysts are usually of no significance to man. Some meat animals such as camels, llamas, water buffalo, and yaks harbor *Sarcocystis* in which the final host is unknown.[48] In animals, chronic sarcocystosis can cause economic losses due to reduced quality and quantity of meat. Additional economic losses occur with those species that form macroscopic cysts resulting in condemnation of whole carcasses or affected parts after slaughter.

Prevention and Control

Although a widespread and common infection in domestic animals, transmission to man through the ingestion of infected meat is a relatively rare occurrence. Species of *Sarcocystis* are highly host-specific, and only two species are known to infect man. The greatest potential risk therefore comes from ingesting raw or uncooked beef or pork. Ensuring that meat is cooked thoroughly or frozen can prevent infection. It is not known if meat fermentation processes will kill the sarcocysts that are present in meat.

To interrupt the life cycle and prevent infection of livestock, carcasses, meat scraps, offal, or other raw or undercooked tissues should not be available to domestic or wild carnivores in areas where livestock are raised. The identification of *Sarcocystis* in meat can be made by direct observation or by a meat digestion method and examination for the presence of bradyzoites.[48]

Toxoplasmosis

Toxoplasma gondii is an obligate intracellular coccidian parasite with a two-host life cycle in which oocyst production occurs in domestic and wild cats.[53] The parasite differs from other coccidia, however, in that it shows a complete lack of specificity for hosts and tissues during its asexual phase and can infect a wide range of animals, including man. The parasite has worldwide distribution and is of both medical and veterinary importance.

Public Health Significance

Infections with *T. gondii* are usually asymptomatic, but can cause mild symptoms of lymphadenopathy through to signs of anemia, enteritis, fever, hepatitis, lymphadenitis, myocarditis, and pneumonitis. Symptoms are worse in immunocompromised individuals, in whom infections can be severe, with cerebral involvement (encephalitis). The parasite is a significant cause of congenital infection and abortion in women.

Life Cycle

The life cycle of *T. gondii* is similar to *Sarcocystis* in that it involves a carnivore (cat) final host but differs in lack of specificity for intermediate hosts that includes virtually every species of warm-blooded animals and birds.

Distribution

Toxoplasma is widespread among man and animals (Table 9–6) and is possibly the most widespread and prevalent protozoan parasite on earth.[53] It has been estimated that more than 500 million people are infected with *Toxoplasma*.[45] In the United States and the United Kingdom, estimated infection rates in man are between 20% and 40%, but are much higher in European countries, with rates ranging from 50% to 80%.[44] Infection appears to be more common in warm climates and low-lying areas, probably relating to climatic conditions favoring oocyst sporulation and survival.[63,177] Encephalitis caused by *Toxoplasma* is a major opportunistic infection in immunocompromised individuals. It is estimated that 30% of AIDS patients seropositive to *T. gondii* will develop encephalitis.[40] In Europe, congenital toxoplasmosis has been reported in 1–6% of newborn children and is probably the most serious form of *Toxoplasma* infection.

Epidemiology and Transmission

Toxoplasmosis is acquired by ingesting sporulated oocysts or viable tissue cysts in meat from infected animals. The extent of human infection resulting from the ingestion of oocysts is not known. The ingestion of oocysts has been incriminated as the cause for the relatively high prevalence of *Toxoplasma* antibody in vegetarians in India.[134] There are few reports of waterborne transmission of *Toxoplasma*.[13] In contrast, food-borne transmission by the ingestion of tissue cysts in raw or undercooked meat from a variety of livestock and game animals has been well documented as a major source of human infection.[45,53] Pigs, sheep, and goats commonly harbor tissue cysts, whereas cattle appear to be relatively resistant to infection.[53] The ingestion of rare or undercooked meat is a common cause of infection in many countries,[44] with cultural habits an important factor in the epidemiology. In France, for example, where raw or undercooked meat is regularly eaten, 84% of French women have antibodies to *T. gondii*. Unpasteurized milk from infected goats has also been implicated as a source of infection.[135,138] Occupational risks of infection may include certain farming practices, and pregnant women and other people in special risk groups should avoid contact with sheep at lambing time.[23]

Prevention and Control

Food-borne transmission of toxoplasmosis through the ingestion of raw or undercooked meat or other animal produce such as milk is well documented. Adequate cooking kills the organism, as does pasteurization. Therefore, to avoid infection, all meat should be cooked thoroughly before eating. Cooking at 60 °C or higher for 3½ minutes or longer renders *Toxoplasma* cysts noninfectious.[47,55] This temperature is necessary at the core of the meat, not just the surface. Microwave ovens should not be used for initial cooking because of the uneven distribution of heat resulting in hot and cold spots.[54] Freezing to –20 °C also kills some strains of *Toxoplasma,* but is less dependable than cooking. Salting, curing, and pickling also kills cysts present in meat,[171] but it is not known if meat fermentation processes render cysts nonviable. The irradiation of meat at 50 kilorads or greater renders *Toxoplasma* cysts noninfectious.[46] The reduction or elimination of oocyst contamination from cat feces also helps prevent toxoplasmosis. Domestic cats should be prevented from hunting birds and rodents and should be fed preferably dry or canned food. Litter trays should be used in the home and emptied daily. Gardeners handling soil should wear gloves and wash their hands thoroughly before eating or handling food. Cats on farms should be prevented from entering feed-storage facilities and animal quarters. Finally, water from lakes, ponds, streams, or other untreated sources should be boiled before drinking or being used to wash food.

Giardiosis

Although many species of *Giardia* (Table 9–5) have been reported in the literature, it is generally considered that there are just three structural types or species. *Giardia* infections occur widely in mammals, birds, reptiles, amphibians, and fish. The species affecting man and the majority of domestic animals is *Giardia duodenalis* (synonymous, *G. intestinalis*, *G. lamblia*, and *Lamblia intestinalis*). Other morphologically different species are *G. muris*, which is identified in rodents, birds, and reptiles, and *G. agilis,* from amphibians.

Public Health Significance

Giardia infections can cause nausea, inappetence, abdominal discomfort, fatigue, foul-smelling watery diarrhea, flatulence, and abdominal distention, lasting from only a few days to several months. Chronic infections lead to malabsorption, weight loss, and debilitation. Mortality has seldom been reported.

Life Cycle

Giardia are flagellate protozoans that are normally found adherent to epithelial surfaces of the small intestine, especially the middle to lower areas of the villi. The life cycle of *Giardia* is simple and direct, the trophozoite stage (motile forms) dividing by binary fission to produce further trophozoites. Intermittently, trophozoites form resistant cyst stages that pass out in the feces of the host.[2]

Distribution

Infection with *G. intestinalis* has been increasingly recognized and has been reported as the most common gastrointestinal protozoan parasite in the world in man[168] (Table 9–6). Very high prevalence rates of between 20% and 60% are common, particularly in children in developing countries.[21,49,52,68,78,100,103,104]

The existence of *Giardia* in domestic animals has been known for many years, but during that time, little information has become available on prevalence, pathogenicity, and the disease caused by these parasites in their hosts. *Giardia* has been reported in cattle, sheep, goats, horses, dogs, cats, rodents, and psittacines, with varying prevalence rates in these hosts.[22,86,87,105,151,155]

Epidemiology and Transmission

Transmission of the resistant cyst stage is via the fecal-oral route through contaminated water, contaminated food, or person-to-person, animal-to-animal contact. Cysts can remain infectious for several months under moist and cool conditions, but lose infectivity under dry and hot conditions.

Infection with *G. intestinalis* is generally associated with conditions of poor hygiene and sanitation, including poor control of water quality.[133,168] Infection is common among children who attend daycare centers, but is also associated with travel.[19,82,168]

Reports of food-borne transmission of *Giardia* have been made only occasionally. Cysts have been detected in salad, fruit, and other food items.[10] Outbreaks of giardiosis have occurred following the consumption of salmon and cream cheese dip,[119] noodle salad,[127] sandwiches,[170] and fruit salad.[130]

Waterborne outbreaks of giardiosis have been reported in the United States and the United Kingdom.[80] In the United States, *Giardia* has been identified more often than any other pathogen in waterborne outbreaks that result in illness. Because *Giardia* is so widespread among domestic and wild animal hosts, it has the potential for zoonotic transmission.

Prevention and Control

The control of giardiosis is generally based on good personal hygiene, proper sanitation, and the treatment of drinking water by a combination of filtration and disinfection. Reports of food-borne infections with *Giardia* are relatively uncommon. To prevent giardiosis, food handlers should ensure good personal hygiene, and fresh fruit and salads should be washed thoroughly. Fermented cheeses may be a source of giardiosis and, wherever possible, milk ingredients should be pasteurized.

Amoebiosis

Entamoeba histolytica is one of six parasitic amoebae of the genus *Entamoeba* known to infect humans[20] (Table 9–5). In man, the primary route of transmission is from human to human by the fecal-oral route, usually associated with poor hygiene or poor water quality. Infections with *E. histolytica* occur worldwide but are more prevalent in the tropics. There are two morphologically identical forms of *E. histolytica*—one nonpathogenic and the other pathogenic.[20] *E. histolytica* has also been recorded in cattle, pigs, dogs, cats, and monkeys.[94]

Public Health Significance

Most human infections with *Entamoeba* are asymptomatic, but approximately 10% of infected individuals develop clinical symptoms of dysentery with bloody or mucoid diarrhea lasting from a few days to several months. Ulceration of the colon may occur, which may progress to ulcerative lesions in the liver, skin, or brain.

Life Cycle

Trophozoites present in the intestine divide by binary fission. Transmission occurs when the trophozoites round up, become smaller, and form a cyst. Nonpathogenic forms of the organism normally live in the lumen of the large intestine. Pathogenic forms invade the mucosa, causing ulceration and dysentery. From there, they may be carried via the portal system to the liver and other organs, where large abscesses may form.

Distribution

It has been estimated that approximately 480 million people, or 12% of the world's population, are infected with *E. histolytica,* with an annual mortality of 40,000 to 110,000 people[69] (Table 9–6). Approximately 10% of those people who are infected every year have clinical symptoms.[164] Five percent of the U.S. population were estimated to be infected, but the prevalence of *E. histolytica* appeared to be declining.[97]

The incidence of *E. histolytica* in animals is unknown. It has been found in naturally infected monkeys of many species throughout the world. Infection in dogs has only been reported sporadically and often through human contacts. Natural infections are apparently rare in pigs and have to be differentiated from other species that occur in the pig.

Epidemiology and Transmission

The ingestion of food and drink contaminated with *E. histolytica* cysts from human feces and direct fecal-oral contact are the most common means of infection. Transmission from human to human by the fecal-oral route is usually associated with poor hygiene or poor water quality. Cysts can be introduced into the mouth by soiled hands of food handlers, family members, hospital personnel, and other close personal contacts; by food contamination via flies; and by water contaminated with sewage. The transmission of *E. histolytica* by water is common in Third World countries where much of the drinking water is untreated. The use of human feces for fertilizer is also an important source of infection.[21] Recognized high-risk groups for infection include travelers, immigrants, migrant workers, immunocompromised individuals, and male homosexuals. Zoonotic infection from animals has not been recorded.

Prevention and Control

The main risk of infection with *E. histolytica* appears to be through drinking untreated or contaminated water or food. Good personal hygiene, especially thorough washing of hands by food handlers, good sanitation, and high-quality water treatment are essential for preventing transmission.

Blastocystosis

Blastocystis was for many years described as a yeast but is now considered to be a protozoan in the subphylum, Blastocysta[81] (Table 9–5). The organism is found in the intestinal tract of man and in many animals, including monkeys, pigs, birds, rodents, snakes, and invertebrates.

Public Health Significance

Infection with *Blastocystis hominis* is generally asymptomatic, but it is becoming increasingly associated with gastrointestinal disease in man.[178] Symptoms include diarrhea, abdominal discomfort, anorexia, flatulence, and other nonspecific gastrointestinal effects.

Life Cycle

Various life cycles have been proposed for *Blastocystis.* Transmission is thought to occur either through the ingestions of a vacuolar or in cyst form (e.g., seen in feces).[154]

Distribution

The prevalence of *B. hominis* in man is very variable (Table 9–6). Studies in man show varying levels of infection in both healthy patients and patients with diarrhea.[101,113] Infection appears to be more common in adults than in children.[43,84,132]

Blastocystis spp. have been reported in a number of animals, but no prevalence figures are available. In pigs, cross-transmission studies have indicated a low host specificity and uncertain pathogenicity.[122]

Epidemiology and Transmission

It is generally presumed that *B. hominis* is transmitted by the fecal-oral route in a similar manner to other enteric protozoa.[66] Waterborne and food-borne transmission are possible routes of infection[66] but have not been investigated. Infections in man have been linked with foreign travel, although one study suggested an association with exposure to pets or farm animals.[43]

Prevention and Control

B. hominis is increasingly reported as an intestinal infection of man, but it is not clear whether it acts as a pathogen. Transmission is by the fecal-oral route, and waterborne and food-borne transmission are potential routes of infection, but have not as yet been investigated to any great extent. Prevention is as with other intestinal protozoa through good hygiene, especially thorough washing of hands by food handlers and the instigation of good food handling practices.

Balantidiosis

Balantidium coli occurs in man, other primates, pigs, and occasionally in dogs, rats, and ruminants[94] (Table 9–5).

The primary host appears to be the pig, in which it is generally regarded as a commensal of the large intestine, where it lives on starch, ingesta, and bacteria.[94] The parasite has occasionally been reported as a food-borne or waterborne zoonotic infection of man.

Public Health Significance

Infection with *B. coli* is usually asymptomatic but may, on occasion, produce superficial to deep ulcers that result in intermittent diarrhea or dysentery. Rare complications include perforation of the bowel wall and extra-intestinal infection of the liver, vagina, ureter, and bladder.

Life Cycle

Trophozoites in the lumen or tissues multiply by binary fission and, under certain conditions, transform into cysts that pass from the body in the feces.

Distribution

The *Balantidium* organism is found worldwide (Table 9–6) in humans, but the prevalence of disease based primarily on case reports appears low. Surveys of fecal samples collected throughout New Guinea revealed prevalence rates of infection with *Balantidium* as high as 29% in some areas in which people lived in intimate contact with pigs.

Epidemiology and Transmission

Transmission via the fecal-oral route is from human to human and possibly from pigs or rats to humans. Transmission is facilitated in confined environments in which hygiene is poor. The cyst form is the source of infection, and cysts can remain viable for days or weeks in moist pig feces. Flies are possible mechanical vectors. Although pigs have been implicated as a major source of human infection, their role is still a source of controversy. In a review of the literature, more than 50% of the people with balantidiosis reported contact with pigs.[166] In one case, a family claimed that infection followed ingestion of raw pork sausage. A waterborne outbreak implicating pigs as the source of infection has been reported.[167]

Balantidiosis in monkeys and primates is normally an endemic infection maintained by the animals themselves.

Prevention and Control

Infection in man is likely to occur through contamination of water or uncooked vegetables. Prevention is the same as for other protozoa. Under severe environmental conditions, potentially contaminated water should be boiled before use. Vegetables should be washed thoroughly before use.

CONCLUSION

Food-borne parasitic diseases are common in many cultures, and human infections can arise following the ingestion of a wide variety of food sources including meats, fish, molluscs, vegetables, or fruit, or products derived from these foods. Transmission is more likely to occur when food is eaten raw or when it has been incompletely cooked, partially pickled or smoked, or poorly preserved. Fermented foods have the potential to be sources of infection with certain parasites, although reports of human infection following the ingestion of fermented food are sparse. Fermentation alone may therefore be insufficient to prevent the transmission of many food-borne parasites, and potentially infected foods should be avoided wherever possible, or alternatively subjected to either freezing or some form of heat treatment prior to processing.

REFERENCES

1. Ackers, J. P. (1997). Gut coccidia—*Isospora, Cryptosporidium, Cyclospora* and *Sarcocystis*. *Sem Gastrointest Dis* 8, 33–44.
2. Adam, R. D. (1991). The biology of *Giardia* spp. *Microbiol Rev* 55, 706–732.
3. Allan, J. C., Avila, G., Garcia Noval, J., Flisser, A. & Craig, P. S. (1990). Immunodiagnosis of taeniasis by coproantigen detection. *Parasitol* 101, 473–477.
4. Anwar, M., Rak, H. & Gyorkos, T. W. (1979). The incidence of *Gongylonema pulchrum* from cattle in Tehran, Iran. *Vet Parasitol* 5, 271–274.
5. Aryeetey, M. E. & Piekarski, G. (1976). Serologische *Sarcocystis*-Studien an Menschen und Ratten. *Z. Parasitenkd* 50, 109–124.
6. Ashford, R. W. (1979). Occurrence of an undescribed coccidian in Paua New Guinea. *Ann Trop Med Parasitol* 73, 497–500.
7. Awad-el-Kariem, F. M., Robinson, H. A., Pety, F., McDonald, V., Evans, D. & Casemore, D. (1998). Differentiation between human and animal isolates of *Cryptosporidium parvum* using molecular and biological markers. *Parasitol Res* 84, 197–201.
8. Azab, M. E. & El-Shennaway, S. F. A. (1992). Investigation of *Sarcocystis* as a causative agent in cardiac disease. *J Egypt Soc Parasitol* 22, 611–616.
9. Badenoch, J. (1990). *Cryptosporidium in water supplies: report of a group of experts.* London: HMSO.
10. Barnard, R. J. & Jackson, G. J. (1984). *Giardia lamblia*: the transfer of human infections by foods. In *Giardia and Giardiasis*, pp. 365–378. Edited by S. L. Erlandsen & E. A. Meyer. New York: Plenum Press.
11. Beaver, P. C., Gadgil, R. K. & Morera, P. (1979). *Sarcocystis* in man: a review and report of five cases. *Am J Trop Med Hyg* 28, 819–844.
12. Beaver, P. C., Jung, R. C. & Cupp, E. W. (1984). *Clinical Parasitology,* 9th edn. Philadelphia, PA: Lea and Febiger.
13. Beneson, M. V., Takafuji, E. T., Lemon, S. M., Greenup, R. L. & Sulzer, A. J. (1983). Oocyst-transmitted toxoplasmosis associated with ingestion of contaminated water. *N Eng J Med* 307, 666–669.
14. Benger, A., Rennie, R. P., Roberts, J. T., Thornley J. H. & Scholten, T. (1981). A human coenurus infection in Canada. *Am J Trop Med Hyg* 30, 638–644.
15. Besser-Weik, J. W., Forfang, J., Hedberg, C. W., Korlath, J. A. & Osterholm, M. T. (1996). Foodborne outbreak of diarrhoeal illness associated with Cryptosporidium parvum—Minnesota, 1995. *J Vet Med Assoc* 276, 1214.
16. Boch, J., Laupheimer, K. E. & Erber, M. (1978). Drei Sarkosporidienarten bei Schlachtrindern in Suddeutschland. *Berl Muench Tieraerztl Wochenschr* 91, 426–431.
17. Bouchier I. (1998). *Cryptosporidium* in water supplies: third report of the group of experts to: Department of the Environment, Transport and the Regions & Department of Health. Norwich, United Kingdom: HMSO.
18. Brennan, M. K., MacPherson, D. W., Palmer, J. & Keystone, J. S. (1996). Cyclosporiasis: a new cause of diarrhoea. *Can Med Assoc J* 155, 1293–1296.
19. Brodsky, R. E., Spencer, H. C. & Schultz, M. G. (1974). Giardiasis in American travellers to the Soviet Union. *J Infect Dis* 130, 319–323.

20. Brucker, G. (1989). Prevalence of amoebiasis and giardiasis in severe intestinal disorders in interpropical countries. In *Secnidazole: A New Approach in 5-Nitroimidazole Therapy*, pp. 3–11. Edited by N. Katz & A. T. Willis. Amsterdam: Excerpta Medica.
21. Bruckner, D. A. (1992). Amoebiasis. *Clin Microbiol Rev* 5, 356–369.
22. Buret, A., den Hollander, N., Wallis, P. M., Befus, D. (1990). Zoonotic potential of giardiasis in ruminants. *J Infect Dis* 162, 231–237.
23. Buxton, D. (1989). Toxoplasmosis. *In Pract* 11, 9–12.
24. Campbell, W. C. (1994). Meatborne helminth infections: Trichinellosis. In *Foodborne Disease Handbook.* Vol. 2, *Diseases Caused by Viruses, Parasites and Fungii,* pp. 255–277. Edited by Y. H. Hui, J. R. Gorham, K. D. Murrell & D. O. Cliver. New York: Marcell Dekker.
25. Carney, W. P., Sudomo, M. & Purnomo. (1980). Echinostomiasis: a disease that disappeared. *Trop Geogr Med* 32, 106–111.
26. Casemore, D. P. (1990). Epidemiological aspects of human cryptosporidiosis. *Epidemiol Infect* 104, 1–28.
27. Casemore, D. P. (1991). Foodborne protozoal infection. In *Foodborne Illness, a Lancet Review*, pp. 108–119. Edited by W. M. Waite & J. P. Arbuthnott. London: Edward Arnold.
28. Casemore, D. P., Jessop, E. G., Douce, D. & Jackson, F. B. (1986). *Cryptosporidium* plus Campylobacter: an outbreak in a semi-rural population. *J Hyg* 96, 95–105.
29. Cebotarev, R. S. & Poliscuk, V. P. (1959). Gongylonematosis of domestic animals under conditions of Ukrainian Polesie and forest-steppe areas. *Acta Parasitol Pol* 7, 549–557.
30. Chen, C. Y., Hsieh, W. C., Shih, H. H. & Chen, S. N. (1987). Evaluation of enzyme-linked immunosorbent assay for immunodiagnosis of chlonorchiasis. *Chung Hua Min Huo Wei Sheng Wu Chi Mien I Hsueh Tsa Chih* 20, 241–246.
31. Childers, A. B., Terrell, R. N., Craig, T. M., Kaufus, T. J. & Smith, G. C. (1982). Effect of sodium chloride concentration, water activity, fermentation method and drying time on the viability of *Trichinella spiralis* in Genoa salami. *J Food Prot* 45, 816–819.
32. Chitwood, M. B., Velázquez, C. & Salazar, N. G. (1968). Capillaria philippensis sp. n. (Nematoda: Trichinellidae), from the intestine of man in the Philippines. *J Parasitol* 54, 368–371.
33. Cross, J. H. (1987). Public health importance of *Angiostrongylus costaricensis* and its relatives. *Parasitol Today* 3, 367–369.
34. Cross, J. H. (1992). Intestinal capillariasis. *Clin Microbiol Rev* 5, 120–129.
35. Cross, J. H. & Basaca-Sevilla, V. (1983). Experimental transmission of *Capillaria philippinensis* to birds. *Trans Ploy Soc Trop Med Hyg* 77, 511–514.
36. Cross, J. H. & Bhaibulaya, M. (1983). Intestinal capillariasis in the Phillipines and Thailand. In *Human Ecology and Infectious Diseases,* pp. 103–136. Edited by N. Cross & J. Cross. New York: Academic Press.
37. Cross, J. H. & Chi, J. H. C. (1982). ELISA for the detection of *Angiostrongylus costaricensis* antibodies in patients with eosinophilic meningitis. *Southeast Asian J Trop Med Public Health* 16, 110–112.
38. Daengsvang, S. (1980). *A Monograph of the Genus Gnathostoma and Gnathostomiasis in Thailand.* Tokyo: Southeast Asian Medical Information Centre.
39. Daengsvang, S. (1982). Gnathostomiasis. In *CRC Handbook Series in Zoonoses.* Section C. Vol. 2, pp. 800–803. Edited by M. G. Schultz. Boca Raton, FL: CRC Press.
40. Dannemann, B. R. & Remington, J. S. (1989). Toxoplasmic encephalitis in AIDS. *Hosp Med* 24, 139–154.
41. Deplazes, P., Eckert, J., Pawlowski, Z. S., Machowska, L. & Gottstein, B. (1991). An enzyme-linked immunosorbent assay for diagnostic detection of Taenia saginata copro-antigens in humans. *Transact R Soc Trop Med Hyg* 85, 391.
42. Donnelly, J. K. & Stentiford, E. I. (1997). The *Cryptosporidium* problem in water and food supplies. *Lebensm -Weiss u-Technol* 30, 111–120.
43. Doyle, P. W., Helgason, M. M., Mathias, R. G. & Proctor, E. M. (1990). Epidemiology and pathogenicity of *Blastocystis hominis. J Clin Micro* 28, 116–121.
44. Dubey, J. P. (1993). *Toxoplasma, Neospora, Sarcocystis* and other tissue cyst-forming coccidia of humans and animals. In *Parasitic Protozoa.* Vol. 6, pp. 1–158. Edited by J. P. Kreir. New York: Academic Press.
45. Dubey, J. P. & Beattie, C. P. (1988). Toxoplasmosis in man (*Homo Sapiens*). In *Toxoplasmosis of Animals and Man,* pp. 41–60. Boca Raton, FL: CRC Press.
46. Dubey, J. P., Brake, R. J., Murrell, K. D. & Fayer R. (1986). Effect of irradiation on viability of *Toxoplasma gondii* cysts in tissues of mice and pigs. *Am J Vet Res* 47, 518–522.
47. Dubey, J. P., Kotula, A. W., Shaar, A., Andrews, C. D. & Lindsay, D. S. (1990). Effect of high temperature on infectivity of *Toxoplasma gondii* tissue cysts in pork. *J Parasitol* 76, 201–204.
48. Dubey, J. P., Speer, C. A. & Fayer, R. (1989). *Sarcocytosis* in humans and other primates. In *Sarcocystosis of Animals and Man*, pp. 143–144. Edited by J. P. Dubey, C. A. Speer & R. Fayer. Boca Raton, FL: CRC Press.
49. Dupont, H. L. & Sullivan, P. S. (1986). Giardiasis: the clinical spectrum, diagnosis and therapy. *Pediatr Infect Dis* 5, 131–138.
50. Duszynski, D. W. & Wattam, A. R. (1988). Coccidian parasites (Apicomplexa: Eimeridae) from insectivores. IV. four new species in *Talpa europea* from England. *J Protozool* 35, 58–62.

51. Eastburn, R. L., Fritsche, T. P. & Terhune, C. A. Jr. (1987). Human intestinal infection with *Nanophyetus salmincola* from salmonid fishes. *Am J Trop Med Hyg* 36, 568–591.

52. Farthing, M. J. G., Mata, L., Urrutia, J. J. & Kronmal, R. A. (1986). Natural history of *Giardia* infection in infants and children in rural Guatemala and its impact on physical growth. *Am J Clin Nutr* 43, 395–405.

53. Fayer, R. (1981). Toxoplasmosis and public health implications. *Can Vet J* 22, 344–352.

54. Fayer, R. (1994). Food-borne and waterborne zoonotic protozoa. In *Foodborne Disease Handbook.* Vol. 2, *Diseases Caused by Viruses, Parasites and Fungi,* pp. 331–362. Edited by Y. H. Hui, J. R. Gorham, K. D. Murrell & D. O. Cliver. New York: Marcel Dekker.

55. Fayer, R. & Dubey, J. P. (1985). Methods for controlling transmission of protozoan parasites from meat to man. *Food Technol* 39, 57–60.

56. Fayer, R., Speer C. A. & Dubey J. P. (1997). The general biology of *Cryptosporidium*. In *Cryptosporidium and Cryptosporidiosis,* pp. 1–41. Edited by R. Fayer. Boca Raton, FL: CRC Press.

57. Fayer, R. & Ungar, B. L. P. (1986). *Cryptosporidium* spp. and cryptosporidiosis. *Microbiol Rev* 50, 458–483.

58. Feachem, R., Bradley, D. J., Garelick, H. & Mara, D. D. (1983). *Sanitation and Disease. Health Aspects of Excreta and Wastewater Management. Taenia, Taeniasis, and Cysticercosis.* World Bank Studies in Water Supply and Sanitation 3. Chichester, England: John Wiley and Sons.

59. Feldheim, W. & Knobloch, J. (1982). Serodiagnosis of *Opisthorchis viverrini* infestation by an enzyme immuno-assay. *Tropenmedizin und Parasitologie* 33, 8–10.

60. Food and Agriculture Organization, World Health Organization, Office International des Epizooties. (1992). *Animal Health Year Book.* Rome: Food and Agriculture Organization/World Health Organization.

61. Ford, P. L. & Duszynski, D. W. (1988). Coccidian parasites (apicomplexa: Eimeriidae) from insectivores. VI. six new species from the eastern mole, *Scalpaus aquaticus*. *J Protozool* 35, 223–236.

62. Ford, P. L. & Duszynski, D. W. (1989). Coccidian parasites (apicomplexa: Eimeriidae) from insectivores. VII. six new species from the hairy-tailed mole, *Parascalops breweri*. *J Protozool* 35, 508–513.

63. Frenkel, J. K., Dubey, J. P. & Miller, N. L. (1970). *Toxoplasma gondii* in cats: faecal stages identified as coccidian oocysts. *Sci* 176, 893–896.

64. Fricker, C. R. (1995). Detection of *Cryptosporidium* and *Giardia* in water. In *Protozoan Parasites and Water*, pp. 91–96. Edited by W. B. Betts, D. Casemore, C. Fricker, H. Smith & J. Watkins. Cambridge, England: The Royal Society of Chemistry.

65. Galbraith, N. S., Barrett, N. J. & Sockett, P. N. (1987). The changing pattern of foodborne disease in England and Wales. *Public Health* 101, 319–328.

66. Garavelli, P. L. & Scaglione, L. (1989). Blastocystosis: an epidemiological study. *Microbiol* 12, 349–350.

67. Gelletlie, R., Stuart, J., Soltanpoor, N., Armstrong, R. & Nichols, G. (1997). Cryptosporidiosis associated with school milk. *Lancet* 350, 1005–1006.

68. Gilman, R. H., Brown, K. H., Visvesvara, G. S., Mondal, G., Greenberg, B., Sack, R. B., Brandt, F. & Khan, M. U. (1985). Epidemiology and serology of *Giardia lamblia* in a developing country: Bangladesh. *Trans Royal Soc Trop Med Hyg* 79, 469–473.

69. Guerrant, R. L. (1986). Amoebiasis: introduction, current status and research questions. *Rev Infect Dis* 8, 218–227.

70. Habeeb, Y. S., Selim, M. A., Ali, M. S., Mahmoud, L. A., Abdel Hadi, A. M. & Shafei, A. (1996). Serological diagnosis of extraintestinal sarcocystosis. *J Egypt Soc Parasitol* 26, 393–400.

71. Hadidjaja, P., Dahri, H. M., Roesin, R., Margono, S. S., Djalins, J. & Hanafish, M. (1982). First autochthonous case of *Fasciolopsis buksi* infection in Indonesia. *Am J Trop Med Hyg* 31(5), 1065.

72. Hall, A., Latham, M. C., Crompton, D. W. T. & Stephenson, L. S. (1981). *Taenia saginata* (Cestoda) in western Kenya: the reliability of faecal examinations in diagnosis. *Parasitol* 83, 91–101.

73. Harrell, L. W. & Deardorff, T. L. (1990). Human nanophyetiasis: transmission by handling naturally infected coho salmon (*Oncorhynchus kisutch*). *J Infect Dis* 161, 146–148.

74. Haswell-Elkins, M. R., Sithithaworn, P. & Elkins, D. (1992). *Opisthorchis viverrini* and *cholangiocarcinoma* in Northeast Thailand. *Parasitol Today* 8, 86–89.

75. Herwaldt, B. L. & Ackers, M. L. (1997). An outbreak of cyclosporiasis associated with imported raspberries. *New Engl J Med,* 336, 1548–1556.

76. Hitchcock, E. R. (1987). Cysticercosis in the United Kingdom. *J Neurol Neurosurg Psychiatry* 50, 1080–1081.

77. Hoge C. W., Shlim D. R., Rajar R., Triplett, J., Shear, M., Rabold, J. G. and Echeverria, P. (1993). Epidemiology of diarrhoeal illness associated with coccidian-like organisms among travellers and foreign residents in Nepal. *Lancet* 341, 1175–1179.

78. Islam, A. (1990). Giardiasis in developing countries. In *Giardiasis,* pp. 235–266. Edited by E. A. Meyer. Amsterdam: Elsevier.

79. Jackson, G. J., Leclerc, J. E., Bier, J. W. & Madden, J. M. (1997). *Cyclospora*: Still another new foodborne pathogen. *Food Technol* 51, 120.

80. Jephcott, A. E., Begg, N. T. & Baker, I. A. (1986). Outbreak of giardiasis associated with mains water in the United Kingdom. *Lancet* i (8453), 730–732.

81. Jiang, J. -B. & He, J. -G. (1993). Taxonomic status of *Blastocystis hominis*. *Parasitol Today* 9, 2–3.

82. Jokipii, A. M. M., Hemitã, M. & Jokipii, L. (1985). Prospective study of acquisition of *Cryptosporidium*, *Giardia lamblia*, and gastrointestinal illness. *Lancet* ii (8453), 487–489.

83. Kacprzak, E., Kurczewska, M. & Stefaniak, J. (1990). Two cases of cryptosporidiosis in adults. *Przegl Epidemiol* 44, 245–248

84. Kain, K. C., Noble, M. A., Freeman, H. J. & Barteluk, R. L. (1987). Epidemiology and clinical features associated with *Blastocystis hominis* infection. *Microbiol Infect Dis* 8, 235–244.

85. Karl, H., Roepstorff, A., Huss, H. H. & Bloesma, B. (1995). Survival of *Anisakis* in marinated herring fillets. *Int J Food Sci Tech* 29, 661–670.

86. Kirkpatrick, C. E. (1989). Giardiasis in large animals. *Compend Contin Ed Pract Vet* 11, 80–84.

87. Kirkpatrick, C. E. & Farrell, J. P. (1982). Giardiasis. *Compend Contin Ed Pract Vet* 4, 367–377.

88. Kliks, M. M. (1983). Anisakiasis in the western United States: four new case reports from California. *Am J Trop Med Hyg* 32, 526–532.

89. Kotula, A. W. (1983). Postslaughter control of *Trichinella spiralis*. *Food Technol* 37, 91–94

90. Kum, P. N. & Nchinda, T. C. (1982). Pulmonary paragonimiasis in Cameroon. *Trans R Soc Trop Med Hyg* 76, 768–772.

91. Laberge, I., Griffiths, M. W. & Griffiths, M. W. (1996). Prevalence, detection and control of *Cryptosporidium parvum* in food. *Int J Food Microbiol* 31, 1–26.

92. Lachberg, S., Thompson, R. C. A. & Lymbery, A. J. (1990). A contribution to the etiology of racemose cysticerocisis. *J Parasitol* 76, 592–594.

93. Lee, S. H., Hong, S. T., Chai, J. Y., Kim, W. H., Kim, Y. T., Song, I. S., Kim, S. W., Choi, B. I. & Cross, J. H. (1993). A case of intestinal capillariasis in Korea. *Am J Trop Med Hyg* 48, 542–546.

94. Levine, N. D. (1985). Chapter 6 - Amebae and Chapter 13 - Ciliophora. In *Veterinary Protozoology*. Ames, IA: Iowa State University Press.

95. Long, E. G., White, E. H., Carmichael, W. W., Quinlisk, P. M., Raga, R., Swisher, B. L., Dangharty, H. & Cohen, M. T. (1991). Morphologic and staining characteristics of a *Cyanobacterium*-like organism associated with diarrhoea. *J Infect Dis* 164, 199–202.

96. Loría-Cortés, R. & Lobo-Sanahuja, J. F. (1980). Clinical abdominal angiostrongylosis: a study of 116 children with intestinal eosinophilic granuloma caused by *Angiostrongylus costaricensis*. *Am J Trop Med Hyg* 29, 538–544.

97. Lushbaugh, W. B. & Pittman, F. E. (1982). Amoebiasis. In *CRC Handbook Series in Zoonoses, Section C: Parasitic Zoonoses*. Vol. 1, pp. 5–13. Edited by L. Jacobs & P. Arambulo. Boca Raton, FL: CRC Press.

98. Malek, E. A. (1980). *Snail-Transmitted Parasitic Diseases*. Vol. 2. Boca Raton, FL: CRC Press.

99. Malek, E. A. (1981). Presence of *Angiostrongylus costaricensis* Morera and Cespedes, 1971, in Colombia. *Am J Trop Med Hyg* 30, 81–83.

100. Martin-Sanchez, A. M., Canut-Blasco, A., Rodrigues-Hernandez, J., Montes-Martinez, I. & Garcia-Rodrigues, J. A. (1992). Epidemiology and clinical significance of *Blastocystis hominis* in different populations groups in Salamanca (Spain). *Eur J Epidemiol* 8, 553–559.

101. Martinez-Cruz, J. M., Bravo-Zamudio, R., Aranda-Patraca, A. & Martinez-Maranon, R. (1989). Gnathostomiasis in Mexico. *Salud Publicica Mex* 31(4), 541–549.

102. Mason, P. & Patterson, B. (1987). Epidemiology of *Giardia lamblia* infection in children: cross-sectional and longitudinal studies in urban and rural communities in Zimbabwe. *Am J Trop Med Hyg* 37, 277–282.

103. McDonald, V. & Awad El-Kariem, F. M. (1995). Strain variation in *Cryptosporidium parvum* and evidence for distinctive human and animal strains. In *Protozoan Parasites and Water*, pp. 104–107. Edited by W. B. Betts, D. Casemore, C. Fricker, H. Smith & J. Watkins. Cambridge, England: The Royal Society of Chemistry.

104. Meloni, B. P., Lymbery, A. J., Thompson, R. C. & Gracey, M. (1988). High prevalence of *Giardia lamblia* in children from a WA aboriginal community. *Med J Aust* 149, 715.

105. Meloni, B. P., Thompson, L. C., Hopkins, R. M., Reynoldson, J. A. & Gracey, M. (1993). The prevalence of *Giardia* and other intestinal parasites in children, dogs and cats from Aboriginal communities in the Kimberley. *Med J Aust* 158, 157–159.

106. Meyer, V. (1965). Marinades. In *Fish as Food*. Vol. III, pp. 165–193. Edited by G Borgstrom. New York: Academic Press.

107. Millard, P. S., Gensheimer, K. F. & Addiss, D. G. (1994). An outbreak of cryptosporidiosis from fresh-pressed apple cider. *J Am Med Assoc* 272, 1592–1596.

108. Morera, P., Pérez, F., Mora, F. & Castro, L. (1982). Visceral larva migrans-like syndrome caused by *Angiostrongylus costaricensis*. *Am J Trop Hyg* 31, 67–70.

109. Murrell, K. D., Stringfellow, F., Dame, J. B., Leiby, D. A., Duffy, C. & Schad, G. A. (1987). *Trichinella spiralis* in an agricultural ecosystem. *J Parasitol* 73, 103–109.

110. Nansen, P. & Henriksen, Sv. Aa. (1986). The epidemiology of bovine cysticercosis (*C. bovis*) in relation to sewage and sludge application on farmland. In *Epidemiological Studies of Risks Associated with the Agricultural Use of Sewage Sludge: Knowledge and Needs,*

pp. 76–82. Edited by J. C. Block, A. H. Havelaar & P. L'Hermite. London: Elsevier.

111. Nawa, Y., Imai, J. I., Ogata, K. & Otsuka, K. (1989). The first record of a confirmed human case of *Grathostoma doloresi* infection. *J Parasitol* 75, 166–169.

112. Nichols, G. & Thom, B. T. (1985). Food poisoning caused by *Cryptosporidium*: a load of tripe. *Commun Dis Rep* 17, 3.

113. Nimri, L. F. (1993). Evidence of an epidemic of *Blastocystis hominis* infections in preschool children in Northern Jordan. *J Clin Microbiol* 31, 2706–2708.

114. Nopparatana, C., Setasnban, P., Chaicumpa, W. & Tapchaisri, P. (1991). Purification of *Gnathostoma spinigerum* specific antigen and immunodiagnosis of human gnathostomiasis. *Int J Parasitol* 21, 677–687.

115. Ona, F. V. & Dytoc, J. N. (1991). *Clonorchis*-associated cholangiocarcinoma: a report of two cases with unusual manifestations. *Gastroenterology* 10, 831–839.

116. Ooi, W. W., Zimmerman, S. K. & Needham, C. A. (1995). *Cyclospora* species as a gastrointestinal pathogen in immunocompetent hosts. *J Clin Microbiol* 33, 1267–1269.

117. Ortega, Y. R., Sterling, C. R., Gilman, R. H., Cama, V. A. & Diaz, F. (1993). *Cyclospora* species: a new protozoan pathogen of humans. *N Engl J Med* 328, 1308–1312.

118. Oshima T. (1972). *Anisakis* and anisakiasis in Japan and adjacent area. In *Progress of Medical Parasitology in Japan.* Vol. 4, pp. 301–393. Edited by K. Morishita, Y. Komiya & H. Matsubayashi. Tokyo: Meguro Parasitological Museum.

119. Osterholm, M. T., Forfang, J. C., Ristinen, T. L., Dean, A. G., Washburn, J. W., Godes, J. R., Rude, R. A. & McCullough, J. G. (1981). An outbreak of foodborne giardiasis. *N Engl J Med* 304, 24–28.

120. Outbreak of cyclosporiasis: Northern Virginia–Washington D.C.–Baltimore, Maryland Metropolitan Area, 1997. (1997). *MMWR* 46, 689–691.

121. Outbreaks of *Esherichia coli* 0157:H7 infection and cryptosporidiosis associated with drinking unpasteurised cider—Connecticut and New York, October 1996. (1997). *MMWR* 46, 4–8.

122. Pakandl, M. (1993). Occurrence of *Blastocystis* sp. in pigs. *Folia Parasitol* 38, 297–301.

123. Pape, J. W., Verdier, R. I., Boncy, M., Boncy, J. & Johnson W. D. Jr. (1994). *Cyclospora* infection in adults infected with HIV: clinical manifestations, treatment, and prophylaxis. *Ann Intern Med* 121, 654–657.

124. Pawlowski, Z. S. (1971). Taeniarhynchosis, a progessive zoonosis in Europe. In: *Comptes-rendus ler Multocolloque Europeen de Parasitologie, Rennes,* 1971, P. 35 (in French).

125. Pawlowski, Z. S. & Schultz, M. B. (1972). Taeniasis and cysticercosis (*Taenia saginata*). *Adv Parasitol* 10, 269–343.

126. Pellérdy, L. (1974). *Coccidia and coccidiosis*, 2nd edn. Berlin, Germany: Paul Parey.

127. Petersen, L. R., Carter, M. L. & Hadler, J. L. (1988). A food-borne outbreak of *Giardia lamblia. J Infect Dis* 157, 846–848.

128. Phamphlett, R. & O'Donoghue, P. (1992). Antibodies against *Sarcocystis* and *Toxoplasma* in humans with the chronic fatigue syndrome. *Aus NZ S Med* 22, 307–308.

129. Poland, G. A., Navin, T. R. & Sarosi, G. A. (1985). Outbreak of parasitic gastroenteritis among travelers returning from Africa. *Arch Int Med* 145, 2220–2221.

130. Porter, J. D. H., Gaffney, C., Heymann, D. & Parkin, W. (1990). Food-borne outbreak of *Giardia lamblia. Am J Public Health* 80, 1259–1260.

131. Punyagupta S., Juttijudata P., & Bunnag T. (1975). Eosinophilic meningitis in Thailand: clinical studies of 484 typical cases probably caused by *Angyostrongylus cantonensis. Am J Trop Med Hyg* 24, 921–931.

132. Qadri, S. M. H., Al-Okaili, G. A. & Al-Dayel, F. (1989). Clinical significance of *Blastocystis hominis. J Clin Microbiol* 27, 2407–2409.

133. Ravdin, J. I., Weikel, C. S. & Guerrant, R. L. (1988). Protozoal enterpathies: cryptosporidiosis, giardiasis, and amoebiasis. In *Clinical Tropical Medicine and Communicable Diseases: Diarrhoeal Diseases*, pp. 503–536. Edited by R. L. Guerrant. London: Balliere Tindall.

134. Rawal, B. D. (1959). Toxoplasmosis: a dye-test survey of sera from vegetarians and meat eaters in Bombay. *Trans R Soc Trop Med Hyg* 53, 61–63.

135. Reimann, H. P., Mayer, M. E., Theis, J. H., Kelso, G. & Behymer, D. E. (1975). Toxoplasmosis in an infant fed unpasteurized goat milk. *J Paediatr* 87, 573–576.

136. Rijpstra, A. C. & Laarman, J. J. (1993). Repeated findings of unidentified small Isopora-like coccidia in faecal specimens from travellers returning to the Netherlands. *Trop Geog Med* 45(6), 280–282.

137. Rommel, M. & Heydorn, A. O. (1972). Beitrage zum Lebenszyklus der Sarkosporidien. III *Isospora hominis* (Railliet I Lucet, 1891) Wenyon, 1923, eine Dauerform der Sarkoporidien des Rindes und des Schweins. *Berl Muench Tieraerztl Wochenschr* 85, 143–145.

138. Sacks, J. J., Roberto, R. R. & Brooks, N. F. (1982). Toxoplasmosis infection associated with raw goat's milk. *J Am Med Assoc* 248, 1728–1732.

139. Sakanari, J. A. & KcKerrow, J. H. (1989). Anisakiasis. *Clin Microbiol Rev* 2, 278–284.

140. Schmutzhard, E., Boongird, P. & Vejjajiva, A. (1988). Eosinophilic meningitis and radiculomyelitis in Thailand caused by CNS invasion of *Gnathostoma*

spinigerum and *Angiostrongylus cantonensis*. *J Neurol Neurosurg Psychiatry* 51, 80–87.

141. Schultz, M. G., Hermos, J. A. & Steele, J. H. (1970). Epidemiology of beef tapeworm infection in the United States. *Public Health Rep* 85, 169–176.

142. Schulze, K. & Zimmermann, T. (1981). Sarkosporidienzysten im Hackfleisch. *Fleischwirtsch* 61, 614–620.

143. Schwartz, D. A. (1980). Review: helminths in the induction of cancer: *Opisthorchis viverrini, Chlonorchis sinensis* and *cholangiocarcinoma*. *Trop Geogr Med* 32, 95–100.

144. Shish, H. H. & Chen, S. N. (1991). Immunodiagnosis of *Angiostrongyliasis* with monoclonal antibodies recognising a circulating antigen of mol. wt. 91,000 from *Angiostrongylus costaricensis*. *Int J Parasitol* 21, 171–177.

145. Sifuentes-Osornio, J., Porras-Cortes, G., Bendall, R. P., Morales-Villarreal, F., Reyes-Teran, G. & Ruiz-Palacios, G. M. (1995). *Cyclospora cayetanensis* infection in patients with and without AIDS: biliary disease as another clinical manifestation. *Clin Infect Dis* 21, 1092–1097.

146. Sirisinha, S., Chawengkirttikul, R., Sermswan, R., Amornpant, S., Mongkolsuk, S. & Panyim, S. (1991). Detection of *Opisthorchis viverrini* by monoclonal antibody-based ELISA and DNA hybridization. *Am J Trop Med Hyg* 44, 140–145.

147. Smith, H. V. (1991). Parasitic protozoa in drinking water. In *UK Health Related Water Microbiology*, pp. 28–43. Glasgow, England: University of Strathclyde.

148. Smith, J. L. (1993). *Cryptosporidium* and *Giardia* as agents of foodborne disease. *J Food Prot* 56, 451–461.

149. Soave, R. (1996). *Cyclospora*: an overview. *Clin Infect Dis* 23, 429–437.

150. Sterling, C. R. (1995). Emerging pathogens: *Cyclospora*, the Microspora, and how many more? In *Protozoan Parasites and Water,* pp. 243–245. Edited by W. B. Betts, D. Casemore, C. Fricker, H. Smith & J. Watkins. Cambridge, England: The Royal Society of Chemistry.

151. Sykes, T. J. & Fox, M. T. (1989). Patterns of infection with *Giardia* in dogs in London. *Trans R Soc Trop Med Hyg* 83, 239–240.

152. Taniguchi, Y., Itashimoto, K., Ichikawa, S., Shimizu, M., Ando, K. & Kotani, Y. (1991). Human gnathostomiasis. *J Cutan Pathol* 18, 112–115.

153. Taylor, M. A. (1995). Protozoan parasites and water: Veterinary aspects. In *Protozoan Parasites and Water,* pp. 246–249. Edited by W. B. Betts, D. Casemore, C. Fricker, H. Smith & J. Watkins. Cambridge, England: The Royal Society of Chemistry.

154. Taylor M. A. (1999). Protozoa. In *The Microbiological Quality and Safety of Food.* Part III, *Foodborne Pathogens,* pp. 1–37. Edited by B. M. Lund, A. C. Baird-Parker & G. W. Gould. Gaithersburg, MD: Aspen Publishers.

155. Taylor, M. A. & Webster, K. A. (1998). Recent advances in the diagnosis of *Cryptosporidium, Toxoplasma, Giardia* and other protozoa of veterinary importance. *Res Vet Sci* 65, 183–193.

156. Taylor, M. A., Catchpole, J., Marshall, R. N. & Green, J. (1993). Giardiasis in lambs at pasture. *Vet Rec* 133, 131–133.

157. Thomson, M. A., Benson, J. W. T. & Wright, P. A. (1987). Two year study of *Cryptosporidium* infections. *Arch Dis Child* 62, 559–563.

158. Tülsner, M. (1978). Die bestimmenden stoffichen Veranderungen im Fisch beim Salzen und Marinieren. *Lebenensmittelindustrie* 25, 169–173.

159. Ungar, B. L. P. (1990). Cryptosporidiosis in humans (homo sapiens). In *Cryptosporidiosis of Man and Animals,* pp. 59–82. Edited by J. P. Dubey, C. A. Speer & R. Fayer. Boca Raton, FL: CRC Press.

160. Upatham, E. S., Viyanant, V., Brockelman, W. Y., Kurathong, S., Lee, P. & Kraengraeng, R. (1988). Rate of re-infection by *Opisthorchis viverrini* in an endemic northeast Thai community after chemotherapy. *Int J Parasitol* 18, 643–649.

161. Outbreaks of cyclosporiasis: United States and Canada, 1997. (1997). *MMWR* 46, 521–523.

162. Van Thiel, P. H. (1976). The present state of Anisakiasis and its causative worms. *Trop Geogr Med* 28, 75–85.

163. Verster, A. (1969). A taxonomic revision of the genus *Taenia* linnaeus, 1758. *Onderstepoort J Vet Res* 36, 3–58.

164. Walsh, J. A. (1986). Problems in recognition and diagnosis of amoebiasis: estimation of the global magnitude of morbidity and mortality. *Rev Infect Dis* 8, 228–238.

165. Walther, M. (1980). *Taenia saginata* cysticercosis: a comparison of routine meat inspection and carcass dissection results in calves. *Vet Rec* 106, 401–402.

166. Walzer, P. D. & Healy G. R. (1982). Balantidasis. In *CRC Handbook Series in Zoonoses, Section C: Parasitic Zoonoses.* Vol. 1, pp. 15–24. Edited by L. Jacobs & P. Arambulo. Boca Raton, FL: CRC Press.

167. Walzer, P. D., Judson, F. N., Murphy, K. B., Healy, G. R., English, D. K. & Schultz, M. G. (1972). Balantidiasis outbreak in Truk. *Am J Trop Med Hyg* 22, 33–41.

168. Warhurst, D. C. & Green, E. L. (1988). Protozoal causes of diarrhoea. *PHLS Microbiol Dig* 5, 31–37.

169. Welch, J. S., Dobson, C. & Campbell, G. (1976). Immunodiagnosis and seroepidemiology of *Angiostrongylus costaricensis* zoonoses in man. *Trans R Soc Trop Med Hyg* 74, 614–623.

170. White, K. E., Haedberg, C. W., Edmondson, L. M., Jones, D. B. W., Osterholm, M. T. & McDonald, K. L.

(1989). An outbreak of giardiasis in a nursing home with evidence for multiple modes of transmission. *J Infect Dis*, 160, 298–304.

171. Work, K. (1968). Resistance of *Toxoplasma gondii* encysted in pork. *Acta Pathol Microbiol Scand* 73, 85–92.

172. World Health Organization. (1979). *Report of a WHO Expert Committee, with the Participation of the FAO.* (Technical Report Series 637).Geneva: Author.

173. World Health Organization. (1983). *Guidelines for Surveillance, Prevention and Control of Taeniasis/Cysticerocosis*. Geneva: Author.

174. Wurtz, R. (1994). *Cyclospora*: a newly identified intestinal pathogen of humans. *Clin Infect Dis* 18, 620–623.

175. Wyllie, A. S. (1984). Cryptosporidiosis. *Br Med J* 289, 1383–1384.

176. Yii, C-Y. (1976). Clinical observations on eosinophilic mengingitis and meningoencephalitis caused by *Angiostrongylus costaricensis* on Taiwan. *Am J Trop Med Hyg* 25, 233–249.

177. Yilmaz, S. M. & Hopkins, S. H. (1972). Effects of different conditions on duration of infectivity of *Toxoplasma gondii* oocysts, *J Parasitol* 58, 938–939.

178. Zierdt, C. H. (1991). *Blastocystis hominis:* past and future. *Clin Microbiol Rev* 4, 61–79.

179. Zinter, D. E. & Migaki, G. (1970). *Gongylonema pulchrum* in tongues of slaughtered pigs. *J Am Vet Med Assoc* 157(3), 301.

CHAPTER 10

Biotechnology and Food Safety: Benefits of Genetic Modifications

T. Verrips

INTRODUCTION

A number of consumer research studies have shown that fermented food products are regarded by consumers as healthy and natural.[9] Factually and from a historic point of view, this is correct. During the approximately 7,000 years that fermented foods have been produced, an enormous selection process has taken place, and the tasty products that are presently on the market are the result of this selection process. It is well known that fermentation of foods and drinks prolongs the shelf life of these products. The historic claims that fermented foods, particularly those derived from milk, prevent diseases are supported by many studies. Those claims related to mucosal health are excellently summarized by Salminen *et al.*[23] Nevertheless, the title of this chapter suggests that at present, there is concern about the safety of fermented food products. This is due to the rapidly growing application of genetically modified food products, including fermented food products. In particular, the penetration on the market of genetically modified plants is remarkable (Figure 10–1 and Table 10–1). In addition, more than 20 food enzymes that are on the market at present are produced using recombinant DNA technology (Table 10–2).

The perception of genetic modification of food products by European consumers is not very positive and varies from country to country. Generally, the attitude of Western consumers is neutral to positive (Figure 10– 2), whereas in most of the developing countries, this technology is seen as an opportunity.[20] However, in spite of the present problems in a number of West European countries, and excluding small groups of dogmatic opponents of genetic engineering, the perception of the consumer will ultimately be determined by the benefits versus risks ratio.

In spite of many efforts, neither scientists nor opinion makers have been able to properly explain the potential risks of genetically modified food products. Therefore, many consumers perceive recombinant DNA technology as an intrinsically dangerous technology, although in the 25 years that this technology has been applied, no unintended dangerous materials have been produced. The main reason for the present perception of this technology by West European consumers is that in many discussions, no clear definitions of the nature of the genetically modified food are used. In an attempt to rationalize the discussions and to avoid inappropriate ethical discussions, let us examine simple decision schemes based on rules of the U.S. Food and Drug Administration (FDA), the United Kingdom, and the Netherlands,[29] and a model for various genetically modified products (Figure 10–3).[30] It should be emphasized that this figure reflects the author's personal view concerning the safety of genetically modified foods that do not contain (antibiotic) resistance markers ("clean"). Genetically modified plants or microorganisms containing antibiotic resistance markers present a very difficult to quantify but realis-

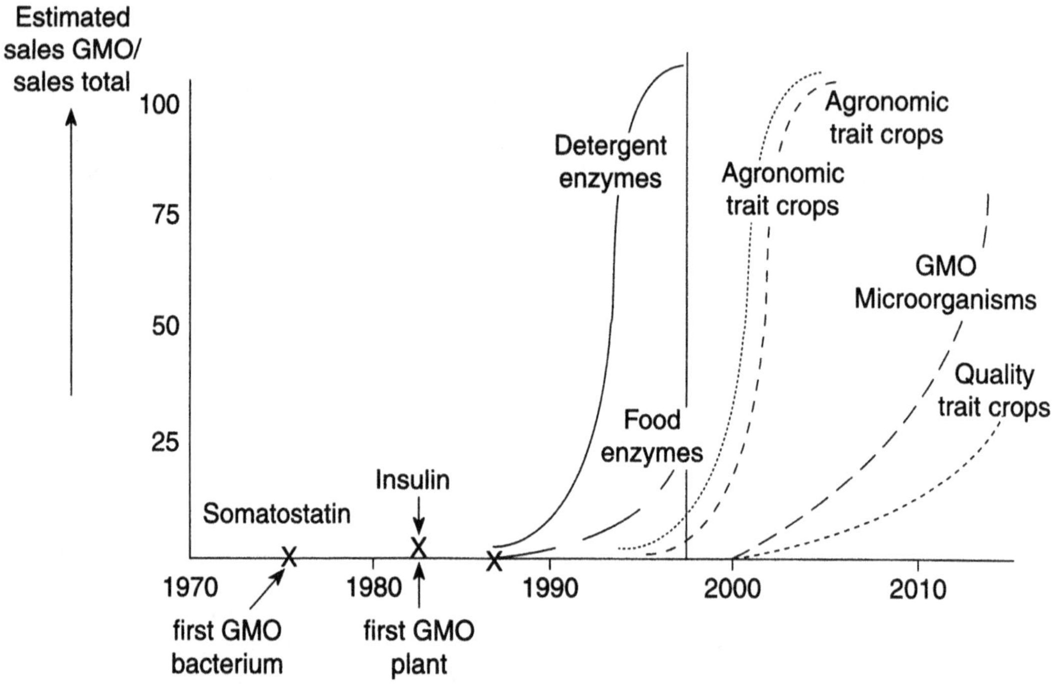

Note: X-axis-penetration as a function of time: The expression of somatostatin in *E. coli* is taken as the starting point and insulin as the first large-scale pharmaceutical product. Y-axis-estimated percent of rDNA product in certain area (Detergent enzymes are nearly 100% rDNA products, about 25% of the food enzymes (e.g., chymosin) are made via rDNA technology).

Figure 10–1 Penetration of rDNA technology in agriculture, pharmaceuticals, and consumer products.

tic risk of spreading genes encoding antibiotic resistance in the environment.[6,10] The chance that these distributed antibiotic-resistant genes are taken up is very low, but not zero, and therefore undermines the argument that genetically modified plants or microorganisms containing a gene that encodes an intrinsically safe protein are safe.

Table 10–1 Rapid Increase of Transgenic Plants

Crop	*1996*	*1997*
Maize	525,000	4,400,000
Soybean	400,000	5,250,000*
Potato/tomato	40,000	500,000†
Oilseed	200,000	1,600,000
Cotton	810,000	1,200,000

*Includes the South American 1996/1997 harvest.
†Includes estimates for China.
Note: In 1994, no transgenic crops were cultivated for usage in the agrofoods industry.

In addition to the lack of proper explanation of the potential risk of genetically modified foods, the benefits for the consumer are either nonexistent or not communicated properly to consumers. This is a major weakness, and only when the benefits to consumers become clear will genetically modified foods be accepted by the majority of consumers. It is obvious that benefits are not absolute values but relative values, strongly influenced by demographic, geographic, and socioeconomic factors.

In this chapter, the supply chains of fermented foods are taken as guides to point out the potential benefits and risks of every step of these chains. The emphasis will be on fermented foods

Table 10–2 Commercial Food Enzymes Made by rDNA Technology (as on the Market at 1.5.1998)

Enzyme	*Commercial Name*	*Producer*	*Application*
α-Galactosidase	α-galactosidase	NOVO N.	Animal feed
Xylanase	Bio-feed wheat	id	id
Lipase	Lipozym	id	Interesterification
ALDC	Maturex	id	Beer
Amylase	Novamyl	id	Bread
1,3 Lipase	Novozym 677	id	Bread
Xylanase	Peptopan	id	Bread
Amylase	Maltogene	id	Starch
Chymosin	Novoren	id	Cheese
Lipase	Novozym 398	id	Interesterification
Lipase	Novozym 435	id	id
Fytase	Pytan	id	Animal feed
Amylase	Termamyl	id	Starch
Transferase	Toryzym	id	Starch
Chymosine	Maxiran	DSM/Gb	Cheese
Phytase	Natuphos	id	Animal Feed
Chymosin	Chymogen	C. Hansen	Cheese
Chymosin	Chymax	Pfizer	Cheese
Xylanase	Biobake 710	Quest	Bread
Invertase		id	Chocolate
Glucanase		id	Beer
α-Galactosidase		id	Guar modification

starting from traditionally bred or genetically modified plants and milk from traditionally bred cattle. The approach of analyzing the safety of fermented products as a function of the supply chain has been adopted, because only with such an approach will all potential risks be included, which is essential to gain the confidence of consumers.

By analogy with microbial risk assessments,[28] risk is defined as the probability of a hazard occurring and hazard as a harmful event. This means that when the newly introduced gene encodes a protein that is intrinsically safe and does not provide a selection benefit for any receiving organism, the hazard is considered as zero. Consequently, the risk is zero and the safety of the food product is 100%. If, on the other hand, the newly introduced gene encodes for a protein that is intrinsically safe but provides the receiving organisms with a selection benefit in the open environment, then the health hazard is zero, but the environmental hazard is not zero. Consequently, the probability of transfer of the gene to the receiver has to be estimated. Genes encoding intrinsically unsafe proteins (i.e., health hazard > 0) are not considered because authorities will not give permission for cloning of such genes in plants and/or microorganisms that enter the food chain; neither will the food industry ever use such organisms.

SUPPLY CHAIN OF FERMENTED FOODS AND ITS IMPORTANCE TO ENSURE SAFE PRODUCTS

Table 10–3 provides a selection of fermented foods based on the raw material(s) that is (or may become in the near future) derived from genetically modified plants.

The steps in the supply chains (including potential beneficial and risk aspects) of fermented foods using traditionally bred plant materials or milk from traditionally bred cows with fermen-

Figure 10–2 Acceptance of rDNA technology for consumer products derived from plants and microorganisms in various countries. Acceptance varies from approximately 90% in Canada, United States, and Portugal to 30 and 45% in Austria and Germany, respectively.

tation processes using genetically modified microorganisms are provided in Figure 10–4. However, Tables 10–1 and 10–3 illustrate that genetically modified plants will become a major source of plant raw material for fermented foods in the near future, and therefore a separate supply chain risk/benefits analysis for fermented products derived from genetically modified plants is provided in Figure 10–5.

It is impossible to deal in this chapter with all of the products in Table 10–3 for both traditional and genetically modified raw materials and/or for traditional and genetically modified microorganisms used in the fermentations. Therefore, only two examples are worked out in more detail.

1. *soy sauce*—a fermented product using genetically modified wheat and/or soybeans, traditional or genetically modified microorganisms, and traditional or recombinant DNA enzymes
2. *cheese*—a fermented product using normal cow's milk, genetically modified functional microorganisms, and/or enzymes made by genetically modified microorganisms

These two examples cover quite well the whole spectrum of fermented food products in which recombinant technology is used at some step in the overall process.

As stated in the introduction to this chapter, fermented foods derived from genetically modified plant material or produced by genetically modified microorganisms will only be accepted if there are clear direct (e.g., significantly better quality or shelf life, more healthy or more convenient) or indirect (e.g., availability, environment) benefits for

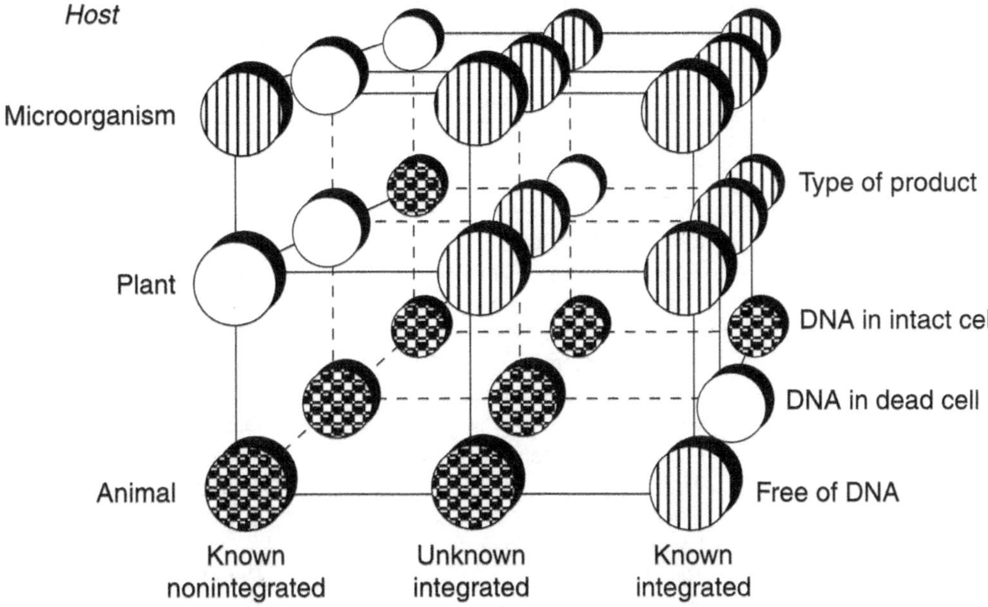

Note: X-axis-physical state of foreign gene(s): "known" means that the location of foreign gene(s) is/are known exactly, "integrated" means stable integration of foreign gene(s) on the chromosome of the host; Y-axis-host organism; Z-axis-whether the consumer product is free of DNA (e.g., cheese made with chymosin) or product contains inactivated cells that produced the rDNA product (i.e., tomato paste) or the rDNA producing cells are "alive" present in the product (i.e., fresh tomatoes). Vertical spheres-no risk; white spheres-risk assessment inconclusive; dotted spheres-either risk assessment inconclusive or ethical objections against such consumer products.

Note: This scheme is based on the assumption that the used constructs do not contain an antibiotic resistance marker. Otherwise, the risk to the environment will increase for all cases except contained fermentation of microorganisms.[29]

Figure 10–3 Summary of risk assessment of rDNA products on basis of three parameters.

the consumer. Some potential benefits will be discussed in the following sections.

Some Consumer Benefits of Genetically Modified Plants

Benefits are relative, and for each group of consumers, benefits will be perceived differently. Consumer studies in Western Europe show that the order of consumer criteria to buy products is quality, health aspects, convenience, environment, and price. It is likely that price together with availability will be on top of such a list in developing countries. In the Western world, there is a surplus of food raw materials and improvement of crop yield is not perceived as a benefit by the consumer; neither is the loss of material during transport seen as a problem. However, the maintenance of quality during transport and at home is considered as a benefit by consumers, as was demonstrated by the initial successful introduction of the Flav Sav tomato (with an antisense polygalacturonase gene that ensured delay of softening of the tomato) by Calgene on the United States market in the mid-1990s.

In many developing countries, however, there is at present a shortage of food raw materials. Therefore, an increase in the yield of crops by genetic modification is seen as a major advantage. There is evidence that shows that in the

Table 10–3 Overview of Most Important Fermented Food Products Derived from Plants of which Genetically Modified Varieties Are on the Market and the Main Functional Microorganisms

Raw Materials	*Functional Microorganisms*	*Product*
Cassava	Bacteria	Chickwangne (Congo)
	Corynebacterium, Geotrichum	Gari (Nigeria)
/ sorghum	Lactic acid bacteria, *Saccharomyces, Candida*	Burukutu (Nigeria)
Maize	*Aspergillus*, lactic acid bacteria, yeast	Chicha (Peru)
	Yeasts, bacteria	Jamin-bang (Brazil)
	Lactic acid bacteria, yeasts, molds	Ogi (Nigeria)
/ cassava	Lactic acid bacteria, yeasts	Banku (Ghana)
Rice	*Monascus*	Ang-kak (China)
	Rhizopus	Lao-chao (China)
	Lactic acid bacteria *Saccharomyces*	Puto (Philippines)
	Aspergillus, Bacillus	Sierra rice (Ecuador)
	Hansenula, Candida, Geotrichum	Torani (India)
/ Black gram	Lactic acid bacteria *Torulopsis*	Idli (India)
/ Carrots	*Hansenula*	Kanji (India)
/ Soybeans	*Aspergillus*, Lactic acid bacteria, yeasts	Miso (China, Japan)
or cassava	Yeasts, molds	Tapé (Indonesia)
Soybean	*Mucor, Aspergillus*	Chee-fan (China)
	Aspergillus, Rhizopus	Meju (Korea)
	Actinomucor	Meitauza (China)
	Bacillus natto	Natto (Japan)
	Actinomucor, Mucor	Sufu (China)
	Rhizopus	Tempeh (Indonesia)
/ rice	*Aspergillus*, yeasts Lactic acid bacteria	Miso (China, Japan)
/ wheat	*Aspergillus*, Lactic acid bacteria	Hamanatto (Japan)
/ wheat	*Aspergillus*, yeasts Lactic acid bacteria	Soy sauce (southeast Asia)
/ wheat	*Aspergillus*	Tao-si (Philippines)
Wheat	*Saccharomyces*	Jalebies (India, Pakistan)
	Molds	Minhin (China)
/milk	Lactic acid bacteria	Kishk (Egypt)

next 10 years, the increase in the supply of agricultural products will be less than the expected increase in world population and food shortage, especially in developing countries, will increase[16] (Figure 10–6). It has also been suggested that only when modern biotechnology results in a second green revolution may this global problem be prevented.[17]

Also, the prevention of the substantial losses during transportation of plant raw materials (up

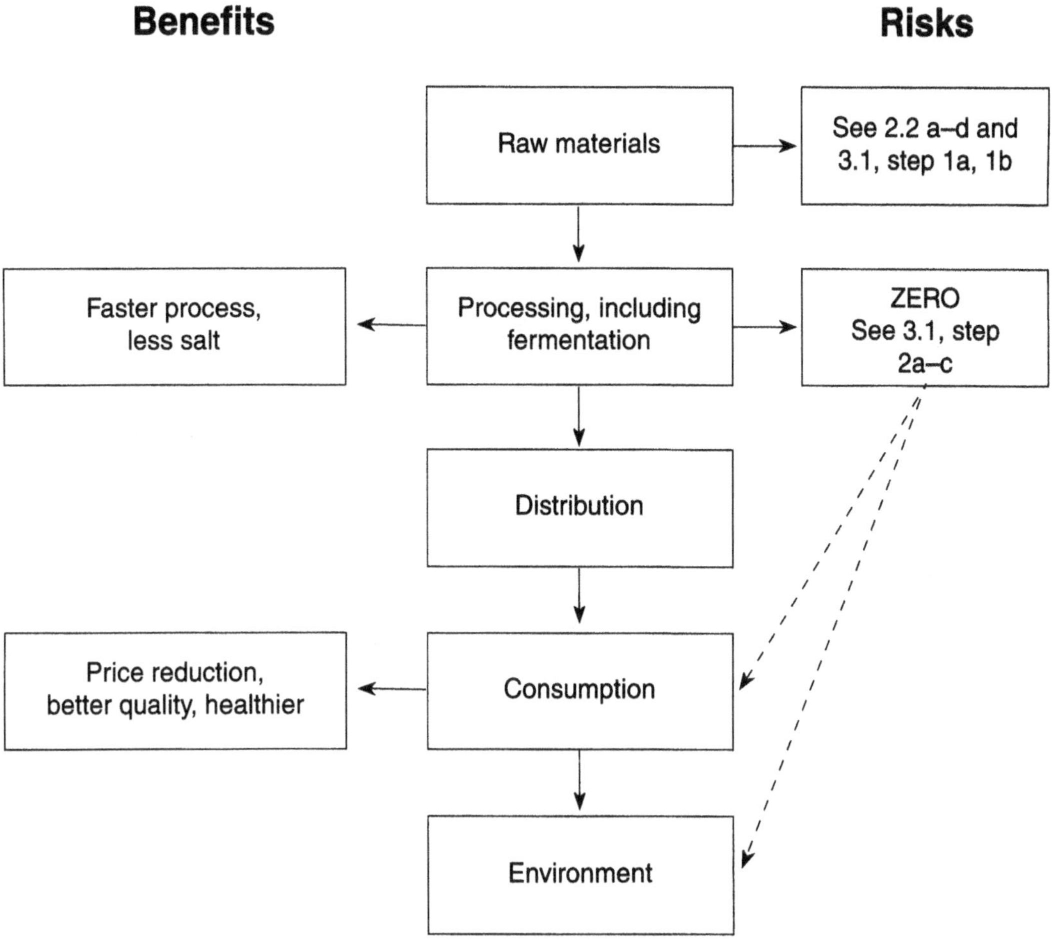

Figure 10–4 Schematic benefit versus risk ratio of rDNA products made by microorganism.

to 30% of harvested plant materials is destroyed due to microorganisms, insects, or uncontrolled endogenous processes) is seen as a major benefit for consumers in that part of the world.

There is evidence that modern biotechnology can significantly contribute to increasing yields or reducing losses of plants and plant products in the first phases of the supply chain. Monsanto's Roundup herbicide-resistant crops need less herbicides (approximately 10% less) and show increased yields (5–15%) (Farmers Organization Argentina, personal communication, November 1997).

Plants producing *Bacillus thuringiensis* BT-protein (preferably more than one variety of this protein so as to minimize the probability of adaptation) are quite well protected against insects, thereby increasing the yield. Modern biotechnology can also contribute to the reduction of losses due to microbial spoilage during distribution. Plants have a rich arsenal to defend themselves against invading microorganisms.[5] They can change the structure of plant tissue by extensive cross linking catalyzed by redox enzymes, produce low molecular mass antimicrobial agents, and degrade cell walls of invading microorganisms using hydrolases they produce. In particular, plant pathogenic fungi can be effectively destroyed by these enzymes. Some companies like Novartis and Zeneca are very active

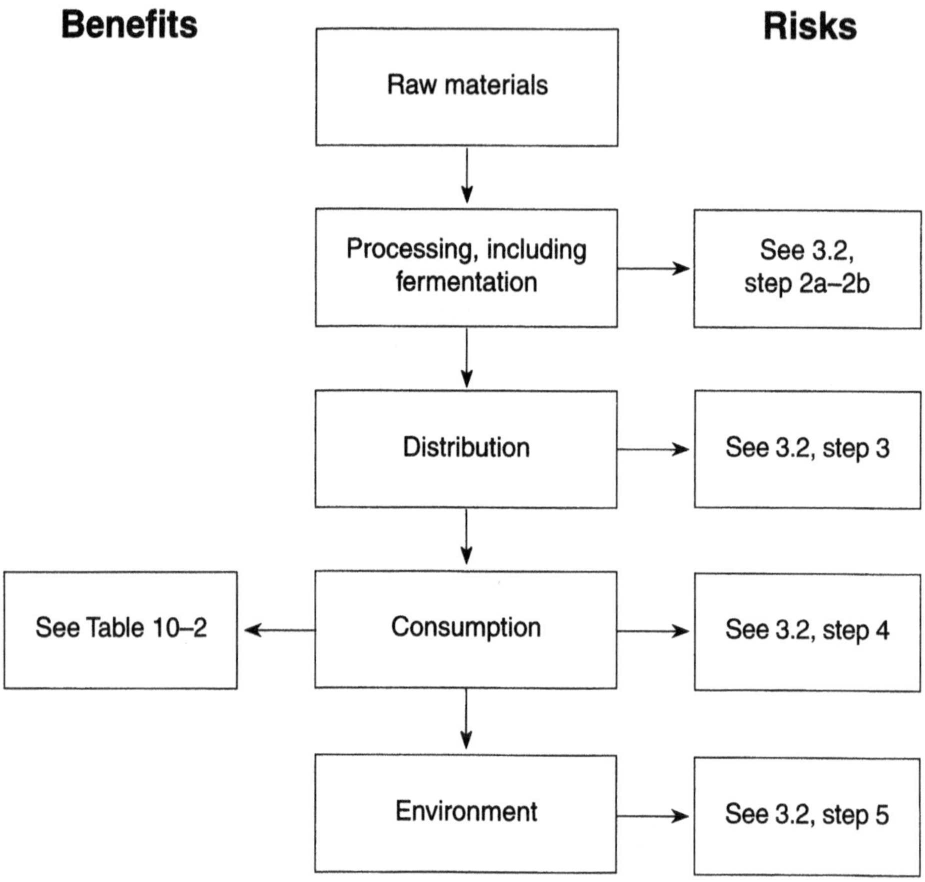

Figure 10–5 Schematic benefit versus risk ratio of rDNA product made by plants.

in this field, as shown by the five and seven patents they filed respectively on this technology between 1980 and 1997. This approach is limited to improving existing plant protection systems. However, recombinant DNA technology can do more. Plant Genetic Systems had a project in the mid-1980s to express potent antimicrobial peptides (e.g., apidaecin)[3] in plants, although with the intention to isolate these antimicrobials from the plant and use them in all type of products. At present, a large number of antimicrobial peptides are known, some of amazing effectivity.[1]

A consumer benefit of currently available genetically modified plants can be a lower price due to higher yield, reduced cost for chemicals, and reduced losses during harvesting and transport. However, an important benefit can be delivered by plants with the right balance of micronutrients that contribute to consumer health. Plants are a well-known and extremely rich source of all kinds of components that may contribute to consumer health, such as antioxidants, organic metal complexes, phytoestrogens, phytosterols, and vitamins. At present, the diet of approximately 2.5 billion people contains insufficient amounts of minerals and vitamins. The consequences of these shortages are dramatic. For example, each year, 2 million young children die and approximately 300,000 children go blind due to a shortage of vitamin A. With the rapid increase in our knowledge of cell and molecular biology, and much better and faster analytical techniques, the number and kind of plant and microbial components with sustain-

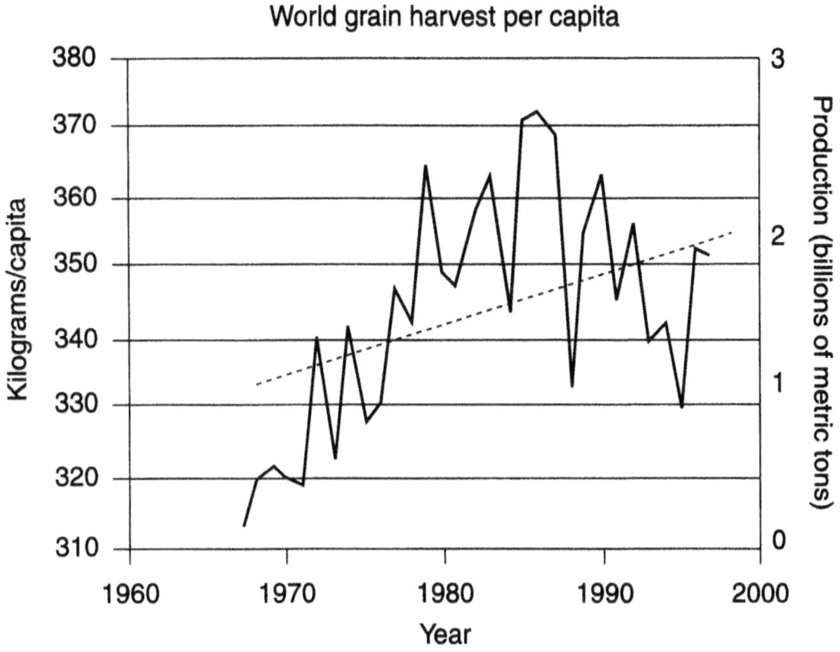

Figure 10–6 World grain harvest during the last 35 years. A steady increase in total production from 1.2 to 2.2 billion of metric tons; a rapid increase in amount per capita from 1966 to 1987, followed by a decrease from 1988 to present.[16] *Source:* Reprinted with permission from FAOSTAT, C.C. Mann, Crop Scientists Seek a New Revolution, *Science,* Vol. 283, pp. 310–314. Copyright 1999 American Association for the Advancement of Science.

able health claims will increase significantly in the near future.

What happens to these components with potential healthy properties during fermentation is not very well known, but it is likely that chemical nature and bioavailability of these components will be influenced (for better or worse) by the various steps in a fermentation process. Some literature is available on the conversion of estrogens and phytoestrogens by the microflora of the gastrointestinal system and show that phytoestrogens can be (extensively) converted into more effective compounds, such as estrone into oestradiol.[21] However, knowledge in this area is too fragmented and limited, and this important aspect has to be researched.

Some Potential Hazards and Risks of Genetically Modified Plants

It is outside the scope of this chapter to go into detail on the potential hazards of genetically modified plants. However, the integral supply chain approach followed requires that some aspects of the potential risk are discussed. At present, there are at least four issues in relation to genetically modified plants.

1. Most of the genetically modified plants contain an antibiotic resistance gene as marker and, although the marker gene as such is not expressed, the potential transfer of this marker gene to other crops in the surrounding area has been studied.[11] Although the frequency of such an event is low, it is measurable; therefore, this approach may contribute to a further spread of antibiotic resistance genes in the environment, thereby contributing to an extremely undesirable further increase of antibiotic resistance of human and animal pathogens.
2. The gene providing the plant with a new desired property, such as herbicide resistance, can also be spread in the environment. This

means that there is a small probability that weeds will pick up this property and become resistant to herbicides. This spread of genes providing herbicide resistance is an issue that has to be addressed.

The spread of genes providing plants with new desired properties related to health benefits of consumers will also occur, but in this case, the gene will most probably not provide the recipient with an ecological benefit; therefore, this risk is so low that it can be neglected.

3. It is still not possible to integrate the new genetic information in a predetermined place in the genome of the plant. The random integration may result in the destruction or enhancement of the expression of genes at the locus of integration and therefore may change the metabolism of the plant. To exclude the introduction of a hazard, one should precisely determine the position of integration and use techniques such as DNA microarrays,[4] proteomics,[13] and gas chromatography coupled to mass spectrometry (GC/MS) analysis of metabolites to determine the effects of this random integration on the metabolism of the plant.
4. It is possible that pieces of plant DNA are taken up by the epithelial cells of the gastrointestinal tract (GIT) of consumers. Hard data on this transfer are scarce, but this is a general phenomenon of digestion and uptake of foods and not just an issue related to genetically modified plant material. However, particularly when antibiotic-resistant markers are used in genetically modified organisms (GMOs), it is necessary to determine the probability of this transfer in order to make a good risk assessment. Of course, it would be much better if antibiotic resistance markers were not present at all.

Some Consumer Benefits of Genetically Modified Microorganisms

As stated in the introduction to this chapter, the popularity of fermented food is based on several aspects, of which the enormous variation in appearance, flavor, and texture is the most important for Western consumers. This enables the consumer to select a product out of this enormous range that is close to personal preference and not an "average" product. When fermented foods are made from plant materials that do not have an optimum composition from a nutritional point of view, such as cassava, fermentation contributes to (partial) supplementation of these components (e.g., amino acids such as methionine, lysine, and tryptophan; vitamins, in particular A and B vitamins; minerals). Fermentation and the accompanying physical processes also contribute to the elimination of undesirable components that are present in plants[2] (see Chapter 4). In general, fermented products also have a better microbiological stability, which is a considerable consumer benefit, especially in developing countries with less well-organized (chilled) distribution chains and often no refrigerator in the homes (see Chapter 2).

On top of the general benefits for the consumer described above, certain microorganisms used in fermentation processes, particularly lactic acid bacteria (LAB), may provide additional health benefits (Exhibit 10–1). Recombinant DNA technology may be able to improve these positive aspects of fermented food products. It is even possible to extend this range of potential health benefits to consumers, for example, by oral immunization against pathogens and/or toxins.[31]

Some Potential Hazards and Risks of Genetically Modified Microorganisms or Their Enzymes Used in the Manufacturing of Fermented Foods

There are several aspects related to the safety of fermented foods. Assuming that extensive research and risk assessment have proved that the newly introduced gene produces an intrinsically safe protein and that this protein does not contribute to the biogenesis of any harmful component, either during fermentation or during digestion of the fermented product, the remaining safety aspects are:

Exhibit 10–1 Potential Consumer Benefits of Fermented Foods

- Improved appearance, flavor, and texture
- Conversion of antinutritional and toxic compounds (e.g., removal of cyanogenic glycosides in cassave)
- Increased microbiological keepability
- Enrichment with amino acids, minerals, and vitamins
- Agonistic action against enteric pathogens*
- Improved lactose utilization (important for developing countries)*
- Conversion of potential precarcinogens into less harmful compounds*
- Stimulation of the mucosal immune system†

*Mainly related to foods fermented with lactic acid bacteria.
†Only a small number of lactic acid bacteria may have this property.

- Can the transfer of newly introduced gene(s) to recipient microorganisms occur in the fermentation process or after consumption of the product in the GIT of consumers?
- Can the transfer of newly introduced gene(s) to recipient cells of the GIT of consumers occur?
- Can the transfer of newly introduced gene(s) to recipient microorganisms in the environment occur?

The risk assessment of the aspects mentioned under the first and third bullet have been described in some detail earlier.[30] As the potential for transfer of genetic information from donor microorganisms to recipient microorganisms is a matter of concern, and new data are available, this will be discussed in some detail using the decision tree provided in Figure 10–7.

Provided that the microorganism used in a fermentation process is intrinsically safe and the recombinant version of this microorganism is substantially equivalent and free of any DNA of the marker free (clean) host organisms, the risk related to the use of these recombinant DNA microorganisms during fermentation processes is zero for both consumer and environment.[30]

When, during the fermentation process, a chemical or physical treatment is applied that lyses the genetically modified microorganism and destroys the functionalities of the genome, it is not necessary to take into account the aspect of potential transfer of genetic material from these microorganisms to recipient cells in the GIT of consumers or safety aspect (third bullet above).

TWO SETS OF EXAMPLES OF SAFETY EVALUATION OF FERMENTED FOODS FOR WHICH RECOMBINANT DNA IS USED

Supply Chain for Soy Sauce Produced from Genetically Modified Soybeans and/or Wheat and/or Using Genetically Modified Microorganisms and/or Enzymes Produced by Genetically Modified Microorganisms

There are a large number of fermented products that are derived from wheat and/or soybean (see Chapter 1). For the supply chain risk/benefits analysis, the well-known soy sauce was used as an example. A general outline of the benefits and risks of genetically modified soybeans was provided on pages 223 and 227, and for microorganisms and enzymes used in the fermentation process, on pages 228. This outline will now be worked out in detail for soy sauce. A simplified scheme of the soy sauce process is depicted in Figure 10–8.

The safety of soy sauce that is derived from genetically modified soybean will be analyzed as a function of the supply chain. As pointed out on page 227, a genetically modified soybean should not have an antibiotic resistance marker because this will increase the risk and is of no benefit to the consumer. For this example, it is assumed that the genetic modification of the soybean results in a faster fermentation process due to the incorporation of enzymes in the soybean and/or wheat that contribute in the first steps of the process. The faster process is assumed to also contribute to the sensory quality of the soy sauce. Therefore, the

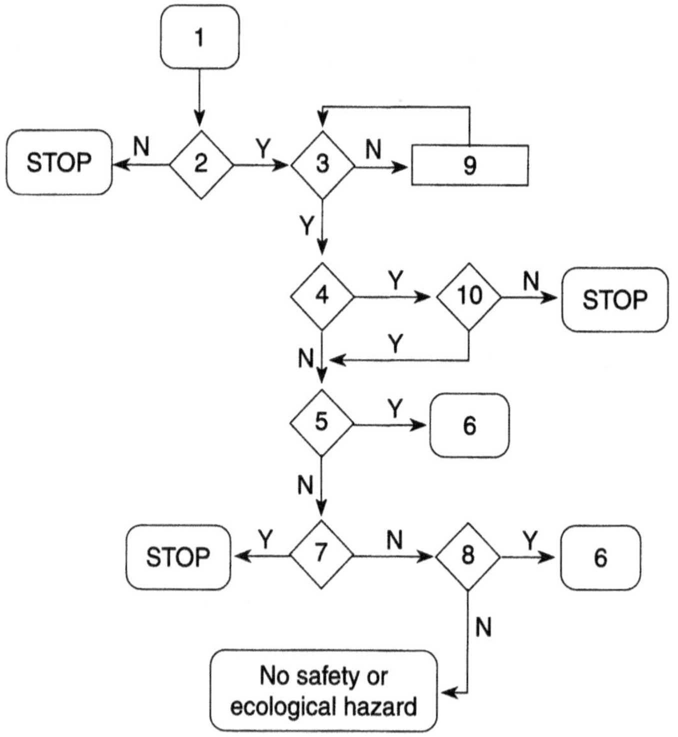

1(E). Generally recognized as safe genetically modified microorganism used in fermentation and still living in the final product
2(Q). Encode the newly introduced gene for an intrinsically safe product?
3(Q). Is the new gene stable integrated in the genome of the host cell?
4(Q). Is an antibiotic resistance marker used during construction of the GMO?
5(Q). Does the newly introduced gene encode for a protein that may result under certain conditions in an ecological advantage for the host cell?
6(A). Carry out full risk assessment.
7(Q). Does the newly introduced gene encode for a protein that after transfer may result under certain conditions in an ecological advantage to recipient mammalian cells?
8(Q). Does the newly introduced gene encode for a protein that after transfer may result under certain conditions in an ecological advantage to recipient microbial cells?
9(A). Integrate new gene, preferably on predetermined locus on the chromosome of the host.
10(Q). Is the antibiotic resistant marker gene eliminated?
(E). Entry or end; Diamonds (Q = questions); Y = yes, N = No

Figure 10–7 General scheme to determine safety to consumers and some environmental consequences of genetically modified microorganisms used in fermentation processes. *Note:* This is a modification of the schemes published in Verrips and Vandenberg.[30]

benefits for the consumer will be better quality and a price reduction.

- **Step 1a (Figure 10–4).** The risk related to the first step of the supply chain (cultivation of the soybeans): Because the newly introduced gene encodes an intrinsically safe protein that does not provide any receiving organism with an ecological advantage, both the environmental and consumer health risks are zero. It should be noted, however, that the environmental hazard is not zero. After studying the frequencies of spontaneous hybridization between oilseed rape (*Brassica napus*) and weedy rape (*B. campestris*) Jorgensen & Anderson[11] concluded that transgenes from oilseed rape may be preserved for many years, and that weedy *B. campestris* with transgenes may present economic (ecological) risks to farmers and plant-breeding companies or biochemical industries. Although their findings of spontaneous hybridization have been confirmed in other studies, unfortunately, they mixed up risk and hazard. Only when the transgene results in a clear ecological advantage for the recipient is there a hazard, and, based on their studies, a realistic probability ($> 10^{-8}$) that the hazard occurs (= risk). Moreover, the transfer of genetic material has happened for millions of years and has not resulted in ecological problems because the transgene did not give the recipient an advantage under a large number of ecological conditions.

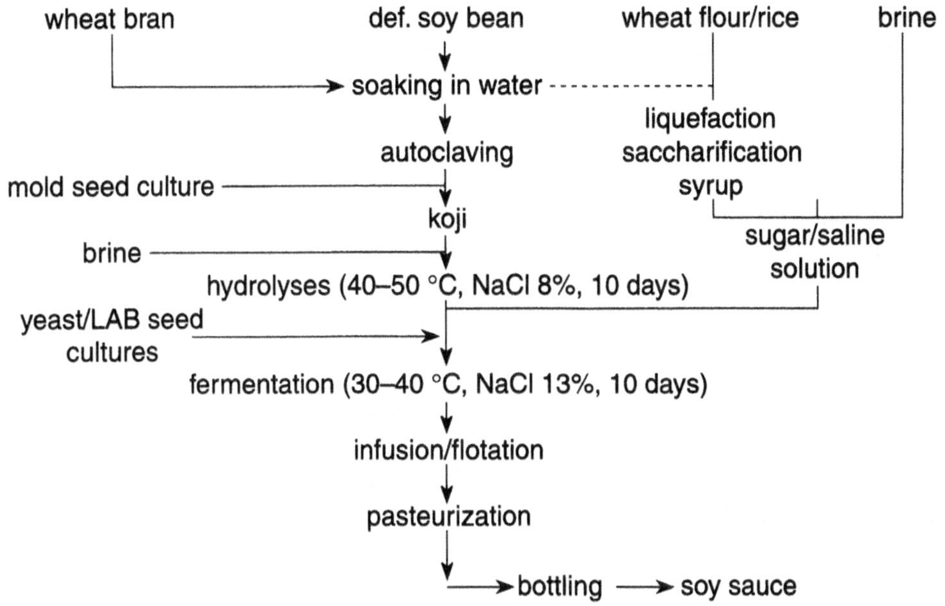

Figure 10–8 Schematic outline of soy sauce process.

It is also important to consider the stability of plant DNA in soil because that DNA can be picked up by microorganisms that are present in the soil. A study showed that "survival" of a neomycin/kanamycin resistance gene from transgenic tobacco was 0.14% after 120 days in soil. Most probably, the persistence is so high because of stabilization of the DNA by soil.[26] Because the stability of DNA in soil is higher than was assumed in the past, the probability that DNA is taken up by soil organisms is also higher. Indeed, various studies have shown that horizontal transfer of plant DNA to microorganisms can occur, but at low frequencies. For instance, based on a number of well-controlled experiments, the transfer of DNA from transgenic potato to the plant pathogenic bacterium *Erwinia chrysanthemi* under natural conditions is calculated to be 10^{-17}, which is much lower than the detection limit of approximately 10^{-12}.[24] Earlier experiments showed that the transfer of hygromycin resistance from various transgenic plants (*B. napus, B. nigra, Datura innoxia, and Vicia narbonensis*) to the fungus *Aspergillus niger* after cocultivation occurred with an unexpectedly high frequency.[10]

- **Step 1b (Figure 10–4).** In cases where the newly introduced gene encodes a protein that can provide an ecological advantage for any recipient organism, the risk is not zero and the risk assessment (Figure 10–7) should be performed before proceeding to step 2.
- **Step 2a (Figure 10–4).** The risk related to the second step of the supply chain (fermentation process): During this process, the soybean is physically denatured; moreover, the chemical composition in certain parts of the process is very hostile for organisms, with the exception of some functional organisms.
- **Step 2b (Figure 10–4).** In the cases that the soy sauce process contains genetically modified microorganisms, it is very likely that the DNA of these microorganisms will also be destroyed during the final steps of the process when the conditions are quite

harsh (but hard evidence is missing). It is again highly recommended to avoid the presence of any nonfunctional foreign genes or genes encoding antibiotic resistance properties in these microorganisms. The hazard of the newly introduced gene product should be analyzed in the same way as described in step 1a for new genes introduced in soybeans and/or wheat.

- **Step 2c (Figure 10–4).** In the case that enzymes produced by genetically modified microorganisms are used in the soy sauce process, the regulations require that the enzyme should be intrinsically safe and free of DNA of the microorganism involved.[29] Practice proves that this requirement can be fulfilled without much difficulty in industry. Even though there is evidence that enzymes can be made free of any DNA, as advocated earlier, it is highly recommended to produce enzymes only with food grade microorganisms that are free from nonfunctional foreign genes, in particular, free from genes encoding antibiotic resistance properties. The enzyme as such should be "substantially equivalent" to the wild type enzyme, another requirement that can be met by industry. Consequently, using recombinant DNA enzymes in the soy sauce process does not pose any hazard and therefore *no risk* to consumer and environment.
- **Steps 3–5 (Figure 10–5).** In all of the subsequent steps of the supply chain, including the step in which consumers introduced the (digested) soy sauce in the environment, the risk related to the use of genetically modified soybean and/or wheat or the use of genetically modified microorganisms or enzymes produced by recombinant DNA technology remains zero.

Supply Chain for Cheeses Produced with Genetically Modified LAB and/or Enzymes Produced by Genetically Modified Microorganisms

As stated in the introduction, fermented products, particularly fermented milk products, have a healthy image. Although a number of studies still have to be carried out to substantiate the potential health benefits of LAB, in particular, the effects of LAB and their products on mucosal health, there are already a large number of publications that support these benefits for consumers.[23] The most important potential health and other benefits are summarized in Exhibit 10–1.

Some of the main aspects of cheese processes are provided in Figure 10–9, and the risk assessment of that process and the other steps in the supply chain will be performed similarly to the assessment for soy sauce.

- **Step 1 (Fig 10–4).** Traditional milk will be used for cheese production; therefore, a risk assessment of GMO-related issues in this step is not necessary.
- **Step 2a (Fig 10–4).** In this step, genetically modified LAB are introduced in the cheese process. As the safety of the end product (cheese) for consumers will be discussed in step 4, only the environmental safety aspects are discussed in this step. It is unrealistic to consider the cheese fermentation process as a completely closed process; therefore, there is a probability that genetically modified LAB will enter the environment. Assuming these LAB do not contain any foreign, nonfunctional genes, and that the newly introduced gene encodes an intrinsically safe protein, the hazard will be zero and consequently the risk will also be zero.

  If, however, the LAB contain a gene-encoding antibiotic resistance or a property providing an ecological advantage to the recipient, the environmental hazard is not zero and a formal risk assessment as discussed in detail in an earlier publication[30] should be performed. More recent data[15,18] support the view that horizontal transfer of genetic material from one microorganism to another, either directly by conjugation or transformation or indirectly via transduction, occurs at a much higher frequency than was assumed in the mid-1980s.
- **Step 2b (Figure 10–4).** In this step, enzymes made by recombinant DNA technol-

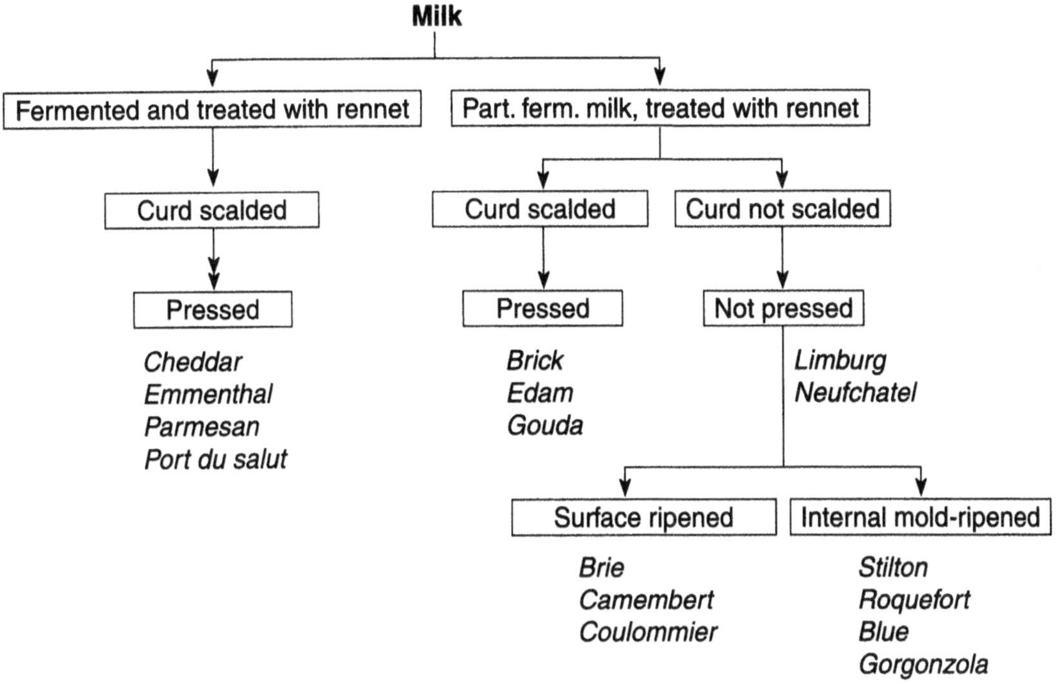

Figure 10–9 Schematic outline of milk fermentation processes and their cheese products.

ogy are introduced.[27] As described in the soy sauce example, an intrinsically safe enzyme produced under proper conditions does not pose a risk to consumer or environment.

- **Step 3 (Figure 10–5).** During distribution, the probability of the escape of LAB into the environment is extremely low, and as horizontal transfer also occurs with low frequencies, this probability is considered to be effectively zero.
- **Step 4 (Figure 10–5).** In this step, the consumer will be in direct contact with the genetically modified LAB and/or the enzyme. For the enzyme, a relatively straightforward procedure has to be followed to prove that the enzyme is intrinsically safe, originates from a well-known source, and is substantially equivalent to the wild type enzyme.[29] If that is the case, the enzyme will not pose a hazard and consequently no risk. For the genetically modified LAB, a formal decision scheme has been developed.[30] Two aspects should be analyzed: the general safety of the fermented food product made with genetically modified LAB and the probability of transfer of genetic information from the LAB to other bacteria in the GIT of consumers (Figure 10–10). In this figure, the following aspects related to a quantitative risk assessment are considered: (1) probability of transfer of genetic material from the GMO to other microorganisms in the GIT as a function of the residence time of the GMO in the GIT, (2) whether the GMO remains alive, (3) whether DNA from lysed GMOs are taken up by normal inhabitants of the GIT, (4) the probability that transfer of genetic information of the GMO to normal inhabitants results in an ecological advantage of the transformed inhabitants and, if so, (5) whether that may pose a health risk to the consumer, or environmental risks. Unfortunately, insufficient data are available to carry out such risk assessment in the proper way.

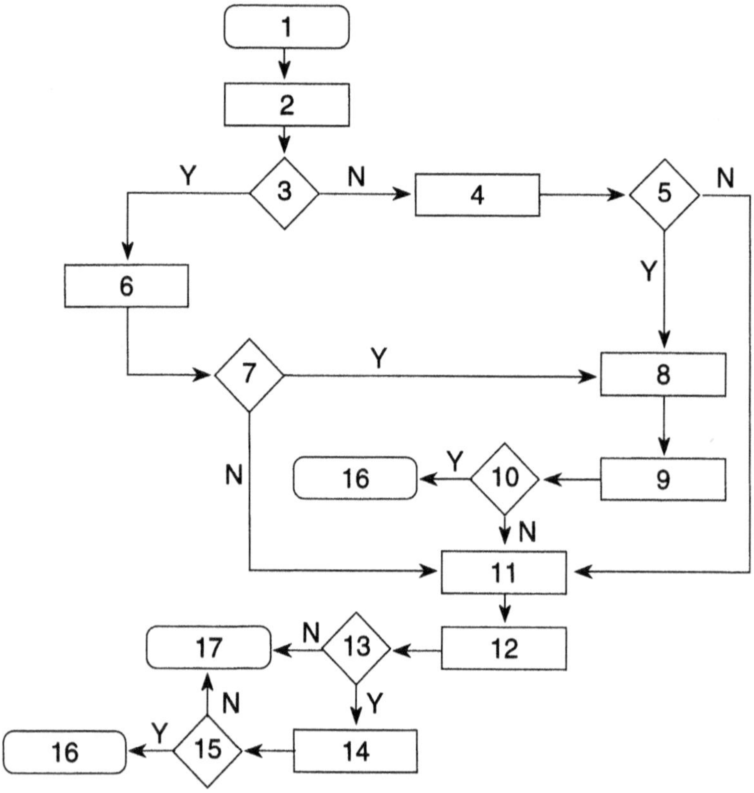

1(E). Genetically modified lactic acid bacterium (GMO)
2(A). Determine the distribution of the residence time of GMO in the gastrointestinal tract (GIT) of consumers. Take the time corresponding with 95% of this distribution curve as *t(r)*.
3(Q). Will the GMO lyse with *P(b)> b* in the GIT within *t(r)*?
4(A). Determine the probability *P(c)* that intact cells of the GMO transfer genetic information to normal inhabitants of the GIT. In these studies, use *t(r)* as contact time and the conditions of the GIT.
5(Q). Is *P(c)>c*?
6(A). Determine the probability *P(f)* that DNA originating from lysed GMO transform normal inhabitants of the GIT (resulting in transformed inhabitants).
7(Q). Is *P(f)>f*?
8(A). Determine whether the transformed inhabitants obtain an advantage over untransformed inhabitants in the GIT: *A(i)*. Define *A(i)* in either faster growth rates *t'(g)*, better adhesion to epithelial cells *h'*, or higher production of certain metabolites *{p(x)', x=1,....}*.
9(A) Determine the probability *P(e)* that any of the events described under *action 8* will result in the formation of a hazardous (=transformed inhabitant produce toxin or that will replace beneficial microorganisms in the GIT) microorganism.
10(Q). Is *{P(c) + P(f)}*P(e) > e*?
11(A). Determine also the probability *P(d)* that the GMO will be transformed by genetic material originating from the common GIT microorganisms (=modified GMO).
12(A). Determine whether the modified GMO gains an advantage over untransformed GMO: *B(i)*. Define *B(i)* in either faster growth rates *t"(g)*, better adhesion to epithelial cells *h"*, or higher production of certain metabolites *{p(x)", x=1,....}*.
13(Q). Is *P(d)> d*?
14(A). Determine the probability *P(g)* that any of the events described under *action 12* will result in the formation of a hazardous GMO.
15(Q). Is *P(d) * P(g) > g*?
16(E). This risk is unacceptable and the GMO should not be released.
17(E). This risk is acceptable and the GMO can be released.
A = action; E = entry or end; Q = question

Figure 10–10 A proposal for a structured assessment of the risk related to the introduction of genetically modified lactic acid bacteria in (or as) food products.[29]

In fact, only fermented products of which the safety has been demonstrated, ecological hazards have been proven to be zero, and consequently, the risk is zero, should be approved. However, assuming that the ecological hazard is not zero, which means that the newly introduced gene confers an ecological benefit on recipient cells, both risk assessments should result in an acceptably low risk. This can only be achieved if the probability of transfer of genetic information from the donor to recipient cells is very low. For horizontal transfer of genetic material between microorganisms, conjugation, transformation, and transduction show a higher probability for plasmid DNA than for chromosomal DNA. Therefore, it is recommended that the foreign DNA be integrated (preferably on a preknown locus) into the chromosome of the microorganism. Integration at preknown loci therefore forms an important component of the risk cube presented in Figure 10–3.

A parameter that is essential for proper risk assessment of events in the human GIT is the residence time of genetically modified microorganisms in the GIT, whether they lyse, whether they conjugate, or whether the DNA liberated during lysis can transform other cells, including human cells. Although the number of reliable data sets have increased during the last decade, they are still not sufficient to solve all of the questions of the decision scheme given in Figure 10–10. Nevertheless, the data available indicate that LAB can survive passage through the GIT.[19] The functional microorganism in many cheese fermentations, *Lactococcus lactis,* survives up to three days, although the survival rate is only 1–2%. Another noteworthy observation in this study was that the PCR method could detect special DNA stretches of this bacterium for up to four days after ingestion,[12] indicating that either the feces contained nonviable *L. lactis* or that these stretches of DNA were liberated during lysis of the bacterium. Experiments with phage M13 DNA ingested by mice showed that a small but measurable percentage of this DNA reached peripheral leukocytes, spleen, and liver via the intestinal wall mucosa and could be covalently linked to mouse DNA.[25] Even taking into account that these experiments are rather artificial, for a proper risk assessment, they have to be included.

From these studies, it can be concluded that there is a probability that genetically modified microorganisms, including LAB, can transfer genetic information to other microorganisms in the GIT. This information is another strong argument to ban genetically modified microorganisms that contain an antibiotic resistance gene. In cases where a nongenetically modified microorganism has such a gene, it should be deleted because further unnecessary spread of antibiotic-resistant genes creates very serious health and environmental problems.

- **Step 5 (Figure 10–5).** The release of feces of consumers into the environment is the final step to be assessed. If the newly introduced gene does not encode for a protein that provides an ecological advantage for the recipient cells, then a risk assessment of this step is not necessary. If this is not the case, then the decision scheme outlined in an earlier publication[30] can be applied for this step. However, this step may become more complex because it should take into account whether the transfer of genetic information from the LAB to recipient microorganisms has occurred in the GIT of consumers and, if so, these modified recipients should be evaluated as well.

In this example, LAB are the functional microorganisms in the fermented product, but for other organisms that are generally recognized as safe (GRAS), such as *Saccharomyces cerevisiae, Kluyveromyces lactis, Penicillium roqueforti, P. camenberti,* and *A. niger,* or *A. awamori,* the same hazard and risk assessment can be applied with similar results.

CONCLUSION

Fermented foods are considered by consumers as natural and healthy.[7,8,14,22] The introduction of genetically modified plants, microorganisms, or enzymes into these products needs careful discussion with authorities and consumer organizations. Many fermented foods contain living microorganisms, and if these organisms are genetically modified, one should consider them as a potential source of DNA for horizontal gene transfer because a number of studies provide circumstantial evidence that, in rare cases, genes have been laterally transmitted among *Eubacteria, Archaea,* and *Eukarya.* However, extensive studies by Puhler's[6] group showed that this transfer occurs at extremely low frequency.

In these discussions, the proven benefits to the consumer should be explained in clear terms. The risk assessment should show that the safety hazard related to the newly introduced gene(s) is zero, and consequently, the risk for consumer is zero. Whenever possible, without losing the consumer benefit, the environmental hazard should also be zero. If that is not possible, the risk assessment should show that the risk is extremely low, preferably less than 10^{-8} per product unit produced. Consumer studies show that provided there is a clear benefit and risk is absent, most consumers will accept these products. To gain the confidence of the consumer for fermented foods in which recombinant DNA technology plays a role, clear and transparent labeling and public relations will be required.

REFERENCES

1. Barra, D. & Simmaco, M. (1995). Amphibian skin: a promising resource for antimicrobial peptides. *Tib Tech* 13, 205–209.
2. Beuchat, L. R. (1983). Indigenous fermented foods. In *Biotechnology.* Vol. 5, pp. 477–528. Edited by H. -J. & G. Reed. Weinheim: Verlag Chemie.
3. Casteels, P. & Ampe, C. (1989). Apidaecins: antibacterial peptides from honeybees. *Embo* 8, 2387–2391.
4. Chervitz, S. A., Aravind, L., Sherlock, G., Ball, C. A., Koonin, E. V., Dwight, S. S., Harris, M. A., Dolinski, K., Mohr, S., Smith, T., Weng, S., Cherry, J. M. & Botstein, D. (1998). Comparison of the complete protein sets of worm and yeast: orthology and divergence. *Science* 282, 2022–2033.
5. Cornelissen, B. J. C., Hooft van Huijsduijnen, R. A. M. & Bol, J. F. (1986). A tobacco mosaic virus-induced tobacco protein is homologous to the sweet-tasting protein thaumatin. *Nature* 321, 531–532.
6. Droge, M., Puhler, A. & Selbitschka, W. (1998). Horizontal gene transfer as biosafety issue: a natural phenomenon of public concern. *J Biotech* 64, 75–90.
7. Gasser, F. (1994). Safety of lactic acid bacteria and their occurrence in human clinical infections. *Bull Inst Pasteur* 92, 45–667.
8. Gilliland, S. E. (1990). Health and nutrition benefits for lactic acid bacteria. *FEMS Microbiol Rev* 87, 175–188.
9. Hamstra, A. M. (1993). Consumer acceptance of food biotechnology. *SWOKA Research Report* 137, The Hague, The Netherlands.
10. Hoffmann, T., Golz, C. & Schieder , O. (1994). Foreign DNA sequences are received by a wild-type strain of *Aspergillus niger* after co-culture with transgenic higher plants. *Curr Genet* 27, 70–76.
11. Jorgensen, R. B. & Andersen, B. (1994). Spontaneous hybridisation between oil seed rape (*Brassica napus*) and weedy *B. campestris*: a risk of growing genetically modified oil seed rape. *Am J Botany* 81, 1620–1626.
12. Klijn, N., Weerkamp, A. H. & de Vos, W. M. (1995). Biosafety assessment of the application of genetically modified *Lactococcus lactis* spp. in the production of fermented milk products. *System Appl Microbiol* 18, 486–492.
13. Koonin, E. V., Tatusov, R. L. & Galperin, M. Y. (1998). Beyond complete genomes: from sequence to structure and function. *Current Opinion in Structural Biology* 8, 355–363.
14. Lee, Y. -K. & Salminen, S. (1995). The coming of age of probiotics. *Trends Foods Sci Technol* 6, 241–245.
15. Lorenz, M. G. & Wackernagel, W. (1994). Bacterial gene transfer by natural gene transformation in the environment. *Microbiol Rev* 58, 563–602.
16. Mann, C. C. (1999). Crop scientists seek a new revolution. *Science* 283, 310–314.
17. Mann, C. C. (1999). Genetic engineers aim to soup up crop photosynthesis. *Science* 283, 314–316.
18. Miller, R. V. (1998, January). Bacterial gene swapping in nature. *Sci Am* 46–51.

19. Pedrosa, M. C., Golner, B. B., Goldin, B. R., Barakat, S., Dallal, G. & Russel, R. (1995). Survival of yoghurt-containing organisms and *Lactobacillus gasseri* (ADH) and their effect on bacterial enzyme activity in the gastrointestinal tract of healthy and hypochlorhydric elderly subjects. *Am J Clin Nutr* 61, 353–358.

20. Ronald, P. C. (1997, November). Making rice disease resistant. *Sci Am* 68–73.

21. Rowland, I., Wiseman, H., Sanders, T., Adlercreuz, H. & Bowey, E. (1999). Metabolism of oestrogens and phytooestrogens: role of the gut flora. *Biochem Soc Trans* 27, 304–308.

22. Salminen, S., Bouley, C., Bouron-Ruault, M. -C., Cummings, J. H., Franck, A., Gibson, G. R., Isolauri, E., Moreau, M. -C., Roberfroid, M. & Rowland, I. (1998). Functional food science and gastrointestinal physiology and function. *Br J Nutr* 80, S147–S171.

23. Salminen, S., Isolauri, E. & Salminen, E. (1996). Clinical uses of probiotics for stabilizing the gut mucosal barrier: successful strains and future challenges. *Ant van Leeuwenhoek* 70, 347–358.

24. Schlüter, K., Futtere, J. & Potrykus, I. (1995). Horizontal gene transfer from a transgenic potato line to a bacterial pathogen (*Erwinia chrysanthemi*) occurs—if at all—at extremely low frequency. *Biotechnol* 13, 1094–1098.

25. Schubbert, R., Renz, D., Schmitz, B. & Doerfler, W. (1997). Foreign (M13) DNA ingested by mice reaches peripheral leukocytes, spleen, and liver via the intestinal wall mucosa and can be covalently linked to mouse DNA. *Proc Nat Acad Sci USA* 94, 961–966.

26. Stiles, M. E. (1996). Biopreservation by lactic acid bacteria. *Ant van Leeuwenhoek* 70, 331–345.

27. Teuber, M. (1990). Production and use of chymosin from genetically altered microorganisms. *Lebensmittel Ind Milchwirtschaft* 35, 1118–1123.

28. Verrips, C. T. (1989). Growth of microorganisms in compartmentalized products. In *Mechanisms of Action of Food Preservation Procedures,* pp. 363–393. Edited by G. W. Gould. London: Elsevier Applied Science.

29. Verrips, C. T. (1995). Structured risk assessment of rDNA products and consumer acceptance of these products. In *Biotechnology.* Vol. 12, pp. 157–196. Edited by H. -J. Rehm & G. Reed. Weinheim, Denmark: VCH Verlaggesellschaft mbH.

30. Verrips, C. T. & Van den Berg, D. J. C. (1996). Barriers to application of genetically modified lactic acid bacteria. *Ant van Leeuwenhoek* 70, 299–316.

31. Wells, J. M., Robinson, K., Chamberlain, L. M., Schofield, K. M. & LePage, R. W. F. (1996). Lactic acid bacteria as vaccine delivery vehicles. *Ant van Leeuwenhoek* 70, 317–330.

CHAPTER 11

Safety Assessment of Probiotics and Starters

Seppo J. Salminen, Atte J. von Wright, Arthur C. Ouwehand, and Wilhelm H. Hozapfel

SAFETY ASPECTS FOR PROBIOTICS AND BIOTECHNOLOGY

Today, milk can be fermented into a wide range of different products with different flavors, consistencies, and structures. Non-dairy fermented products such as fermented vegetables and cereals are increasingly popular in different parts of the world, and they put new demands on starter cultures. Additionally, the focus on probiotics has increased the demand on starter culture properties. This poses great demands on the starter cultures and their characterization, including their activity, stability, and resistance to bacteriophages. Specific species and strains of starter cultures are used and selected for novel applications. Thus, it is also important to consider the safety aspects of new and especially novel starter cultures. Thus far, the selection of general starter cultures has focused on improved technological properties and more effective natural strains, which are easily applied for production technology, industrial processes, and starter culture storage.

The selection of current and more effective probiotic strains has focused on natural strains with specific target properties such as adhesion to different target mucosal models, immunogenic properties, and the production of antimicrobial substances or competitive exclusion of pathogens. Fighting pathogens with other microbes has been recommended earlier.[66] It is likely that in the future, more advanced biotechnological selection and modification systems will be applied. Thus, it is important to define the safety assessment standards and assess the differences between natural and genetically modified strains of starters. Genetically modified strains of lactobacilli or bifidobacteria have not been introduced into the field of probiotics or starters yet. However, it is likely that they will appear in the future as a result of work presently being conducted. In this respect, the European Union has a suggested procedure and decision tree for the safety assessment, as indicated by Jonas and coworkers.[42] This procedure should be followed (see Chapter 10).

Probiotic products provide a variety of documented health benefits to all consumers.[4] There are no apparent safety concerns for immunocompetent subjects. However, today, one has to consider the growing number of immunodeficient subjects. Specific probiotics used in foods have been shown to block the adhesion of pathogenic bacteria to human intestinal mucosa and to prevent enteroinvasive pathogens from invading the mucosal cells by competitive exclusion, antimicrobial production, or cell aggregation.[4,11] Other studies have provided good evidence of stabilizing and normalizing the gut mucosal barrier, thus promoting the health of the host from infants to the elderly.[4]

In the case of probiotic foods, the application of probiotics is similar to any fermented product containing organisms that are already present in the human gastrointestinal tract (GIT). Thus, no specific concerns appear in relation to the consumption of any fermented foods or foods containing viable bacterial cultures, such as probiotics. However, as disease-specific applications increase, it may be necessary to further evaluate the safety aspects of novel probiotics. In addition, high concentrations of microorganisms that are well adapted to the human GIT have probably not been deliberately used in fermented foods before.

The use of starter cultures in milk products is based on the development of lactic acid and other acids to decrease the pH resulting in microbiologically stable and safe fermented products. These principles have also been applied to meat, vegetable, and cereal products. In broad terms, starter cultures enhance the safety of these products, but in the future, selection of novel strains will need safety assessment mechanisms when new species and even new genera may be used.

Safety Assessment of Probiotics and Starters

Four major approaches may be used to assess the safety of probiotic and starter strains (Exhibit 11–1). Several groups such as the European Union Probiotic Demonstration Working Group have suggested safety assessment procedures.[2,26,67,68] In general, the safety assessment of microbial food supplements is not well developed, and often a case-by-case approach has to be taken. A review on past safety studies has been presented by Donohue *et al.*[19]

Exhibit 11–1 Approaches for Assessing the Safety of Probiotic and Starter Strains

1. Characterization of the genus, species, and strain and its origin will provide an initial indication of the presumed safety in relation to known probiotic and starter strains
2. Studies on the intrinsic properties of each specific strain and its potential virulence factors
3. Studies on adherence, invasion potential, and the pharmacokinetics of the strain
4. Studies into interactions between the strain, intestinal and mucosal microflora, and the host

Suggested Virulence Factors and Other Harmful Properties

Several potential virulence factors have been proposed for lactic acid bacteria (LAB) (Figure 11–1). Many of these are known virulence factors for "real" pathogens; their importance for generally non-pathogenic organisms such as lactobacilli is uncertain. No significant risk factors or virulence factors have so far been identified for LAB or bifidobacteria used as starters or probiotics.

Strong adhesion has been suggested as a virulence factor because it may facilitate translocation and platelet aggregation, and thus may be connected with endocarditis.[26,31,55] In a study comparing clinical *Lactobacillus* isolates from bacteremia patients with current probiotic and dairy strains, no indication that clinical bacteremia strains would have stronger adhesion than common probiotics was found.[44] Clinical bacteremia isolates also showed no general ability to aggregate platelets; in fact, most of them were nonaggregating strains. Also, the platelet-aggregating strains showed relatively low adhesion ability in intestinal mucosal models. Thus, the relation of platelet aggregation and adhesion may be involved in some cases, but it does not appear to be a strict virulence factor for LAB. However, when more potent specific probiotic microbes are selected or modified from current probiotics, it may be important to assess both platelet aggregation properties and adhesion properties in more than one model. The adhesion models may include the Caco-2 cell model simulating the intestinal epithelium and the fecal or intestinal mucus model to simulate the mucosal barrier.[43,69,76,77]

The ability to adhere to extracellular matrix (ECM) components such as fibronectin, collagen, laminin, and so forth, has been suggested to be of importance for pathogenic microorganisms. ECM components form the major constituents of wounded tissues and can thus serve as receptors for invading microorganisms. Interaction with these components is therefore consid-

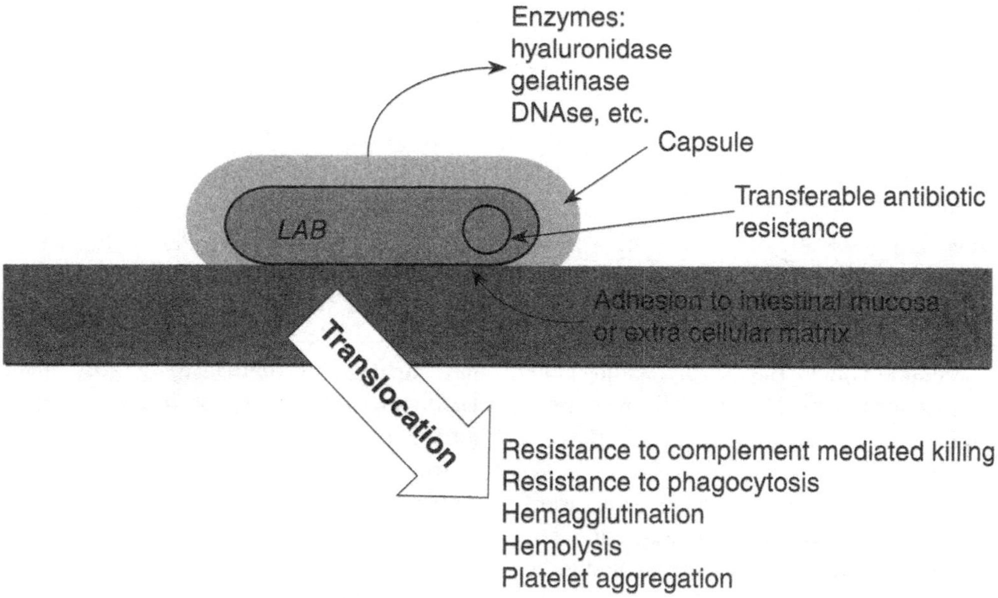

Figure 11–1 Potential virulence factors suggested for lactic acid bacteria

ered to play a role in the earliest steps of bacterial pathogenesis.[32] Such interaction has also been suggested to promote passage to the circulation.[31] For probiotics, it has also been suggested to be a beneficial property because the probiotics may colonize the damaged tissue and prevent subsequent colonization by invasive pathogens.[75] In fact, the healing of damaged gastric mucosa has been observed to be enhanced after colonization by lactobacilli.[22] Hemolysis and hemagglutination have been suggested as potential virulence factors; hemolysis may lead to anemia and edema. The ability to cause hemolysis, through the production of α-hemolysin, has been found to be general among *Lactobacillus* isolates from food, clinical, and fecal origin,[5,36] and may therefore not be a relevant property for potential virulence. Hemagglutination, on the other hand, was found to be rare.[5]

The ability to form capsules has been suggested to play a role in pathogenesis because it may protect from phagocytosis.[72] This property was found to be rare among *L. rhamnosus* strains from clinical, food, and fecal origin.[5]

Translocation, the passage of viable bacteria from the intestine to other sites in the body, is likely to be a normal event because viable intestinal organisms can be recovered from the mesenteric lymph nodes of a small percentage of healthy animals. LAB may also enter the circulation through other routes such as dental manipulation, complications from delivery in postpartum women, and patients with gastrointestinal complications.[36] In the blood, bacteria will come in contact with the complement system that has bacteriolytic activity. Resistance to complement-mediated killing may enhance the survival in the blood and thus increase virulence.[59] Likewise, resistance to phagocytosis may contribute to increased virulence.

Regarding the enterococci, several unresolved issues deserve special attention. Presently, it may be concluded that their virulence is still not well understood, but adhesins, hemolysin, hyaluronidase, aggregation substance, gelatinase, and their ability to transfer antibiotic resistance may be considered as putative virulence factors.[24]

SAFETY OF CURRENT PROBIOTICS

The Role of the Host

Current probiotics have mostly been derived from studies on the genera and species in the healthy human GIT. Taxonomy forms the first basis for safety assessment. Lactobacilli and bifidobacteria have a natural association with the infant and human mucosal surfaces of the mouth, GIT, and genitourinary tract. This has been considered as an indicator that they are safe, common commensals of the healthy normal human microflora. Data available on current probiotics and starters attest to their safety in food use.

Lactobacillus species are rarely isolated from bacteremia cases. The incidence of *Lactobacillus*-caused bacteremia has been estimated to be 0.1–0.3%.[26,70] Surveillance studies have not reported any current probiotic or starter strains to be involved in septicemia.[70] However, two case studies have been reported where probiotic organisms may have been involved in the etiology of disease—a liver abscess caused by *L. rhamnosus* indistinguishable from *Lactobacillus* GG[60] and a case of *Lactobacillus* endocarditis possibly caused by probiotic *L. rhamnosus*.[50] The most likely source of the infection is often considered to be the intestine. Subjects with *Lactobacillus* bacteremia usually have severe underlying diseases, predisposing them to infections. Factors considered to be predisposing to infections of intestinal origin are diabetic state, poor nutritional status, compromised immune system (e.g., due to AIDS or transplantation), traumatic injury, and peritoneal inflammation. Dental manipulation has also been suggested to be a potential risk factor.[36,46,82]

In immunocompromised hosts, current probiotics offer protective mechanisms including immune stimulation and colonization resistance to pathogens, which may give incentives for their use to protect immunodeficient patients from pathogenic microbes. Studies with congenitally immunodeficient mice, however, showed an increased mortality in newborn pups to mothers that had been mono-associated with *Lactobacillus* GG and *L. reuteri*.[80] This indicates that immunocompromised neonates may be a subpopulation deserving special attention, not only for possible infection by LAB, but also by other microorganisms.

The presence of predisposing factors suggests that not only are bacterial properties important for the development of *Lactobacillus* infection, but also that the host properties play a significant role.

It is important to first verify potential virulence factors for probiotics in the immunodeficient host. Biotechnology may offer advantages here for the development of disease-specific health microbes for the treatment of specific target populations.[81]

Antibiotic resistance may be a property of some LAB, but it is generally not transferable, as indicated in studies with one probiotic.[78] However, the risk of transfer of antibiotic resistance may be a significant problem for enterococci, other novel probiotics, and starter strains if the transfer potential is not included in the selection of novel strains (see Chapter 10).

Safety Assessment Procedures

In a European Demonstration project on probiotic properties, the safety assessment of probiotic microbes has been a special target. Recommendations for safety assessment procedures have been made by the group, outlining the targets and steps for safety assessment of both natural and novel probiotics[19,56,68] (Exhibit 11–2).

RELEVANCE OF A SOUND TAXONOMIC BASIS

Strong competition in the probiotic market, on the one hand, and unresolved questions around the regulatory situation on the other, indicate the complexity of the present situation with regard to newly selected functional strains of organisms. Species level identification alone does not provide sufficient transparency required by industry, responsible scientists, regulatory bodies, and also the modern consumer. It is a well-established fact that the species and even genus designation may provide a strong indication of typical habitats and possible origin of

Exhibit 11–2 Suggested Steps for the Assessment of the Safety of Probiotic and Starter Strains[67,68]

1. Strains with a long history of safe use will be safe as starters or probiotic strains.
2. Strains that belong to species for which no pathogenic strains are known but which do not have a history of safe use may be safe as probiotic or starter strains, but should be treated as novel foods.
3. Strains that belong to species for which pathogenic strains are known should be treated as novel foods.
4. Characterization of the intrinsic properties of new strains. Strains carrying transferable antibiotic resistance genes should not be marketed.
5. Strains that are not properly taxonomically described (DNA-DNA hybridization, rRNA sequence determination) should not be marketed.
6. Assessment of the acute and subacute toxicity of ingestion of extremely large amounts of the probiotic and starter strains should be carried out.
7. Estimation of the infective properties *in vitro*, using human cell lines and intestinal mucus, and in animal models, using sublethally irradiated or immunocompromised animals.
8. Assessment of the effects of the strains on the composition and activity of the human intestinal microflora
9. Epidemiological and postmarketing surveillance

an organism (Table 11–1). In addition, the generally accepted safety and technical applicability of a species and especially of "new" strains may strongly be indicated by the species or even genus to which it belongs (Table 11–2). Particularly, strains selected for their specific functional properties have to be clearly characterized below the species level. This is in the interest of the manufacturer, after considerable investment in screening and selection procedures and in clinical studies, and also for postmarketing surveillance. Molecular fingerprinting methods may be applied as reliable and highly discriminatory tools in order to meet these challenges. In this way, requirements of legislative bodies and governmental control organs for exact information on physiological features and nomenclature of "new" strains may be satisfied.

The long tradition of "safe and acceptable" application of traditional starter cultures (e.g., for fermented milk products) does not exist for most probiotic strains on the present-day market. The longest history of proven health benefits and "safe use" is probably exemplified by the *L. paracasei* strain "Shirota" and some strains of the so-called *L. acidophilus* "group." Presently, strains of these species, that is, the *L. acidophilus* group (thereby including the "related" species such as *L. crispatus* and *L. johnsonii*) (see Table 11–1), the *L. casei/paracasei* "group," and also some *Bifidobacterium* spp., predominate in probiotic yogurts.[34,45,71] Most of these strains originate from the human GIT and have confirmed much of the pioneering work of Reuter and coworkers in the 1960s[48,61–63] on lactobacilli autochthonous to the human GIT. The "biotypes" suggested by Lerche & Reuter[48] for differentiating within the *L. acidophilus* group have largely been supported by the DNA homology groups reported by other workers in 1980.[41,47] This finally resulted in six phylogenetically related species presently considered to belong to this "group." Although independent species, they cannot be readily distinguished by phenotypic features.[34,51,71] Also in confirmation of earlier work by Reuter *et al.*, representatives of two further homofermentative groups, *L. salivarius* and the *L. casei* group, involving strains of *L. paracasei* and *L. rhamnosus,* are presently used in probiotic products. Among the heterofermentative lactobacilli also shown by Reuter's group to be part of the normal microbial population of the human GIT, especially strains presently classified as *L. reuteri,* show potential as both human and animal probiotics.[84]

As with the *L. acidophilus* group, confusion also exists concerning the correct nomenclature of the *L. casei* group. Following the suggestion of Collins *et al.*,[13] most *L. casei* strains were transferred to the new species *L. paracasei,* while the former *L. casei* ssp. *rhamnosus* was

Table 11–1 Typical Properties of Species of the So-called "Acidophilus Group" (modified acc. to ref. 35, 51 and Reuter (1997): personal communication).

Species	*Habitat**	*mol % G + C in the DNA*	*"Biotypes" acc. to*	*DNA Homology groups acc. to*	
Reference			*(48)*	*(47)*	*(41)*
L. acidophilus	Most ?	32–37	I,/II	I a	A-1
L. amylovorus	P/C	40	IV (III)	I b	A-3
L. crispatus	H/G	35–38	III	I c	A-2
L. gallinarum	G	33–36	—	I d	A-4
L. gasseri	H/C	33–35	I	II a	B-1
L. johnsonii	H/P/G	32–38	I, II	II b	B-2

*H = humans; P = pigs; C = cattle; G = poultry.

elevated to species status. Dicks *et al.*,[18] however, suggested the reclassification of *L. casei* ATCC 393 (one of the few remaining strains in this species) and *L. rhamnosus* ATCC 15820 as *L. zeae*, and the designation of strain ATCC 334 as the neotype strain of *L. casei*, simultaneously with the rejection of the name *L. paracasei*. Subsequently, this suggestion was strongly supported by phylogenetic data from Mori *et al.*[53] and phenotypic data from Klein *et al.*[45] Yet, because of nomenclature rules, the full suggestion of Dicks *et al.*[18] has not yet been accepted by the judicial commission. Therefore, practically all strains presently designated as *L. casei* are officially in fact representatives of *L. paracasei*. This has been shown by a recent investigation on strains from novel-type yogurts on the market.[71]

Functional properties and safety of particular strains of *L. casei, L. rhamnosus, L. acidophilus, L. crispatus, L. gasseri,* and *L. johnsonii* are in-

Table 11–2 General Classification of Starter Cultures in Potential Risk Groups

"Safe"	*Doubtful*	*Risk Group*
Bifidobacterium	*Bif. dentium*	*Peptostreptococcus* spp.
Carnobacterium (most spp.)	*Carnobacterium piscicola* (?)*	*Streptococcus* spp. (most)
Enterococcus casseliflavus	*Enterococus* spp. (most)†	
Ent. malodoratus	*Lb. catenaforme*	
Ent. mundtii	*Lb. rhamnosus*‡	
Ent. saccharolyticus	*Lc. garvieae*	
Lactobacillus (most spp.)	*Vagococcus*	
Lactococcus		
Leuconostoc		
Oenococcus		
Pediococcus		
Streptococcus thermophilus		
Weissella		

*Some strains are pathogenic to fish.
†Some strains of *Ent. faecium* and *Ent. faecalis* with a history of safe use are known.
‡Several strains of *Lb. rhamnosus* with a "safe history" are known.

creasingly being studied. Species such as *L. rhamnosus, Enterococcus faecium,* and *E. faecalis* are presently classified as "risk group 2" organisms (i.e., potential pathogens) in Germany.[7] Some strains, however, find application in probiotic products, and are considered not to constitute any health risk because of their history of safe use (Table 11–2). Differentiation of these strains from those of, for example, clinical, environmental, or animal origin, poses a special challenge that may be solved by modern molecular typing techniques. These genotyping techniques are valuable tools for the identification of LAB, either on species level or for differentiation of strains to the clonal level.[35] Advantages include their universal applicability, whereas the high discriminatory power of particular techniques such as pulsed-field gel electrophoresis (PFGE) may enable differentiation among closely related strains of a similar phenotype. Depending on the particular situation, molecular methods such as restriction enzyme analysis (REA), PFGE, randomly amplified polymorphic DNA (RAPD-PCR), and ribotyping may be applied on chromosomal DNA for the typing of probiotic strains. Plasmid profiling, by contrast, is not considered suitable for the typing of individual strains due to the instability of extrachromosomal DNA.

REA may be applied for differentiation among strains of *L. acidophilus*,[64] *L. casei/L. rhamnosus*,[3] and *L. reuteri*.[3,73] As for other methods relying on the separation of DNA fragments by electrophoresis, the complexity of the banding patterns necessitates the use of computer-aided multivariate analysis.[9] In addition, the selection of the appropriate restriction enzyme(s) is essential in order to obtain revealing discriminatory patterns. PFGE has been used successfully to differentiate strains of important probiotic bacteria such as bifidobacteria,[65] *L. casei*,[23] and *L. acidophilus*.[64] Enabling differentiation among different clones of a particular species, PFGE was shown, for example, to distinguish among individual strains of *E. faecalis* on the basis of 25 patterns obtained, as compared to only 7 patterns obtained with ribotyping.[28] PFGE may therefore enable intraspecies differentiation between probiotic and clinical strains of enterococci.[57,58] Ribotyping, on the other hand, is mainly applicable on species level, and reveals fingerprinting patterns that are more stable and more easily interpretable than those obtained by REA.[9] In addition, this technique is highly reproducible and allows the possible use of a universal probe for all species because of the similarity of conserved regions of ribosomal genes.[29,58] Whereas it has been found appropriate for studying the diversity of strains of *L. reuteri* and *L. fermentum* from the mouse ileum,[54] ribotyping may also be applied to characterize strains of different *Lactobacillus* species,[54,64] particularly of the *L. acidophilus* group.[64] In a similar fashion, soluble cell protein fingerprintings revealed by polyacrylamide gel-electrophoresis also enable species-level distinction within the *L. acidophilus* group.[45,58]

The RAPD-PCR technique often reveals poorly reproducible patterns and, similar to PFGE, it needs to be performed under carefully controlled conditions. This technique has, however, been used for species-level distinction among strains of *Bifidobacterium*[65] and of the *L. acidophilus* group.[21] Using three different arbitrary primers, RAPD-PCR produced revealing information concerning the genetic diversity among dairy lactococcal strains.[15] Applying a combination of two decamer primers in a single PCR reaction, a multiplex RAPD-PCR technique enabled differentiation of *Lactobacillus* strains from the GIT of mice.[15] Oligonucleotide probes complementary to rRNA gene targets may be used for *in situ* detection of strains in mixed populations of potentially probiotic lactobacilli,[6,30,33,58] enterococci,[6] and bifidobacteria.[25]

Amplified fragment length polymorphism (AFLP), another more recent fingerprinting technique, uses the selective amplification of restriction fragments from a digest of total genomic DNA. It is a powerful discriminatory fingerprinting technique, and, with the exception of *Listeria monocytogenes*,[1] it has thus far been applied mainly for species-level differentiation of Gram-negative species such as *Xanthomonas* spp.,[39] *Aeromonas* spp.,[38,39] *Acinetobacter* spp.,[40] *Pseudomonas* spp.,[27] and *Campylobacter* strains[20] from poultry. It also provided a sound basis for classification of *Alcaligenes faecalis*-like isolates as a new species, *Ralstonia gilardii*.[12]

TRANSFER OF ANTIBIOTIC RESISTANCE

The roles of different physiological gene transfer mechanisms—transduction, conjugation, and transformation—in *Lactobacillus*, *Lactococcus*, *Pediococcus*, and *Streptococcus thermophilus* have been reviewed recently.[79] In transduction, DNA is conveyed from one host to another by the mediation of an infecting bacteriophage; in conjugation, the DNA transfer occurs by direct cell-to-cell contact; and in transformation, the DNA is conveyed through direct uptake of free soluble DNA by the organism. In principle, transduction is limited by the phage host range and is thus confined within bacterial strains that are very closely related to each other. In conjugation and transformation, the DNA exchange can occur across species and even genera. Although both transduction and conjugation seem to be relatively common phenomena among LAB, there has been little evidence of physiological transformation in these genera and species. Conjugation, often accompanied by the production of pheromone type of proteins, is also common among enterococci.[10]

With the notable exception of the enterococci, antibiotic resistance plasmids and transposable genetic elements are relatively rare among LAB,[79] and apparently no antibiotic transferable resistance determinants have, so far, been detected among the bifidobacteria.[49] Among the lactobacilli, certain strains of *L. fermentum, L.acidophilus, L. reuteri,* and *L. plantarum* have been shown to harbor resistance plasmids against erythromycin (and related macrolides), tetracycline, and chloramphenicol.[68] Thus, the possibility of antibiotic resistance spreading from the LAB common in fermented foods seems to be remote. However, the prevalence of antibiotic resistance in enterococcal strains used or prevalent in many indigenous fermented foods needs to be evaluated.

From a safety point of view, the crucial question is whether antibiotic resistance determinants conveying resistance to clinically important drugs could be transferred in the intestine from harmless commensals to actual or opportunistic pathogens. Although experimental evidence from real-life situations either in humans or in conventionally reared animals is difficult to produce, some data obtained by using gnotobiotic animals indicate the presumable conjugational *in vivo* transfer of antibiotic resistance markers. For example, the transfer of an enterococcal wide host-range plasmid pAMß1 has been observed between different genera of LAB in the gut of germ-free mice.[14,52] Thus, the spread of antibiotic resistance among the intestinal microflora remains a possibility, and care should be taken in choosing strains, either for conventional fermented foods or for probiotic purposes, to avoid strains carrying transferable drug resistance determinants and strains that have the ability to facilitate plasmid transfer.

When evaluating the antibiotic resistance profiles among different species and strains, it is crucial to differentiate between intrinsic resistance and resistance that is mediated by special genetic elements. Vancomycin resistance is a particularly illustrative example. Vancomycin is one of the last resort antibiotics against multiresistant staphylococci. Therefore, transferable vancomycin resistance among the enterococcal strains is a serious clinical problem, both because of the increasing incidence of enterococcal nosocomial infections and also in view of the possible transfer of the resistance to staphylococcal strains.[4] By contrast, the intrinsic resistance in several genera and species of LAB (i.e., *L. rhamnosus*) due to the special structure of the cell wall does not pose any risk.[8,30,78] These bacteria are not pathogens, they do not transfer the resistance to other species or strains, and in the rare occasions where they might be involved in opportunistic infections, there are plenty of effective alternative antibiotics to use.

GENETICALLY MODIFIED ORGANISMS

Questions concerning the safety of genetically modified organisms (GMOs) have increasing relevance for LAB because recombinant DNA techniques now find general application to many of these species and strains.[79] Although no genetically modified probiotics or starters are yet on the market, extensive research is concentrated on the development of LAB-based oral vaccines.[74,83] This application is very close to the probiotic concept and should it become legally acceptable and gain the confidence of the public, it is conceivable that recombinant-DNA techniques could also be used to improve the health-promoting or technological properties of human or animal probiotic strains. Genetic modification may also improve the properties of starter strains (e.g., by shortening the ripening process or through increased production of aroma compounds).

When food-associated microorganisms are to be genetically modified, the safety of the resulting GMO should be exceptionally well guaranteed. The current EU-legislation classifies foods containing GMOs automatically as novel foods and requires a rigorous safety evaluation according to the multistep procedure defined in the EU Novel Food Regulation.[37] In the United States, the Food and Drug Administration does not make a rigorous distinction between traditional and genetically modified foods, expecting similar safety standards of both and requiring a specific approval of the latter only in cases where they differ significantly from their conventional counterparts.[16]

It is generally considered unacceptable that food-associated GMOs contain antibiotic resistance markers in the final genetic constructs. Different food-grade selection systems have been developed for the genetic modification of LAB, mainly lactococci. These can be based on sugar fermentation genes, components of the lactic acid bacterial proteolytic systems, and bacteriocin resistance.[17,79] Although these markers are often more difficult to apply than the traditional antibiotic resistance used in the genetic studies, they are being constantly developed. It is therefore to be expected that similar constructs will also be available for the lactobacilli.

In cases where the genetic modification of starter or probiotic strains is limited to the transfer of well-known traits from strains of the same or closely related species, the procedure can be regarded as self-cloning and hardly entails any risks greater than those already connected with the use of unmodified parental strains. If, however, genetic constructs containing genes from completely different organisms are used, and the resulting GMOs have properties that are not naturally occurring among strains of the species, the safety of the strain has to be carefully considered, also taking into account the environmental consequences, in addition to the well-being of the host. The latter might become important if the genetic modification intentionally or adventitiously could alter or expand the ecological niche of the organism. From the point of view of the host, the harmlessness of new gene products has to be established in terms of the amounts likely to be present in foods or feeds or in the intestine. The safety depends naturally on the physiological activity and fate of the substances in question.

At present, the above considerations are hypothetical because no such novel starter or probiotic GMOs have been introduced to markets yet. Knowing the rapid developments in this field, however, the situation may change at any moment. The enterprises that are first to market genetically modified starters or probiotics will, undoubtedly, create the administrative precedent cases that will be critical for future evaluation and approval procedures.

CONCLUSION

Foods containing LAB have an important nutritional value and foods containing probiotic LAB may also have direct health effects. In most cases, these microorganisms will be ingested alive, and the assurance of safety of the consumer is therefore of major importance.

Starter strains often have a long history of safe use. For probiotics, this is often not the case. In general, lactobacilli are rarely associated with disease; thus far, only two cases are known where probiotic lactobacilli may have been involved in disease. However, potential risk groups can be identified, mainly immunocompromised subjects. No common virulence factors have so far been identified among lactobacilli; the main potential risk factor so far is antibiotic resistance and the ability to transfer this trait. This is also one of the main reasons why care should be taken when enterococci are to be used as probiotics or starter cultures.

In conclusion, there is no indication that the general public is at risk from the consumption of lactobacilli or bifidobacteria used as probiotics or starters.

REFERENCES

1. Aarts, H. J. M., Hakemulder, L. E. & Van Hoef, A. M. A. (1999). Genomic typing of *Listeria monocytogenes* strains by automated laser fluorescnce analysis of amplified fragment length polymorphism fingerprinting patterns. *Int J Food Microbiol* 49, 95–102.
2. Adams, M. R. & Marteau, P. (1995). On the safety of lactic acid bacteria from food. *Int J Food Microbiol* 27, 263–264.
3. Ahrné, S. & Molin, G. (1997). Restriction endonuclease analysis of total chromosomal DNA of *Lactobacillus*. *Microecol Ther* 26, 27–30.
4. Arthur, M., Reynolds, P. & Courvalin, P. (1996). Glycopeptide resistance in enterococci. *Trends Microbiol* 4, 401–407.
5. Baumgartner, A., Kueffer, M., Simmen, A. & Grand, M. (1998). Relatedness of *Lactobacillus rhamnosus* strains isolated from clinical specimens and such from foodstuffs, humans and technology. *Lebensm Wiss Technol* 31, 489–494.
6. Betzl, D., Ludwig, W. & Schleifer, K. H. (1990). Identification of lactococci and enterococci by colony hybridisation with 23S rRNA-targeted oligonucleotide probes. *Appl Environ Microbiol* 56, 2927–2929.
7. BG Chemie (Berufsgenossenschaft der chemischen Industrie). Sichere Biotechnologie. Eingruppierung biologischer Agenzien: Bakterien. Merkblatt B 006, 12/95; ZH 1/346. Jedermann-Verlag Dr. Otto Pfeffer oHG, Heidelberg.
8. Billot-Kelin, D., Gutman, L., Sablé, S., Guittet, E. & Van Heijenoort, J. (1994). Modification of Peptidoglycan precursors is a common feature of the low-level Vancomycin resistant species of *Lactobacillus casei*, *Pediococcus pentosaceus*, *Leuconostoc mesenteroides*, and *Enterococcus gallinarum*. *J Bacteriol* 176, 2398–2405.
9. Charteris, W. P., Kelly, P. M., Morelli, L. & Collins, J. K. (1997). Selective detection, enumeration and identification of potentially probiotic *Lactobacillus* and *Bifidobacterium* species in mixed bacterial populations. *Int J Food Microbiol* 35, 1–27.
10. Clewell, D. B. (1981). Plasmids, drug resistance, and gene transfer in the genus *Streptococcus*. *Microbiol Rev* 45, 409–436.
11. Cocconier, M. H., Bernet, M. F., Chauvière, G. & Servin, A. L. (1993). Adhering heat-killed human *Lactobacillus acidophilus*, strain LB, inhibits the process of pathogenicity of diarrhoeagenic bacteria in cultured human intestinal cells. *J Diarrhoeal Dis Res* 11, 235–242.
12. Coenye, T., Falsen, E., Vancanneyt, M., Hoste, B., Govan, J. R. W., Kersters, K. & Vandamme, P. (1999). Classification of *Alcaligenes faecalis*-like isolates from the environment and human clinical samples as *Ralstonia gilardii* sp. nov. *Int J Syst Bacteriol* 49, 405–413.
13. Collins, M. D., Ash, C., Farrow, J. A. E., Wallbanks, S. & Williams, A. M. (1989). 16S ribosomal ribonucleic acid sequence analysis of lactococci and related taxa. Description of *Vagococcus fluvialis* gen. nov., sp. nov. *J Appl Bacteriol* 67, 453–460.
14. Connel, M. A., Mercer, A. A. & Tannock, G. W. (1991). Transfer of plasmid pAMß1 between members of the normal microflora inhabiting the murine digestive tract and modification of the plasmid in a *Lactobacillus reuteri* host. *Microb Ecol Health Dis* 4, 343–355.
15. Daud Khaled, A. K., Neilan, B. A., Henriksson, A. & Conway, P. L. (1997). Identification and phylogenetic analysis of *Lactobacillus* using multiplex RAPD-PCR. *FEMS Microbiol Lett* 153, 191–197.
16. Department of Health and Human Services, Food and Drug Administration. (1992). Statement of policy: foods derived from new plant varieties. *Federal Register* 57(104).
17. De Vos, W. M. (1999). Safe and sustainable systems for food-grade fermentations by genetically modified lactic acid bacteria. *Int Dairy J* 9, 3–10.
18. Dicks, L. M. T., Du Plessis, E. M., Dellaglio, F. & Lauer, E. (1996). Reclassification of *Lactobacillus casei* ATCC 3493 and *Lactobacillus rhamnosus* ATCC 15820 as *Lactobacillus zeae* nom. rev., designation of ATCC 334 as neotype of *L. casei* subsp. *casei*, and rejection of the name *Lactobaclllus paracasei*. *Int J Syst Bacteriol* 46, 337–340.
19. Donohue, D., Salminen, S. & Marteau, P. (1998). Safety of probiotic bacteria. In *Lactic Acid Bacteria: Microbiology and Functional Aspects*, pp. 369–384. Edited by S. Salminen & A. Von Wright. New York: Marcel Dekker.

20. Duim, B., Wassenaar, T. M., Rigter, A. & Wagenaar, J. (1999). High-resolution genotyping of *Campylobacter* strains isolated from poultry and humans with amplified fragment length polymorphism fingerprinting. *Appl Environ Microbiol* 65, 2369–2375.
21. Du Plessis, E. M. & Dicks, L. M. T. (1995). Evaluation of random amplified polymorphic DNA (RAPD)-PCR as a method to differentiate *Lactobacillus acidophilus*, *Lactobacillus crispatus*, *Lactobacillus amylovorus*, *Lactobacillus gallinarum*, *Lactobacillus gasseri*, and *Lactobacillus johnsonii*. *Curr Microbiol* 31, 114–118.
22. Elliott, S. E., Buret, A., McKnight, W., Miller, M. J. S. & Wallace, J. L. (1998). Bacteria rapidly colonize and modulate healing of gastric ulcers in rats. *Am J Physiol* 275, G425–G432.
23. Ferrero, M., Cesena, C., Morelli, L., Scolari, G. & Vescovo, M. (1996). Molecular characterisation of *Lactobacillus casei* strains. *FEMS Microbiol Lett* 140, 215–219.
24. Franz, C. M. A. P., Holzapfel, W. H. & Stiles, M. E. (1999). Enterococci at the crossroad of food safety? *Int J Food Microbiol* 47, 1–24.
25. Frothingham, R., Duncan, A. J. & Wilson, K. H. (1993). Ribosomal DNA sequences of bifidobacteria: implications for sequence-based identification of the human colonic flora. *Microbiol Ecol Health Dis* 6, 23–27.
26. Gasser, F. (1994). Safety of lactic acid bacteria and their occurrence in human clinical infections. *Bull Inst Pasteur* 92, 45–67.
27. Geornaras, I., Kunene, N. F., Von Holy, A. & Hastings, J. W. (1999). Amplified fragment length polymorphism fingerprinting of *Pseudomonas* strains from a poultry processing plant. *Appl Environ Microbiol* 65, 3828–3833.
28. Gordillo, M. E., Singh, K. V. & Murray, B. E. (1993). Comparison of ribotyping and pulsed-field gel electrophoresis for subspecies differentiation of strains of *Enterococcus faecalis*. *J Clin Microbiol* 31, 1570–1574.
29. Grimont, F. & Grimont, P. D. A. (1986). Ribosomal ribonucleic acid gene restriction as potential taxonomic tools. *Ann Inst Pasteur/Microbiol* 137B, 165–175.
30. Handwerger, S., Pucci, M. J., Volk, J. K., Liu, J. & Lee, M. S. (1994). Vancomycin-resistant *Leuconostoc mesenteroides* and *Lactobacillus casei* synthesize cytoplasmic peptidoglycan precursors that terminate in lactate. *J Bacteriol* 176, 260–264.
31. Harty, D. W. S., Oakey, H. J., Patrikakis, M., Hume, B. B. H. & Knox, K.W. (1994). Pathogenic potential of lactobacilli. *Int J Food Microbiol* 24, 179–189.
32. Hasty, D. L., Courtney, H. S., Sokurenko, E. V. & Ofek, I. (1994). Bacteria-extracellular matrix interactions. In *Fimbriae: Adhesion, Genetics, Biogenesis, and Vaccines*, pp. 197–211. Edited by P. Klemm. Boca Raton, FL: CRC Press.
33. Hertel, C., Ludwig, W., Pot, B., Kersters, K. & Schleifer, K. H. (1993). Differentiation of lactobacilli occurring in fermented milk products by using oligonucleotide probes and electrophoresis protein profiles. *Syst Appl Microbiol* 16, 463–467.
34. Holzapfel, W. H., Haberer, P., Snel, J., Schillinger, U. & Huis in't Veld, J. H. J. (1998). Overview of gut flora and probiotics. *Int J Food Microbiol* 41, 85–101.
35. Holzapfel, W. H., Haberer, P., Geisen, R., Björkroth, J. & Schillinger, U. (2000). Taxonomy and important features of probiotic microorganisms in food and nutrition. *Am J Clin Nutr* (in press).
36. Horwitch, C. A., Furseth, H. A., Larson, A. M., Jones, T. L., Olliffe, J. F. & Spach, D. H. (1995). Lactobacillemia in three patients with aids. *Clin Infect Dis* 21, 1460–1462.
37. Hugget, A. C. & Conzelmann, C. (1997). EU regulation on novel foods: consequences for the food industry. *Food Sci Technol* 8, 133–139.
38. Huys, G., Coopman, R., Janssen, P. & Kersters, K. (1996). High-resolution genotypic analysis of the genus *Aeromonas* by AFLP-fingerprinting. *Int J Syst Bacteriol* 46, 572–580.
39. Janssen, P., Coopman, R., Huys, G., Swings, J., Bleeker, M., Vos, P., Zaubeau, M. & Kerstens, K. (1996). Evaluation of the DNA fingerprinting method AFLP as a new tool in bacterial taxonomy. *Microbiology* 142, 1881–1893.
40. Janssen, P., Maquelin, K., Coopman, R., Tjernberg, I., Bouvet, P., Kersters, K. & Dijkshoorn, L. (1997). Discrimination of *Acinetobacter* genomic species by AFLP fingerprinting. *Int J Syst Bacteriol* 47, 1179–1187.
41. Johnson, J. L., Phelps, C. F., Cummins, C. S., London, J. & Gasser, F. (1990). Taxonomy of the *Lactobacillus acidophilus* group. *Int J Syst Bacteriol* 30, 53–68.
42. Jonas, D. A., Antignac, E., Antoine, J. M., Classen, H. G., Huggett, A., Knudsen, I., Mahler, J., Ockhuizen, T., Smith, M., Teuber, M., Walker, R. & De Vogel, D. (1996). The safety assessment of novel foods: guidelines prepared by the ILSI Europe novel foods task force. *Food Chem Toxicol* 34, 931–940.
43. Kirjavainen, P. V., Ouwehand, A. C., Isolauri, E. & Salminen, S. J. (1998). The ability of probiotic bacteria to bind to human intestinal mucus. *FEMS Microb Lett* 167, 185–189.
44. Kirjavainen, P. V., Tuomola, E. M., Crittenden, R. G., Ouwehand, A. C., Harty, D. W. S., Morris, L. F., Rautelin, H., Playne, M. J., Donohue, D. C. & Salminen, S. J. (1998). *In vitro* adhesion and platelet aggregation properties of bacteremia-associated lactobacilli. *Infect Immun* 67, 2653–2655.
45. Klein, G., Pack, A., Bonaparte, C. & Reuter, G. (1998). Taxonomy and physiology of probiotic lactic acid bacteria. *Int J Food Microbiol* 41, 103–125.

46. Larvol, L., Monier, A., Besnier, P. & Levecq, H. (1996). Abcês hépatique à *Lactobacillus acidophilus*. *Gastroenterol Clin Biol* 20, 193–195.

47. Lauer, E., Helming, C. & Kandler, O. (1980). Heterogeneity of the species *Lactobacillus acidophilus* (Moro) Hansen and Moquot as revealed by biochemical characteristics and DNA-DNA hybridisation. *Zbl Bakt Hyg I Abt Orig* C1, 150–168.

48. Lerche, M. & Reuter, G. (1962). Das Vorkommen aerob wachsender gram-positiver Stäbchen des Genus *Lactobacillus* Beijerinck im Darminhalt erwachsener Menschen. *Zbl Bakt I Abt Orig* 185, 446–481.

49. Lim, K. S., Huh, C. S. & Baek, Y. J. (1993). Antimicrobial susceptibility of Bifidobacteria. *J Dairy Sci* 76, 2168–2174.

50. Mackay, A. D., Taylor, M. B., Kibbler, C. C. & Hamilton-Miller, J. M. T. (1999). *Lactobacillus* endocarditis caused by a probiotic organism. *Clin Microbiol Infect* 5, 290–292.

51. Mitsuoka, T. (1992). Intestinal flora and ageing. *Nutr Rev* 50, 438–446.

52. Morelli, L., Sdarra, P. G. & Bottazzi, V. (1988). *In Vivo* transfer of pAMß1 from *Lactobacillus reuteri* to *Enterococcus faecalis*. *J Appl Bacteriol* 65, 371–375.

53. Mori, K., Yamazaki, K., Ishiyama, T., Katsumata, M., Kobayashi, K., Kawai, Y., Inoue, N. & Shinano, H. (1997). Comparative sequence analysis of the genes encoding for 16S rRNA of *Lactobacillus casei*-related taxa. *Int J Syst Bacteriol* 47, 54–57.

54. Ning, W., Mackie, R. I. & Gaskins, H. R. (1997). Biotype and ribotype diversity among *Lactobacillus* isolates from mouse ileum. *Syst Appl Microbiol* 20, 423–431.

55. Oakey, H. J., Harty, D. W. S. & Knox, K. W. (1995). Enzyme production by lactobacilli and the potential link with infective endocarditis. *J Appl Bacteriol* 78, 142–148.

56. Ouwehand, A. C. & Salminen, S. J. (1998). The health effects of viable and non-viable cultured milks. *Int Dairy J* 8, 749–758.

57. Piehl, I., Bülte, M. & Reuter, G. (1995). Pulsfeld-Gelelektroforese zur Feindifferenzierung von Mikroorganismen – ein Beitrag zur Charakterisierung von Probiotika-Stämmen. In *Proc 36th Arbeitstagung des Arbeitsgebietes "Lebensmittelhygiene,"* Vol. 2. pp. 226–233. Gießen: Deutsche Veterinärmedizinische Gesellschaft eV.

58. Pot, B., Hertel, C., Ludwig, W., Descheemaker, P., Kersters, K. & Schleifer, K. H. (1993). Identification and classification of *Lactobacillus acidophilus*, *L. gasseri* and *L. johnsonii* strains by SDS PAGE and rRNA-targeted oligonucleotide probe hybridisation. *J Gen Microbiol* 139, 513–517.

59. Prescott, L. M., Harley, J. P. & Klein, D. A. (1993). *Microbiology.* Dubuque, IA: Wm. C. Brown Publishers.

60. Rautio, M., Jousimies-Somer, H., Kauma, H., Pietarinen, I., Saxelin, M., Tynkkynen, S. & Koskela, M. (1999). Liver abscess due to a *Lactobacillus rhamnosus* indistinguishable from *L. rhamnosus* strain GG. *Clin Infect Dis* 28, 1159–1160.

61. Reuter, G. (1965). Das Vorkommen von Laktobazillen in Lebensmitteln und ihr Verhalten im menschlichen Intestinaltrakt. *Zentralbl Bakt Hyg I Abt Orig A* 197, 468–487.

62. Reuter, G. (1965). Untersuchungen über die Zusammensetzung und Beeinflußbarkeit der menschlichen Magen- und Darmflora unter besonderer Berücksichtigung der Laktobazillen. *Ernährungsforsch* 10, 429–435.

63. Reuter, G. (1969). Zusammensetzung und Andwendung von Bakterienkulturen für therapeutische Zwecke. Arzneim-Forsch *Drug Res* 50, 951–954.

64. Roussel, Y., Colmin, C., Simonet, J. M. & Decaris, B. (1993). Strain characterisation, genome size and plasmid content in the *Lactobacillus acidophilus* group (Hansen and Mocquot). *J Appl Bacteriol* 74, 549–556.

65. Roy, D., Ward, P. & Champagne, G. (1996). Differentiation of bifidobacteria by use of pulsed-field gel electrophoresis and polymerase chain reaction. *Int J Food Microbiol* 29, 11–29.

66. Saavedra, J. M. (1995). Microbes to fight microbes: a not so novel approach to controlling disease. *J Pediatr Gastroenterol Nutr* 21, 125–129.

67. Salminen, S. & Von Wright, A. (1998). Current probiotics-safety assured? *Microb Ecol Health Dis* 10, 68–77.

68. Salminen, S., Von Wright, A., Morelli, L., Marteau, P., Brassard, D., De Vos, W., Fondén, R., Saxelin, M., Collins, K., Mogensen, G., Birkeland, S. -E. & Mattila-Sandholm, T. (1998). Demonstration of safety of probiotics: a review. *Int J Food Microbiol* 44, 93–106.

69. Salminen, S., Bouley, C., Boutron-Rualt, M. -C., Cummings, J. H., Franck, A., Gibson, G., Isolauri, E., Moreau, M. -C., Roberfroid, M. & Rowland, I. (1998). Functional food science and gastrointestinal physiology and function. *Br J Nutr Suppl* 1, 147–171.

70. Saxelin, M., Chuang, N. H., Chassy, B., Rautelin, H., Mäkelä, P. H., Salminen, S. & Gorbach, S. L. (1996). Lactobacilli and bacteremia in Southern Finland: 1989–1992. *Clin Infect Dis* 22, 564–566.

71. Schillinger, U. (1999). Isolation and identification of lactobacilli from novel-type probiotic and mild yoghurts and their stability during refrigerated storage. *Int J Food Microbiol* 47, 79–87.

72. Sims, W. (1964). A pathogenic *Lactobacillus*. *J Pathol Bacteriol* 87, 99–105.

73. Stahl, M. & Molin, G. (1994). Classification of *Lactobacillus reuteri* by restriction endonuclease analysis of chromosomal DNA. *Int J Syst Bacteriol* 44, 9–14.

74. Steidler, L., Robinson, K., Chamberlain, L., Schofield, K. M., Reault, E., Le Page, R. W. & Wells, J. M. (1998). Mucosal delivery of murine interleukin-2 (IL-2) and IL-6 by recombinant strains of *Lactococcus lactis* coexpressing antigen and cytokine. *Infect Immun* 66, 3183–3189.

75. Toba, T., Virkola, R., Westerlund, B., Björkman, Y., Sillanpää, J,. Vartio, T., Kalkkinen, N. & Korhonen, T. K. (1995). A collagen-binding S-layer protein in *Lactobacillus crispatus*. *Appl Environ Microbiol* 61, 2467–2471.

76. Tuomola, (née Lehto) E. M. & Salminen, S. J. (1998). Adhesion of some probiotic and dairy *Lactobacillus* strains to Caco-2 cell cultures. *Int J Food Microbiol* 41, 45–51.

77. Tuomola, E., Ouwehand, A. C. & Salminen, S. (1999). Human ileostomy glycoproteins as a model for small intestinal mucus to investigate adhesion of probiotics. *Lett Appl Microbiol* 28, 159–163.

78. Tynkkynen, S., Singh, K. & Varmanen, P. (1998). Vancomycin resistance factor of *Lactobacillus rhamnosus* GG in relation to enterococcal vancomycin resistance (*van*) genes. *Int J Food Microbiol* 41, 195–204.

79. Von Wright, A. & Sibakov, M. (1998). Genetic modification of lactic acid bacteria. In *Lactic Acid Bacteria: Microbiology and Functional Aspects,* pp. 161–210. Edited by S. Salminen & A. Von Wright. New York: Marcel Dekker.

80. Wagner, R. D., Warner, T., Robers, L., Farmer, J. & Balish, E. (1997). Colonization of congenitally immunodeficient mice with probiotic bacteria. *Infect Immun* 65, 3345–3351.

81. Wagner, R. D. & Balish, E. (1998). Potential hazards of probiotic bacteria for immunodeficient patients. *Bull Int Pasteur* 96, 165–170.

82. Wells, C. L., Maddaus, M. A. & Simmons, R. (1988). Proposed mechanisms for the translocation of intestinal bacteria. *Rev Infect Dis* 10, 958–979.

83. Wells, J. M., Wilson, P. W., Norton, P. M., Gasson, M. J. & Le Page, R. W. F. (1993). *Lactococcus lactis*: high level expression of tetanus toxin fragment C and protection against lethal challenge. *Mol Microbiol* 8, 1115.

84. Wolf, B. W., Garleb, K. A., Ataya, D. G. & Casas, I. A. (1995). Safety and tolerance of *Lactobacillus reuteri* in healthy adult male subjects. *Microb Ecol Health Dis* 8, 41–50.

CHAPTER 12

Practical Applications: Prospects and Pitfalls

Yasmine Motarjemi, A. Asante, Martin R. Adams, and M. J. Robert Nout

INTRODUCTION

Fermentation is one of the oldest technologies used for food preservation. Over the centuries, it has evolved, been refined, and diversified until today, when a large variety of foods are derived from this technology, both in the industrialized and the developing countries, in households, small-scale food industries, and large commercial enterprises. Fermented foods form a major part of the human diet all over the world. In some regions, mainly in African countries, fermentation plays an important role in the nutrition of infants and young children because it is used for the preparation of complementary foods.†

Advances in food science and technology have given rise to a wide range of new food technologies. Nevertheless, fermentation has remained one of the most important food processing techniques throughout human history. Many benefits are attributed to fermentation. It can preserve food (i.e., increase shelf life), improve digestibility, enrich food, and enhance taste and flavor. It is also an affordable technology, and thus is accessible to all populations. Furthermore, fermentation has the potential to enhance food safety by controlling a great number of pathogens in foods. Thus, it makes an important contribution to human nutrition, particularly in developing countries, where economic problems are a major barrier to ensuring food safety.

In general, fermented foods, particularly those produced under controlled conditions, have a good record of safety and are implicated in outbreaks of diseases relatively infrequently. There are, however, more concerns with traditional and artisanal productions, where the application of fermentation is based largely on experience and knowledge gained through trial and error by generations of food producers and households. Such an empirical approach presents a major pitfall. Depending on the process, ingredients, raw material, and environmental conditions, the process may lead to unsafe products.

This chapter reviews the importance of fermentation from a public health and food safety point of view. Specific emphasis is put on problems in developing countries where risks of food contamination and disease are greater, and the potential contribution of fermentation for processing and storage of foods more important. The chapter examines the risks and benefits of fermentation for human nutrition, with a particular focus on the fermentation of complementary

---

Note: This chapter is partly based on the report of a Joint Food and Agriculture Organization(FAO)/World Health Organization (WHO) Workshop on fermentation as a household technology to improve food safety held in Pretoria, South Africa, 11–15 December 1995.[18]

†The term *complementary foods* refers to any nutrient-containing foods, be they solids or liquids, other than breast milk, that are given to infants and young children during the period of complementary feeding. This period of complementary feeding corresponds to the period during which other foods are provided along with breast milk.

Source: Portions of this chapter are adapted from *Fermentation: Assessment and Research. Report of a Joint FAO/WHO Workshop on Fermentation as a Household Technology to Improve Food Safety,* Pretoria, South Africa, December 11–15, 1995, © 1996, World Health Organization.

foods, and looks into the prospects for promoting the technology and improving the safety and nutritional quality of the derived products. It also presents two applications of the Hazard Analysis and Critical Control Point (HACCP) concept to fermented foods as examples.

BENEFITS OF LACTIC ACID FERMENTATION

Depending on the organism used (i.e., molds, yeasts, bacteria), there are different types of fermentation processes. Fermentation mediated by lactic acid bacteria (LAB) is the process that is presently of greatest interest to food safety and public health. Other types of fermentation are also important to public health because they may contribute to food security and overcoming malnutrition. The focus of this chapter is on lactic fermentation.

There is considerable evidence to show that lactic acid fermentation inhibits the growth, survival, and toxin production of a number of pathogenic bacteria[1] (see also Chapter 2). The inhibitory effect of lactic acid fermentation on pathogenic bacteria as well as the spoilage organisms is due to the rapid growth and acid production by LAB, the consequent decrease in pH, and the formation of other antimicrobial factors associated with LAB, such as bacteriocins, hydrogen peroxide, ethanol, and diacetyl.[1,8] The extent to which pathogens are inhibited depends on the organism concerned, the temperature, the amount of acid produced, and the properties of the food (e.g., the buffering capacity). For instance, in cereals and vegetable products that are weakly buffered, an efficient lactic acid fermentation will produce a pH of 4 or less, at which the growth of bacterial pathogens is inhibited and many bacteria die. This aspect of fermentation, as seen in the following paragraphs, is of great importance to public health, particularly in developing countries where safe food storage by cold or hot holding is difficult because of socioeconomic constraints.

Lactic acid fermentation has also been associated with the reduction of certain naturally occurring toxins in plant foods, notably a reduction of cyanogenic glycosides that are present in cassava. Cassava is a major staple in Africa and, due to the presence of these compounds, it has been associated with several health problems.[51] Although the degradation of the cyanogenic glycosides is mediated by endogenous plant enzymes, microbial activities carried out during fermentation contribute to the detoxification process by softening the plant tissues (see Chapter 4).

Nutritional benefits have also been attributed to fermentation. Some fermentation processes lead to enhanced digestibility of carbohydrates as a result of degradation of oligosaccharides and dietary fiber, particularly prevalent in foods of plant origin. The use of amylase-rich flour (ARF) combined with a small amount of a lactic acid starter culture can increase the nutrient density of starchy foods while keeping a semi-liquid consistency. This makes the fermented food particularly interesting as complementary food for infants and young children. Fermentation also provides an optimum pH for the activity of phytase and the degradation of phytate. The degradation of phytate increases the amount of soluble iron several fold, thereby increasing its bioavailability. Lactic fermentation also decreases the tannin content of cereals and, in this way, has a positive impact on mineral absorption and the protein digestibility of cereals.

Over and above the influence on pathogenic bacteria and toxic compounds leading to improved food safety and quality, specific beneficial health effects such as prophylaxis against *Escherichia coli* and other pathogens and hypocholesterolemic and anticarcinogenic effects have also been attributed to some selected strains of LAB. A number of such organisms with claimed health-related properties are already used in food production and are available in many countries. The use of probiotics raises two fundamental questions: (1) Are the organisms used safe? and (2) Are the health claims valid? The former has been discussed in Chapter 11 of this book. The latter is the subject of extensive research. A number of studies have shown some specific health benefits with regard to lactose intolerance and gastrointestinal disorders. So far, there is no conclusive explanation for the underlying molecular mechanisms. Also, there are a number of claims that have not been substantiated and need further investigation.[13,22]

Advances in genetic modification open new doors to fermentation technology.[34] Genetic modification may be carried out on the substrate (e.g., plants) or on the microorganisms used for fermentation. Fermentation may also be used for the production of enzymes with enhanced properties and improved quality (see Chapter 10). These can have significant benefits in terms of improving the fermentation process and the quality and safety of the final product.

Genetic modification can be used in different ways to improve the fermentation process and to enhance the quality and safety of the final product. For instance, starter cultures may be improved with regard to their rate of growth, potential for lactic acid production, salt tolerance, inability to metabolize organic acids, resistance to bacteriophage, production of flavor compounds, and so on.[6,8] Genetic modification may also be used to remove or reduce hazards in the raw material, as in the production of a variety of cassava containing less cyanogenic glycosides.

As discussed in Chapter 10, the prospect of using genetic modifications for better control of fermentation processes and to enhance the safety and quality of derived products has two preconditions: (1) an adequate method for assessing the safety and nutritional implications of these products and (2) consumer acceptance. FAO, WHO, and other organizations have developed strategies to assess the safety of such foods,[17,50] and further work in this area is planned. To gain consumer acceptance, genetic modification needs to be considered in the broader perspectives of risks and benefits for the consumer, environment, socioeconomic implications (in both developing and industrialized countries), and, in general, the need for such technology. The prospect of using genetic modification technology for improving fermentation processes relies on an appropriate risk communication with consumers.

It should be mentioned that genetic modification can also occur spontaneously in nature and lead to new starter strains. Conventional methods of screening and selection of new starter strains with desirable properties have been used for the production of products with enhanced quality.[34]

PITFALLS OF FERMENTATION

As simple as it may seem, fermentation is also a sensitive and complex technology. The safety and quality of the final product depend on a number of factors, which need to be carefully controlled. These factors are

- quality and safety of raw materials, including the initial level of contamination
- levels of environmental hygiene and sanitation
- quality and safety of starter culture
- safety of metabolites
- processing conditions and degree of acidity achieved

In certain applications, particularly in the context of cottage industry or household application of fermentation, it may be difficult to control these factors adequately. Not infrequently, fermented products have been incriminated in infections or intoxications. In certain settings, inadequate fermentation has been a major cause of food-borne intoxications. For instance, in China, 2,861 cases of botulism (745 outbreaks) were reported in the year 1989, causing some 421 deaths. The major implicated foods were homemade fermented bean products. Fermented fish and salmon eggs are a significant cause of botulism occurring among the native Alaskan Inuit population in the North American continent. Fermented Inuit products such as white whale meat, seal flippers, or salmon eggs lack fermentable carbohydrates and therefore do not undergo a sufficient pH reduction to prevent toxigenesis.[20] Cheeses of different categories have also been implicated in outbreaks of various types of food-borne diseases.[24] Examples of other documented outbreaks of illness are listed in Table 12–1. It is likely that outbreaks of food-borne disease caused by fermented foods produced under unhygienic conditions or with faulty handling of the products occur more frequently than are officially reported. However,

Table 12–1 Examples of Food-Borne Disease Outbreaks Associated with Fermented Foods

Implicated Food	*Causative Agent*	*Number of Cases*	*References*
Vegetables			
Paste of soybeans and wax gourds	*Clostridium butyricum*	6	32
Sauerkraut	Histamine		30
Milk products			
Curd (yogurt)	*Campylobacter jejuni*	160	7
Yogurt	*Clostridium perfringens*	167	33
Hazelnut yogurt (hazelnut puree was contaminated)	*Clostridium botulinum*	27	36
Sour milk	*Clostridium botulinum*	11	41
Yogurt	*Escherichia coli* O157	16	35
Meat products			
Semi-dry sausages	*Escherichia coli* O111:NM	23	10
Pork (salami)	*Salmonella anatum*	52	4
Pork (labh, raw and nahm, fermented)	*Trichinella*	27	26
Salami stick	*Salmonella typhimurium*	85 (including 13 secondary cases)	12
Salami	*Escherichia coli* O157	23	11
Sausages (Lebanon bologna)	*Salmonella typhimurium*	26	38
Fish			
Fish (seal flipper)	*Clostridium botulinum*	1	39
Fish (beaver tails)	*Clostridium botulinum*	7	40
Fish (salmon fish heads)	*Clostridium botulinum*	8	39
Salmon eggs	*Clostridium botulinum*	15*	20
Cheeses			
Soft cheese (Mexican-style)	*Listeria monocytogens*	142	24, 27
Soft cheese	*Salmonella berta*	82 (including 3 secondary cases	15
Cheese	*Salmonella enteritidis*	≅700	9
Soft cheese	*Salmonella dublin*	42	28
Goat milk cheese	*Salmonella paratyphi*	273	14
Cheddar cheese	*Salmonella heidelberg*	339† (28000–36000 est.)	19
Mozzarella cheese	*Salmonella typhimurium*	321	3
Cheese	*Escherichia coli* O157	22	44
Cheese (Brie, Camembert, Coulommiers)	*Escherichia coli* O124:B17	387 est.	29
Cheese (Brie, Camembert)	*Escherichia coli* O27:H20	170	3
Cheese (Brie, Camembert)	*Clostridium botulinum*	27	37
White processed cheese	*Streptococcal pharyngitis*	197	5
Mexican-style soft cheese	*Streptococcus zooepidemicus*	16	16
Mexican-style soft cheese	*Brucella melitensis*	31	3
Hand-pressed direct set cheese	*Staphylococcus aureus*	16	3
Cheese	*Staphylococcus aureus* and *Shigella sonnei*	≥ 50**	40
Swiss cheese	Histamine	6	43

*15 cases involved in 7 outbreaks between 1971–1984.

†339 cases where reported in Colorado. From the attack rates noted (28–36%) and the amount of cheese presumably consumed (2830 kg), it is estimated that between 28,000 and 36,000 persons were affected in total.

**Bacillary dysentery, which affected numerous persons who had eaten various French cheeses purchased at a Paris airport in 1982, was reported by several Scandinavian countries.

weaknesses in the food-borne disease surveillance system, particularly in regard to household food preparation, do not provide for identification and reporting of these outbreaks.

Although LAB can inhibit the growth of certain food-borne pathogens, care should be taken to ensure that the raw material is of high hygienic quality and the risk of contamination from the environment is minimized. A number of food-borne hazards are not controlled by lactic acid fermentation and so present a serious threat to health if they persist in food. For instance, enterohemorrhagic *E. coli* has shown patterns of acid resistance and may survive certain fermentation processes. Yogurt and fermented meat have been recognized as potential vehicles of enterohemorrhagic *E. coli* infection. Food and waterborne viruses, a relatively frequent cause of gastroenteritis, may survive high levels of acidity. For instance, Simian rotavirus has been shown to survive a high level of acidity during 24 hours of storage in model fermented foods[53] (see Chapter 8). There is little information on the effect of fermentation on parasites, such as *Cryptosporidium, Giardia lamblia,* and food-borne trematodes. The cysts or metacercariae of these organisms often show resistance to adverse conditions (see Chapter 9). Thus, the possibility that they may survive fermentation should not be excluded. Most toxins produced by algae, bacteria, and molds are also unaffected by fermentation (see Chapter 5).

In view of this, it is important not to rely only on the fermentation step *per se* to eliminate or reduce hazards to safe levels. To ensure safety, it is important to design the process in such a way as to (1) prevent contamination, (2) eliminate or reduce the hazards as much as possible, and (3) control the growth of pathogens. For this reason, fermentation steps frequently need to be combined with other process operations such as soaking, cooking, and so forth. Lactic acid fermentation also has a limited effect on antinutritional factors, such as protease inhibitors and lectins, and other process operations such as heat treatment should be used to destroy these antinutritional factors.

There are also some concerns with the organic acids that are produced during lactic acid fermentation. Microbially produced lactic acid is usually a mixture of the optical isomers L(+) and D(-) lactic acid. D(-) lactic acid cannot be metabolized by humans. Excessive production and intake of D(-) lactic acid may result in acidosis, a disturbance of the acid-alkali balance in the blood. However, very little is known about the "toxicity" of D(-) lactic acid for malnourished or sick children, and little data are available on the D(-) lactic acid content of fermented foods prepared at the household level. Further studies in this area are needed to elucidate the role of D(-) lactic acid and methods to control its production.[1,2,45,54]

Histamine poisoning is also a potential problem with fermented foods. Although histamine poisoning has been commonly associated with the consumption of scrombroid-type fish such as tuna and mackerel, certain fermented foods such as wine, soy sauce, and, in particular, cheese, may also present a risk. Poisoning due to histamine or other biogenic amines is caused by an accumulation of histamine or other type of biogenic amines (e.g., tyramine, phenylethylamine) following metabolic activity of decarboxylase-positive LAB such as *Lactobacillus büchneri.* However, the problem can be controlled to a large extent by observing good manufacturing practice during production. For instance, during the manufacture of cheese, the control of factors such as hygienic quality of milk and temperature of storage minimize the risk of histamine formation[42] (see Chapter 6).

IMPORTANCE OF FOOD FERMENTATION IN PUBLIC HEALTH

Food-borne diseases are a major public health problem all over the world, in both industrialized and developing countries. The industrialized countries, benefiting from a higher standard of living, good water supply and sanitation, and technologies for processing and preserving foods, have succeeded in combating many food-borne infections. This has resulted in a reduction of a great number of food-borne infections such as typhoid fever, cholera, and shigellosis. Nevertheless, food-borne diseases remain a widespread public health problem, and statistics in these countries indicate that possibly up to 30%

of the population may suffer from a food-borne illness annually.[31]

It is the developing world that bears the brunt of the problem. Although statistics on the incidence of food-borne diseases are not available, the high prevalence of diarrheal diseases in these parts of the world, particularly in infants and young children, is an indication of an underlying food safety problem. In 1997, 4,000 million cases of diarrhea were estimated to occur in the world.[52] Approximately 1,500 million episodes of diarrhea occur annually in children under the age of five, and more than 1.8 million children die as a result. Indirectly, diarrheal diseases kill many more children because they are one of the major underlying factors of malnutrition. It is estimated that annually, some 13 million children under the age of five die from the associated effect of malnutrition.

The etiological agents responsible for food-borne diseases are broad and include bacteria, viruses, and parasites, all of which have been discussed here individually in their respective chapters.

The sources of contamination of food are diverse and include polluted water, night soil, dust, flies, domestic animals, dirty utensils, and food handlers. Raw foods themselves may also be a source of contaminants because many foods harbor pathogens or come from infected animals. Moreover, during the preparation process, there is an added risk of cross-contamination. However, one of the major factors leading to food contamination is time-temperature abuse during food preparation and storage with the result that pathogens survive, grow, and produce toxins (Figure 12–1).

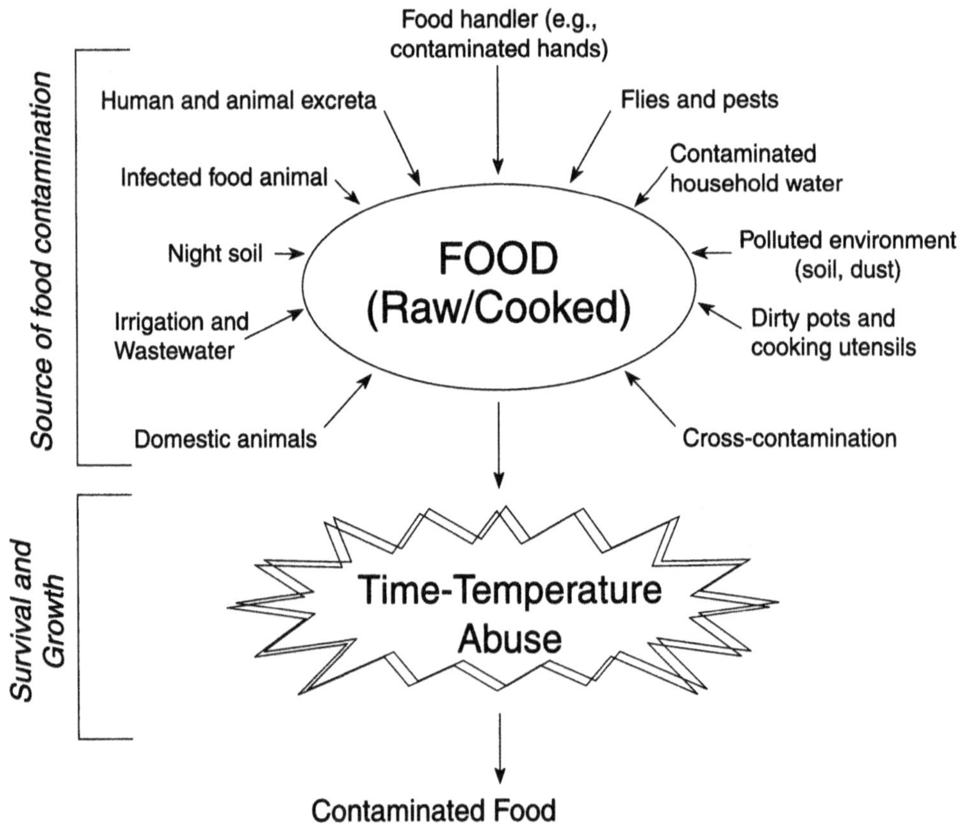

Figure 12–1 Sources of food contamination

In addition to agents of diarrheal diseases, food may also be a vehicle for chemical hazards, whether naturally present or contaminating the food as a result of poor agricultural practices or environmental pollution as dealt with in Chapters 4, 5, and 6. Depending on the dose, chemical hazards may lead to acute intoxication or long-term health problems such as cancers and other chronic diseases. Food may also contain antinutritional factors such as enzyme inhibitors, phytates, lectins, and polyphenols, which interfere with the digestion, absorption, or other aspects of metabolism of nutrients in foods.

Food processing technologies are applied for a variety of reasons. They may be applied to render food more digestible or edible, to retain or enhance sensory quality, to increase shelf life, to improve nutritional quality, and/or to render food safe. Food fermentation has proved to possess many of these features; in this way, it is an important food technology. However, its role in improving the nutritional quality of foods, particularly complementary foods, and preserving foods and preventing growth of most pathogenic organisms make this technology particularly important from a public health point of view.

Fermentation is particularly important for food safety and for the prevention of diarrheal diseases in the developing regions. Although the application of basic rules of food hygiene can prevent a great proportion of diarrheal diseases, in the developing countries, the application of these rules is sometimes hampered by socioeconomic constraints such as inadequate supply of safe water, lack of knowledge or facilities for safe preparation and storage of food (e.g., refrigeration, fuel for hot holding or thorough reheating), and lack of time to prepare food properly before each meal. As a result, some households, particularly low-income ones, are simply not able to apply essential food safety principles, such as feeding infants with freshly prepared foods, chill storage, hot storage, reheating of stored foods, and so forth.

Thus, fermentation provides an economic means of preserving food and inhibiting the growth of pathogenic bacteria even under conditions where refrigeration or other means of safe storage are not available. At the same time, it can enhance the nutritional quality of some foods. In several African countries, the technology is used in particular in the preparation of foods for infants and young children.[2] For instance, in Kenya, Nigeria, the United Republic of Tanzania and Uganda, it is customary to give infants fermented cereals, or root-crop products. Fermentation is also used to produce beverages. Again, in areas where the safety of the water cannot be ensured, fermentation processes contribute to reducing the risk of waterborne diseases.

PRACTICAL INTERVENTION TO ENHANCE SAFETY OF FERMENTED FOODS

To take advantage of the benefits that fermentation offers and, at the same time, to minimize its risks, it is important to examine the fermentation process carefully and to develop a plan where hazards associated with the different production steps are considered and controlled. In Chapter 3, the concept of Hazard Analysis and Critical Control Point (HACCP) as a method of food safety assurance was explained, and a plan for a fermented indigenous food that is common in Africa and prepared at household level was presented. Other examples of the application of the HACCP concept to fermented African foods have also been published elsewhere.[46–48] In this section, a second example of an indigenous African food, *togwa,* is presented to illustrate how the concept of HACCP can be used to evaluate a food production or preparation process in order to identify possible safety problems and how they may be controlled.

This study illustrates that the process of togwa preparation, as described here, leads to a high-risk product. The risk lies in the fact that "power flour" (an ARF obtained from germinated seeds), which may be contaminated, is added after cooking. If fermentation fails, proliferation of contaminant bacteria may occur and acid-resistant pathogens may survive. Possible options to reduce risks include accelerated fermentation by back-slopping or use of starter culture.[21] Extreme care should be taken to prevent postcooking contamination. The HACCP study of this product identifies numerous critical control points (CCPs) reflecting the high degree of concern. The reasons for concern are that (1) the food is intended primarily for infants and young children, who, due to their susceptibility to infections, require great care in the preparation of their food; (2) a wide range of hazards have been considered in this study; and (3) the preparation is envisaged in a household setting with rudimentary environmental conditions. Because the risk of contamination in this setting is higher, several steps in the preparation likely to prevent or control contamination need to be carefully controlled.

A second example shows an HACCP study for the production of a traditional cheddar cheese. Cheddar cheese has been implicated in outbreaks of salmonellosis and *Staphylococcus aureus* intoxication.[19,25,49] In one outbreak, the problem was due to improper application of the HACCP system (i.e., inadequate pasteurization and a lack of corrective action when the monitoring results indicated failure in reaching critical limits).[49]

Both examples also underline the importance of good hygienic practice (GHP) as a prerequisite for the production of fermented foods, particularly with regard to the prevention of contamination. In the case of cheddar cheese, it can be seen that after the pasteurization step, there is no other step that can eliminate pathogens. Therefore, GHP is extremely important to prevent recontamination.

HACCP Study of Togwa*†

1. **Product description**—Togwa is prepared by many tribes in Tanzania, and there are variations in its production. The preparation involves mixing cooked cereals (e.g., maize, sorghum) with germinated cereals (e.g., finger or bulrush millet or sorghum). Fermentation takes approximately 9–12 hours, depending on whether or not an old batch is used as a starter culture. If an old batch or power flour is used, the fermentation is completed in approximately 6–9 hours.
2. **Intended use**—Most fermented gruels in Tanzania remain edible for one to two days. Beyond this period, the product is too sour, or gives off an unpleasant odor. Fermented gruels are traditionally given to children less than five years of age.
3. **Flow diagram**—Figure 12–2 shows the flow diagram of togwa.
4. **Hazards of concern**—Hazards considered in this context include biological (e.g., bacteria, viruses, parasites), chemical (e.g., contaminants, mycotoxins), and physical agents.
5. **Identification of hazards, control measures, and CCPs**—Table 12–2 shows hazards associated with each step in the preparation of togwa and of power flour, which is an essential ingredient in this gruel.
   a. *Raw material:* Major hazards in maize, millet, and sorghum are toxins (e.g., aflatoxin produced by molds, and agrochemicals). For prevention of the activity of toxigenic molds during storage, the raw material should, as far as possible, be stored under appropriate conditions. When the ambient temperature and humidity are high, storage time should be limited. Insofar as agrochemicals are concerned, households cannot do much except to get assurance from suppliers concerning the safety of the products. The possibility of accidental contamination of grains during storage with agrochemicals should be prevented.

      The grains may also contain foreign matter such as stones and insect fragments. These will not be eliminated at a later step, so it is important that households thoroughly clean the raw material.

*Application of the HACCP system has been simplified and adapted to household conditions. Although the same approach can be used for production on a cottage and industrial scale, the requirements in terms of critical control points, critical limits, and monitoring procedures may be different and more severe.

†Model HACCP plans are not appropriate for use until validated for a specific food and food process.

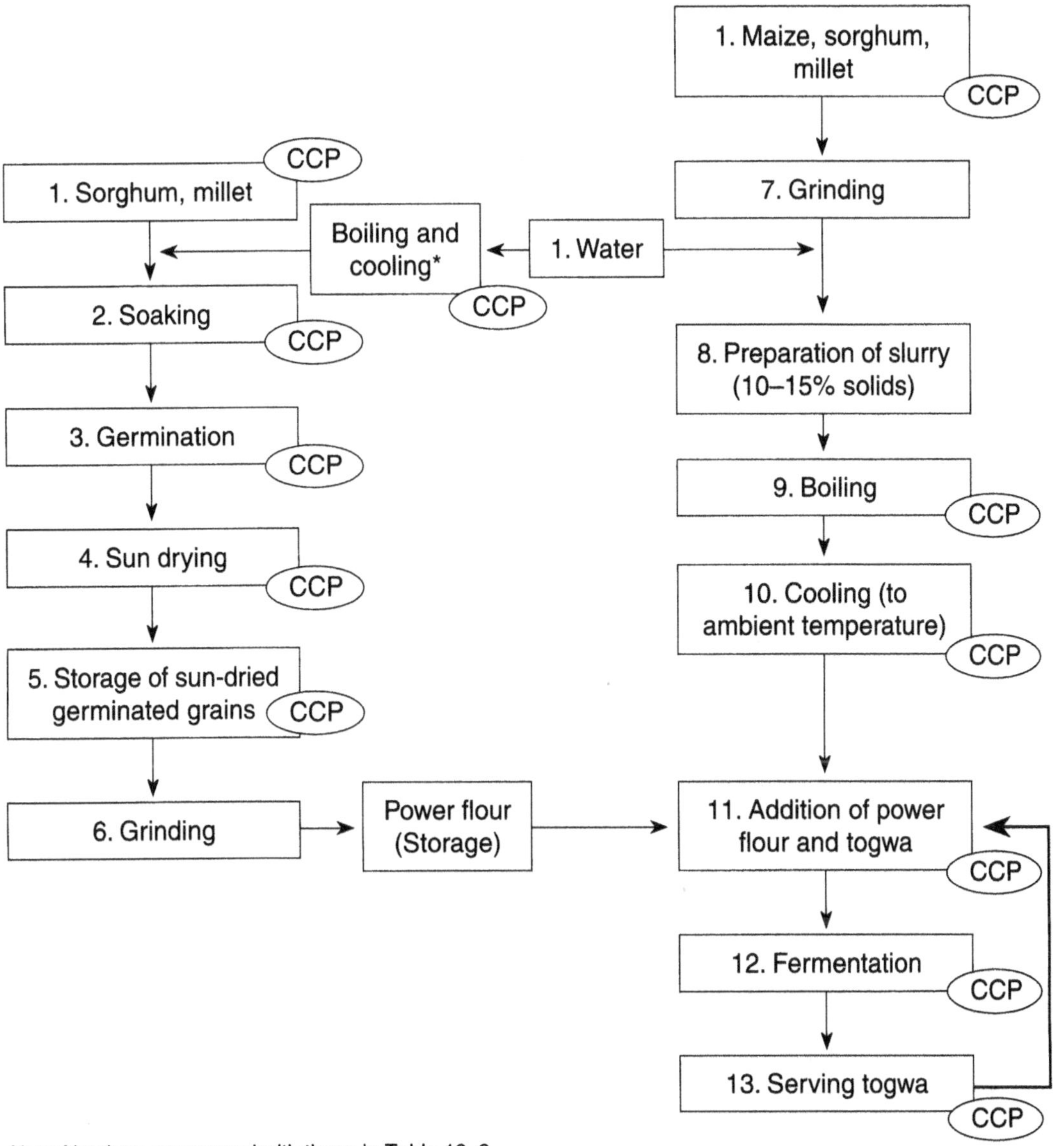

Note: Numbers correspond with those in Table 12–2.
*Although this step was not part of the original togwa preparation, it was added after the HACCP study because boiling water was identified as being essential and needed to ensure the hygienic quality of power flour.

Figure 12–2 Flow diagram of togwa. The high number of CCPs identified in this study is due to the fact that the study is considered in poor hygienic conditions.

Water used in the preparation of slurry may be contaminated. Boiling the slurry (see step 9) will eliminate eventual pathogens. Therefore, water used for the preparation of slurry is not a CCP, although as part of a good hygienic practice, safe water should be used as far as possible in the preparation of togwa and for washing hands and utensils. On the other hand, the use of safe water in the preparation of power flour is essential because this

Table 12–2 HACCP Study of the Preparation of Togwa in Households

Step	*Hazards*	*Control Measures*	*CCPs*	*Critical Limit**	*Monitoring Procedure*	*Corrective Actions*
1. Raw material i) Maize Sorghum Millet	a. Mycotoxins	a. i) Obtain assurance from supplier of adequate preharvest and postharvest handling of grains	a. Yes	a. i) No moldiness, good smell	a. Observation, smelling	a. Discard the raw material and change supplier
		ii) Store grains in dry (and if possible cool) area, limit storage time		ii) Short storage time, adequate temperature humidity of storage area	Time keeping (measurement of temperature or humidity if possible)	Utilize the raw material as quickly as possible.
	b. Agrochemicals	b. Obtain assurance from supplier of adequate preharvest and postharvest handling of grains	b. No			
	c. Pathogens: *Bacillus cereus, Salmonella, Escherichia coli*	c. Heat treatment, fermentation	c. No			
	d. Physical: insects and stones	d. Manual cleaning	d. Yes	d. No visible stones	d. Observation	d. Reclean.
1. Raw material ii) Water	a. Chemical contaminants, depending on the source	a. Obtain assurance about the source of water; use only safe water	a. Yes	a. Clear, free of odor and off taste	a. Observation, smelling, and tasting	a. Use another source of water.
	b. Pathogens (e.g., *Escherichia coli, Campylobacter, V. cholerae, Salmonella, Cryptosporidium, Giardia lamblia, Entamoeba histolytica*) Rotavirus	b. If safe water (i.e., filtered and disinfected) is not available, boil the water	b. Yes for step 2; No for step 8	b. Bubbles	b. Observation	b. Reboil.

*In this study, the preparation of togwa is considered in a household environment; therefore, the critical limits are expressed in qualitative values.

continues

Table 12–2 continued

Step	*Hazards*	*Control Measures*	*CCPs*	*Critical Limit**	*Monitoring Procedure*	*Corrective Actions*
2. Soaking	Growth of microorganisms	As far as possible at low temperatures	Yes	Water should remain free from odor or foam	Observation, smelling	Refresh water.
3. Germination	Growth of microorganisms (e.g., toxigenic molds)	As far as possible at low temperature	Yes	Grains should remain free of moldiness	Observation	Remove moldy grains.
4. Sun drying	a. Contamination and introduction of foreign matter	a. Protect the sprouts	Yes	a. No foreign matter	a. Observation	a. Clean if possible. If not, discard
	b. Inadequate drying may lead to growth of microorganisms during storage	b. Ensure thorough and fast drying		b. Sufficient time, adequate exposure to sun, dry ambient conditions, adequate air circulation, no mold	b. Time keeping, observation	b. As long as there is no mold growth, redry under proper conditions; otherwise, discard.
5. Storage of sun-dried germinated grains	a. Contamination/introduction of foreign matter	a. Protect the grains	Yes	a. No foreign matter	Observation	a. Clean if possible. If not, discard
	b. Growth of toxigenic molds, if the moisture content is high	b. Keep dry		b. Dry conditions of storage, no mold		b. Discard
6. Grinding to power flour	Introduction of filth, dirt and foreign matter	Use clean and properly maintained equipment	No			
7. Grinding	same as step 6					

continues

Table 12–2 continued

Step	*Hazards*	*Control Measures*	*CCPs*	*Critical Limit**	*Monitoring Procedure*	*Corrective Actions*
8. Slurry preparation	Contamination with pathogens through utensils and/or water	Use clean utensils and safe water	No			
9. Boiling	Survival of pathogens	Thorough boiling	Yes	Bubbles	Observation	Reboil.
10. Cooling	a. Growth of bacterial spores	a. Ensure rapid cooling	Yes	a. Short time, room temperature within four hours	a. Time keeping	a. Reboil.
	b. Contamination	b. Protect the porridge during the cooling process		b. No foreign matter	b. Observation	b. Depending on the nature of contamination, either clean, reboil or discard
11. Addition of:						
a. power flour	a. Contamination with pathogens by power flour	a. Ensure hygienic quality of power flour	Yes	a. Power flour of high hygienic quality	a. & b. Observation	a. & b. Use another power flour or togwa.
b. Togwa	b. Contamination with acid-resistant pathogens by togwa	b. Ensure the safety of previously prepared togwa		b. Absence of disease upon consumption of previously prepared togwa		
12. Fermentation	a. Growth and formation of toxin by *Staphylococcus aureus*	a. Rapid fermentation	a. Yes	a. Acid taste and characteristic odor within 24 hours	a. Observation	a. Discard the material.
	b. Survival of acid-tolerant pathogens	b. Minimize contamination with acid-tolerant pathogens (see step 11)*	b. No			
13. Serving	Recontamination with pathogens by hands, utensils, and environment	Wash hands and use clean utensils	Yes	Hands properly washed with soap and clean water	Observation	Reheat the food thoroughly.

*As there is no critical control point in the subsequent steps that would ensure the killing of acid-tolerant pathogens surviving the fermentation step, the present process of togwa preparation may lead to a high-risk product.

process does not include a step that would ensure the elimination of pathogens introduced through the raw materials (i.e., water, sorghum, or millet). In addition, pathogens introduced through the raw material may proliferate during the soaking and germination periods and may also survive the sun-drying stage. Therefore, the safety of water used in power flour preparation is critical for minimizing contamination. The hygienic quality of the power flour is particularly important for the safety of the final product because there is no step that would ensure the killing of acid-resistant pathogens after the power flour has been added. Therefore, after the HACCP study, it was suggested to add a boiling step in the preparation of power flour to ensure safety of water and minimize the contamination of power flour.

b. *Soaking:* During the soaking period, bacterial growth will occur. The use of safe water may minimize the final bacterial load. If at all possible, soaking should take place at low temperatures in order to minimize the growth of microorganisms.
c. *Germination:* This usually takes place in layers a few centimeters thick, spread on mats or leaves, and covered with leaves, mats, or gunny sacks to avoid excessive dehydration. From time to time, the material must be aired and mixed while checking and adjusting the degree of grain humidity. Further contamination by mats and microbial growth can occur at this stage. Microbial growth, particularly with respect to toxigenic molds, can also occur. As far as possible, germination should be carried out in cool conditions in order to minimize microbial growth.
d. *Sun drying:* During this step, the cover is removed and the sprouted grains are spread on mats or bamboo trays to dry in the sun. This may take a few days. All kind of contaminants and foreign matter can fall into the sprouted grains if they are not well protected. In addition, insufficient drying may lead to a microbially unstable product during subsequent storage. Thorough drying is critical for the stability of the product.
e. *Storage of sun-dried germinated grains:* Hazards associated with this step are contamination (e.g., by rodents and other animals during storage) and microbial growth, particularly toxigenic molds, if the germinated grains are not properly dried or are kept in humid conditions.
f. *Grinding to power flour:* This step may introduce dirt and foreign matter into the product. As part of a good hygienic practice, this should be avoided as far as possible by using clean and properly maintained equipment. However, it is unlikely that this step will introduce any major health hazard.
g. *Grinding:* Same as step f.
h. *Preparation of slurry:* Except for the safety of water and cleanliness of utensils, no other major hazard is associated with this step. Because the subsequent boiling step will kill the pathogens that may have been introduced at this step, it is not considered a CCP. Nevertheless, as part of a GHP, households should use clean utensils and safe water as much as possible.
i. *Boiling:* Boiling should be thorough in order to gelatinize all the starch. This step is also essential to kill nearly all pathogens (bacterial spores may survive).
j. *Cooling:* The pot should be covered to protect against dirt or other foreign matter falling in. It is important that cooling is carried out as fast as possible. Prolonged cooling may present an opportunity for bacterial spores to grow. When large quantities of togwa are prepared, the cooling time can be

reduced by dividing the togwa into small portions.

k. *Addition of power flour and togwa:* Two ingredients may be added here to initiate the fermentation, that is, power flour and previously fermented togwa. The contamination brought about by power flour is diverse and difficult to control. Viruses and other acid-tolerant agents are of particular concern here. The togwa starter culture is quite acidic, and thus will contain only those contaminants that are acid tolerant.

   During the fermentation that follows, acid-sensitive pathogens may be killed. However, acid-tolerant pathogens may survive. A power flour prepared under hygienic conditions may minimize the contamination of togwa. However, in the absence of a final killing step such as thorough reheating, the presence of acid-tolerant pathogens in the final product may not be excluded.

l. *Fermentation:* During fermentation, a rapid dominance of LAB may be expected. This is supported by the short fermentation time to reach the required acidity. A rapid fermentation is critical for killing acid-sensitive pathogens and for preventing bacterial growth and the production of toxin. The addition of togwa enhances fermentation and may be beneficial provided it does not introduce acid-resistant pathogens.

m. *Serving:* It is important to ensure that pathogens are not re-introduced into togwa by dirty hands and utensils. Therefore, these have to be washed carefully with safe water. Depending on the hygienic measures taken to prepare togwa, the final product could be more or less contaminated because there is no final CCP that would ensure the killing of acid-resistant pathogens. Thorough reheating would greatly contribute to the safety of the final product. However, implications in terms of textural and other changes should be considered because the final product may become unacceptable to the consumer.

HACCP Study of Traditional Cheddar Cheese*[23]

1. **Product description**—Cheddar cheese is a hard-pressed cheese that is originally from the United Kingdom. It has a firm body and closed texture. It is composed of approximately 37% water, 33% fat, 25% protein,1% lactose, and 4% ash. The curd is textured after cutting and scalding. It is then milled and salted and is pressed in a mold. The cheese is wrapped in firm blocks and stored at approximately 4–6 °C. The cheese is stored from a few weeks to several months, during which time it ripens and develops its flavor (mild cheese up to 3 months; medium cheese 6 months, and mature cheese 8–12 months). Starter culture used for the production of cheddar cheese consists of *Lactococcus lactis* subsp. *lactis* and *Lc. lactis* subsp. *cremoris.* Other organisms that also play a role during ripening are *Lactobacillus casei, Lb. plantarum,* and *Lb. brevis.*
2. **Intended use**—General population, including vulnerable groups.
3. **Flow diagram**—Figure 12–3 shows the flow diagram of traditional cheddar cheese manufacture.
4. **Hazards of concern**—Major hazards are of bacterial origin. Several pathogens such as *Mycobacterium tuberculosis, Brucella abortus, Salmonella, Campylobacter,* and *Listeria monocytogenes* may survive in cheddar cheese. Other hazards of concern are antibiotic residues and histamine.
5. **Identification of hazards, control measures, and CCPs**—Table 12–3 shows

*Model HACCP plans are not appropriate for use until validated for a specific food and food process.

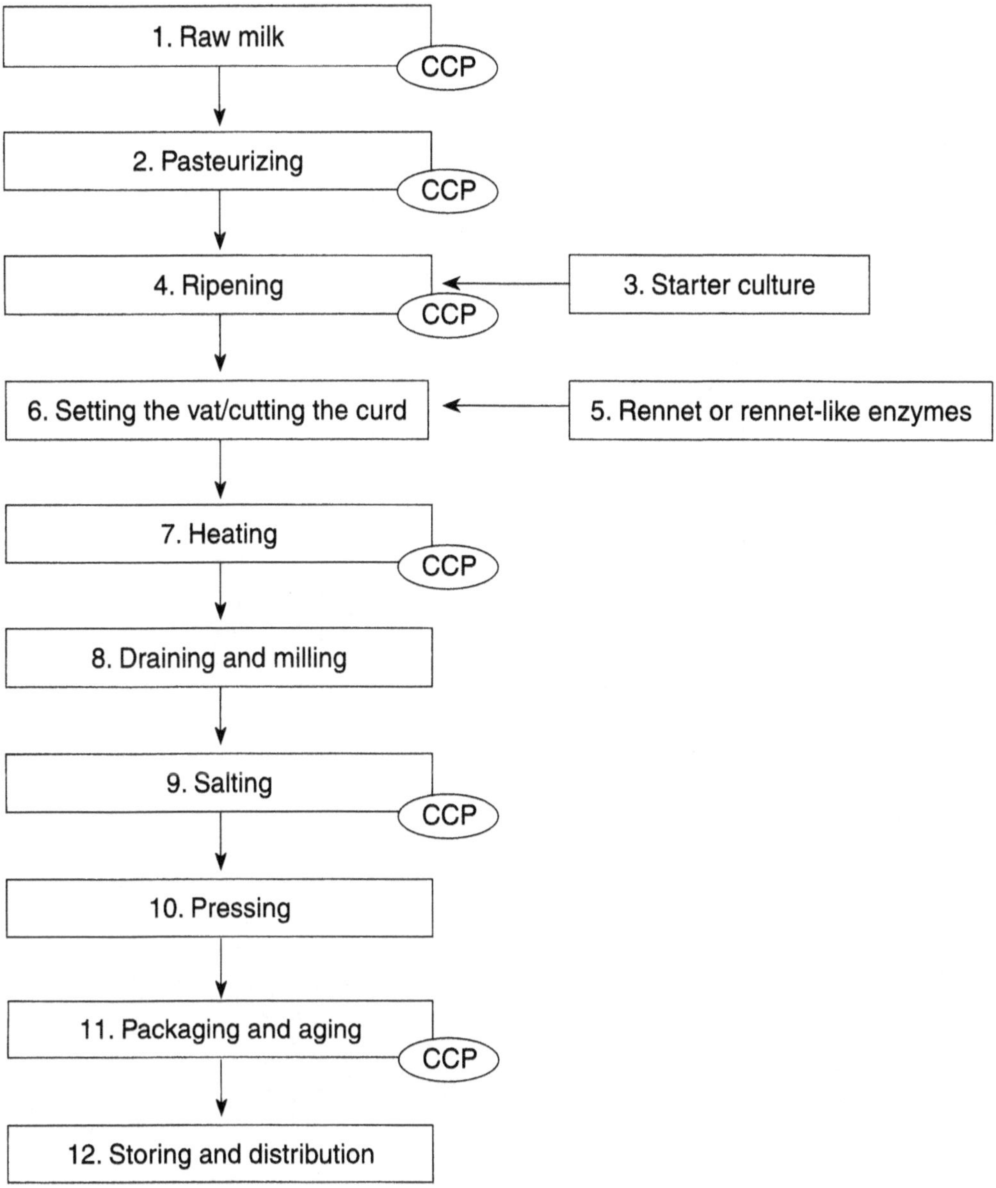

Note: Numbers correspond with those in Table 12–3.

Figure 12–3 Flow diagram of traditional cheddar cheese

hazards associated with each step of the production of cheddar cheese.

- *Raw material:* Raw milk when received by a dairy may be contaminated with a wide range of pathogens, as mentioned above, as well as spoilage organisms. In addition, milk may contain antibiotic residues. Hygienic handling of milk at the farm and during transport is essential to prevent con-

Table 12–3 HACCP Study of the Preparation of Traditional Cheddar Cheese in Industrial Setting

Step	*Hazards*	*Control Measures*	*CCPs*	*Critical Limit*	*Monitoring Procedure*	*Corrective Actions*
1. **Raw material: milk**	a. Pathogenic bacteria: *Mycobacterium* spp., *Brucella* spp, *Campylobacter, Listeria, Salmonella*	Hygienic collection at the farm and transport Refrigeration (maximum temperature 7 °C)	a. No			
	b. Antibiotic residues	b. Control of incoming milk	b. Yes	b. According to Codex Alimentarius Commission	b. Chemical or microbiological tests	b. Reject the milk.
2. **Pasteurization**	Survival of pathogens	Heating	Yes	72 °C for 15 seconds or equivalent	Temperature, time	Repasteurize.
3. **Starter culture**	Presence of hazards in the starter culture	Assurance by the supplier of the starter culture on the quality	No			
4. **Ripening** (addition of the starter culture)	a. Contamination with pathogens	a. Cleaning and disinfection of vat and other equipment	a. No			
	b. Insufficient fermentation	b. Follow instructions of manufacturer of starter culture; use appropriate proportion of starter culture and milk; keep milk warm	b. Yes	b. 30–31 °C, development of acidity of 0.14% before step 8. Proportion of starter culture and milk according to the instructions of the manufacturer	b. Temperature titratable acidity	b. Correct the temperature and the amount of starter culture.
5. **Rennet or rennet like-enzymes**	Presence of pathogens	Assurance of quality by the supplier	No			

continues

Table 12–3 continued

Step	*Hazards*	*Control Measures*	*CCPs*	*Critical Limit*	*Monitoring Procedure*	*Corrective Actions*
6. **Setting the vat and cutting the curd**	Contamination with pathogens, of utensils used for adding the rennet	Cleaning and disinfecting of cutting knives and hygienic practice	No			
7. **Heating**	Arresting the fermentation	Control temperature of heating	Yes	37 °C	Temperature	Readjust the temperature.
8. **Cheddaring (draining and milling)**	Contamination of utensils and environment with hazards	Hygienic practice	No			
9. **Salting**	Growth of *S.aureus* at subsequent steps	Adjusting the salt level and its distribution	Yes	1.7 %	Amount of salt added	Readjust the proportion of salt and curd.
10. **Pressing**	Contamination of the press with pathogens	Strict hygiene of the equipment and personnel	No			
11. **Aging**	Growth of *Staphylococcus aureus;* growth of decarboxylase positive lactic acid bacteria and formation of histamine	Adequate storage (see also step 8)	Yes	5 °C	Temperature	Correct the temperature of storage. Consider removal of the product if the loss of temperature control has exceeded several hours.
12. **Storing and distributing**	Mold growth	Adequate packaging, refrigeration, and rapid distribution	No			

Note: This table was adapted from ICMSF.[23]

tamination and keep the level of pathogens possibly present to a minimum. To ensure that raw material is of adequate hygienic quality, the milk may be tested for its bacterial load with methylene blue or by some other rapid test. The receipt of the raw milk is a CCP for antibiotic residues because these will not be eliminated by further processing of the milk into cheese.

- *Pasteurization:* The pasteurization step is a CCP because it is essential for the elimination of pathogens that may be present in the raw milk. It will also reduce the number of decarboxylase-positive LAB able to produce histamine.
- *Starter culture:* The contamination of starter culture has in the past caused food-borne disease outbreaks. Therefore, the possibility that the starter culture may be contaminated or be of poor quality (leading to poor fermentation) should be considered, and assurance should be obtained from the supplier of its quality.
- *Ripening of the milk:* At this stage, the milk is cooled to 30–31 °C and poured into the cheese vat, and the starter culture is added. It is important that the starter culture is of adequate quality and is added in sufficient quantity to ensure rapid fermentation. The temperature of the milk should also be kept at an optimal level to ensure adequate starter culture activity. Starter culture activity may be measured by pH changes. The cheese vat and other equipment should be cleaned and disinfected before use. To prevent postpasteurization contamination, strict hygiene should be applied at this stage and later stages of production.
- *Rennet or rennet-like enzymes:* As for starter culture, it is important to obtain assurance from the supplier that the rennet is of adequate quality.
- *Setting the vat and cutting the curd:* The rennet or rennet-like enzymes are added and the milk is left undisturbed to form a gel. The coagulated milk is then cut into cubes. To prevent contamination with cutting utensils, hygienic practices are strictly observed.
- *Heating:* The curd is heated to 37–39 °C and the curd and whey are stirred. The curd shrinks as a result of heating and the whey is released. It is important to have control of temperature in order to maintain the activity of the starter culture. At the end of this step, the whey should have a titratable acidity of 0.14–0.16% and the curd pH should be approximately 6.
- *Draining and milling:* The whey is drained from the vat. The curd develops a plastic consistency as a result of continued acid development. The curd is cut into sections and passed through a mill to form ribbons. No major hazard is associated with this step. However, hygienic practices should be strictly observed to prevent recontamination.
- *Salting:* Salt is added to shrink the curd and further separate the whey. The amount of salt should be carefully controlled because it will have an impact on subsequent acid development and favor growth of *Staphylococcus aureus*. The salt content should be approximately 1.7%.
- *Pressing:* After salting, the curd is placed in a box and pressed. In this way, the remaining whey is removed. To prevent recontamination, this step should be carried out under strict hygienic conditions.
- *Aging:* The cheese is vacuum packed and aged at approximately 5 °C for a few weeks to few months. The control of temperature at this stage is important and critical for the prevention of growth of possibly present *Staph. aureus* and/or activity of decarboxylase-positive bacteria.
- *Storing and distribution:* To ensure safety and quality, storage should be at

refrigeration temperature and the product should be distributed rapidly.

RESEARCH NEEDS

Research needs in the area of fermentation are extensive and relate to several areas.

- identification of new organisms and technological developments leading to new products with enhanced nutritional or organoleptic quality
- assessment of safety of starter cultures, be they traditional organisms selected for their specific features or genetically modified organisms
- interaction of LAB with pathogens in the intestinal tract and risks for transfer of pathogenicity and virulence
- potential of fermentation technology to control pathogens
- identification and selection of organisms with potential health benefits
- transfer of technology to settings where the technology can be of specific benefit to health
- implementation and evaluation of health education interventions carried out to ensure safe application of fermentation.

Additional guidance on research needs and priorities in this area are provided elsewhere.[6,18]

CONCLUSION

Fermentation is and will continue to remain an important technology. For foods for infants and young children, it presents clear nutritional and safety advantages. In the industrialized countries, the technology may not be important as a preservation technique, but advances in areas such as molecular biology will help in the development of new products, with diversified taste and aroma and perhaps health benefits. The challenge will be to ensure that organisms used as starter culture are safe and that the process is designed and applied in such a way that potential hazards, in particular acid-resistant pathogens, are controlled.

The greatest benefits but also risks are currently in the developing countries and in places where the technology is used in an artisanal way. In the developing world, where refrigeration facilities are lacking, the technology can make a great contribution to public health by preventing the growth of pathogens. However, substantial health education campaigns are necessary to ensure that the technology is applied in an appropriate manner and leads to safe and nutritious products.

REFERENCES

1. Adams, M. R. & Nicolaides, L. (1997). Review of the sensitivity of different pathogens to fermentation. *Food Control* 8, 227–239.
2. Adams, M. R. (1998). Fermented weaning foods. In *Microbiology of Fermented Foods.* Vol. 2, 2nd edn., pp. 790–811. Edited by B. J. B. Wood. London: Blackie Academic and Professional.
3. Altekruse, S. F., Babagaleh, T. B., Mowbray, J. C. et al. (1998). Cheese-associated outbreaks of human illness in the United States, 1973 to 1992: sanitary manufacturing practices protect consumers *J Food Prot* 61, 1405–1407.
4. Baldwin, I., Hobson, P., Weinstein, P. et al. (1992). *Salmonella anatum* foodborne outbreak, South Australia. *Commun Dis Intell* 16, 94–98.
5. Bar-Dayan, Y., Bar-Dayan, Y. et al. (1996). Foodborne outbreak of streptococcal pharyngitis in an Israeli base air force. *Scandavian J Infect Dis* 28, 563–566.
6. Buckenhüskes, H. J. (1997). Fermented vegetables. In *Food Microbiology: Fundamentals and Frontiers.* Edited by M. P. Doyle, L. Beuchat & T. J. Montville. Washington, DC: American Society of Microbiology Press.
7. BGVV/RKI. (1997). *Abschlussbericht der Untersuchung eines Ausbruchs von Gastroeneritis im landkreis Anhalt-Zerbst, 1997. Internal Report from the Bundesinstitut für gesundheitlichen Verbraucherschutz und Veterinärmedizin.* Berlin, Germany.
8. Caplice, E. & Fitzerald, G. F. (1999). Food fermentation: role of microorganisms in food production and preservation. *Intl J Food Microbiol* 50, 131–149.
9. CCDR. (1999). *Salmonella enteritidis* outbreak due to contaminated cheese: Newfoundland. *Canada Comm Dis Report*, 25–3.
10. CDC. (1995). Community outbreak of hemolytic uremic syndrome attributable to *Escherichia coli* O111:NM: South Australia. *MMWR* 550–551, 557.

11. CDC. (1995). *Escherichia coli* O157:H7 outbreak linked to commercially distributed dry-cured salami: Washington and California. *MMWR* 44, 157–160.

12. Cowden, J. M., O'Mahony, M. et al. (1989). A national outbreak of *Salmonella typhimurium* DT 124 caused by contaminated salami sticks. *Epidemiol and Infect* 103, 219–225.

13. De Roos, N. M. & Katan, M. B. (2000). Effects of probiotic bacteria on diarrhoea, lipid metabolism, and carcinogenesis: a review of papers published between 1988 and 1998. *Am J Clin Nutr* 71, 405–411.

14. Desenclos, J. -C., Bouvet, P. & Benz-Lemoine, E. (1996). Large outbreak of *Salmonella* serotype *paratyphi* B infection caused by a goat's milk cheese, France, 1993: a case finding and epidemiological study. *Br Med J* 312, 91–94.

15. Ellis, A., Preston, M., Borczyk, A. et al. (1998). A community outbreak of *Salmonella berta* associated with a soft cheese product. *Epidemiol and Infect* 120, 29–35.

16. Espinosa, F. H. et al. (1983). Group C Streptococcal infections associated with eating homemade cheese: New Mexico. *MMWR* 32, 510–516.

17. FAO/WHO. (1996). *Biotechnology and Food Safety. Report of a Joint FAO/WHO Consultation, Rome Food and Agriculture Organization of the United Nations.* Rome, Italy.

18. FAO/WHO. (1996). *Fermentation: Assessment and Research. Report of a Joint FAO/WHO Workshop on Fermentation as Household Technology To Improve Food Safety.* WHO Document WHO/FNU/FOS/96.1. Geneva: World Health Organization.

19. Fontaine, R. E., Mitchell, L. C. et al. (1980). Epidemic salmonellosis from cheddar cheese: surveillance and prevention. *Am J Epidemiol* 111, 247–253.

20. Hauschild, A. H. W. & Gauvreau, L. (1985). Foodborne botulism in Canada, 1971–84. *Canadian Med Assoc J* 133, 1141–1146.

21. Holzapfel, W. (1997). Use of starter cultures in fermentation on a household scale. *Food Control* 8, 241–258.

22. Hove, H., Norgaard, H. & Brobech, M. (1999). Lactic acid bacteria and the human gastrointestinal tract. *Eur J Clin Nutr* 53, 339–350.

23. ICMSF. (1988). *Microorganisms in Foods 4: Application of the Hazard Analysis Critical Control Point (HACCP) System To Ensure Microbiological Safety and Quality.* London: Blackwell Scientific Publications.

24. James, S. M., Fannin, S. L., Agee, B. A. et al. (1985). Listeriosis outbreak associated with Mexican-style cheese. *MMWR* 34, 357–359.

25. Johnson, E., Nelson, J. & Johnson, M. (1990). Microbiology safety of cheese made from heat treated milk, part II: microbiology. *J Food Prot* 53, 519–540.

26. Khamboonruang, C. & Nateewatana, N. (1975). Trichinosis: a recent outbreak in Northern Thailand. *Southeast Asian J Trop Med and Public Health* 6, 74–78.

27. Linnan, M. J., Mascola, L., Lou, X. D. et al. (1998). Epidemic listeriosis associated with Mexican-style cheese. *N Engl J Med* 319, 823–828.

28. Maguire, H., Cowden, J., Jacob, M. et al. (1992). An outbreak of *Salmonella dublin* infection in England and Wales associated with unpasteurized cows' milk cheese. *Epidemiol and Infect* 109, 389–396.

29. Marier, R., Wells, J. G., Swanson, R. C. et al. (1973). An outbreak of enteropathogenic *Escherichia coli* foodborne disease traced to imported French cheese. *Lancet* Dec. 15, 1376–1378.

30. Mayer, K. & Pause, G. (1972). Biogene amines in Sauerkraut. *Leben.-Wiss.-Technol.* 5, 108–109.

31. Mead, P. S. et al. (1999). Food related illness and death in the United States. *Emerging Infect Dis* 5, 607–625.

32. Meng, X., Karasawa, T., Zou, K. et al. (1997). Characterization of a neurotoxigenic *Clostridium butyricum* strain isolated from the food implicated in an outbreak of foodborne type E botulism. *J Clin Microbiol* 35, 2160–2162.

33. MOH (1993). *Clostridium perfringens* food poisoning at Tabuk Camp. *Saudi Epidemiol Bull* 1, 2.

34. Mollet, B. (1999). Genetically improved starter strains, opportunities for the dairy industry. *Int Dairy J* 9, 11–15.

35. Morgan, D., Newman, C. P., Hutchinson, D. N. et al. (1993). Verotoxin producing *Escherichia coli* O157 infections associated with the consumption of yoghurt. *Epidemiol and Infect* 111, 181–187.

36. O'Mahony, M., Mitchell, E. et al. (1990). An outbreak of foodborne botulism associated with contaminated hazelnut yoghurt. *Epidemiol and Infect* 104, 389–395.

37. Pourshafie, M. R., Saifie, M., Shafiee, A. et al. (1998). An outbreak of food-borne botulism associated with contaminated locally made cheese in Iran. *Scan J Infect Dis* 30, 92–94.

38. Sauer, C. J., Majkowski, J., Green, S. et al. (1997). Foodborne illness outbreak associated with a semi-dry fermented sausage product. *J Food Prot* 60, 1612–1617.

39. Shaffer, N., Wainwright, R. B., Middaugh, J. P. et al. (1990). Botulism among Alaska natives: the role of changing food preparation and consumption practices. *West J Med* 153, 390–393.

40. Sharp, J. C. M. (1987). Infections associated with milk and dairy products in Europe and North America, 1980–85. *Bull WHO* 65, 397–406.

41. Smith, D. H., Timms, G. L. & Refai, M. (1979). Outbreak of botulism in Kenyan nomads. *Annals of Trop Med and Parasitol* 4, 2–7.

42. Stratton, J. et al. (1991). Biogenic amines in cheese and other fermented foods: a review. *J Food Prot* 54, 460–470.
43. Taylor, S. L. et al. (1982). Outbreak of histamine poisoning associated with consumption of Swiss cheese. *J Food Prot* 45, 455–457.
44. The Pennington Group. (1997). Report on the circumstances leading to the 1996 outbreak of infection with *E. coli* O157 in Central Scotland: the implication for food safety and the lessons to be learned. April, 9–11.
45. Uribarri, J., Oh, M. S. & Carroll, H. J. (1998). D-lactic acidosis: a review of clinical presentation, biochemical features, and pathophysiolocial mechanisms. *Medicine* 72, 73–82.
46. World Association of Industrial and Technological Research Organizations. (1998). *HACCP System for Traditional African Fermented Foods: Kenkey.* Taastrup, Denmark: Danish Technological Institute.
47. World Association of Industrial and Technological Research Organizations. (1998). *HACCP System for Traditional African Fermented Foods: Soumbala.* Taastrup, Denmark: Danish Technological Institute.
48. World Association of Industrial and Technological Research Organizations. (1998). *HACCP System for Traditional African Fermented Foods: Soy-ogi.* Taastrup, Denmark: Danish Technological Institute.
49. Wall, P. (1998). Food-borne human disease: is it a veterinary problem? *Vet J* 51, 34–35.
50. World Health Organization. (1991). *Strategies for Assessing the Safety of Foods Produced by Biotechnology: Report of Joint FAO/WHO Consultation.* Geneva: Author.
51. World Health Organization. (1996). Konzo, a distinct type of upper motoneuron disease. *Weekly Epidemiolog Rec* 71, 225–228.
52. World Health Organization. (1997). *World Health Report.* Geneva: Author.
53. Wood, G. F. & Adams, M. R. (1992). Effects of acidification, bacterial fermentation and temperature on the survival of rotavirus in a model weaning food. *J Food Prot* 55, 52–55.
54. Yusof, R. M., Morgan, J. B. & Adams, M. R. (1993). Bacteriological safety of a fermented weaning food containing L-lactate and nisin. *J Food Prot* 56, 414–417.

List of Sources

CHAPTER 1

Figure 1-2 Adapted with permission from H. Uhlig, *Enzyme Arbeiten Fur Uns. Technische Enzyme und Ihre Anwendung*, p. 37, © 1991, Carl Hanser Verlag.

Figure 1-3 Reprinted with permission from A.G.J. Voragen and L.A.M. Van den Broek, Fruit Juices, in *Biotechnological Innovations in Food Processing*, Biotol Team, ed., p. 189, © 1991, Butterworth-Heinemann, Ltd.

Figure 1-4 Reprinted with permission from N.C. Carpita and D.M. Gibeaut, Structural Models of Primary Cell Walls in Flowering Plants: Consistency of Molecular Structure with the Physical Properties of the Walls During Growth, *The Plant Journal*, Vol. 3, p. 7, © 1993, Blackwell Science Ltd.

Figure 1-5 Adapted from R.J. Sturgeon, *Advances in Macromolecular Research Vol. 1*, Copyright 1997, pp. 1-64, with permission from Elsevier Science.

CHAPTER 2

Figure 2-1 Reprinted with permission from M.R. Adams and C.J. Hall, Growth Inhibition of Food-Borne Pathogens by Lactic and Acetic Acids and Their Mixtures, *International Journal of Food Science Technology*, Vol. 23, pp. 287-292, © 1988.

Table 2-2 Data from A. Ouwehand, Antimicrobial Components From Lactic Acid Bacteria, in *Lactic Acid Bacteria: Microbiology and Functional Aspects*, S. Salminen and A. Von Wright, eds., pp. 139-159, © 1998, Marcel Dekker, Inc.

Table 2-3 Data from S. Saono, R.R. Hull, and B. Dhamcharee, eds., *A Concise Handbook of Indigenous Fermented Foods in the ASCA Countries*, p. 157, © 1986, The Government of Australia.

CHAPTER 3

Figure 3-1 Reprinted with permission from Y. Motarjemi and M. van Schothorst, *HACCP: Principles and Practice*, © 1999, World Health Organization.

Figure 3-2 Reprinted with permission from *Assessment of Fermentation as a Household Technology for Improving Food Safety, A Report of the WHO Consultation*, © 1996, World Health Organization.

Table 3-1 Reprinted with permission from *Assessment of Fermentation as a Household Technology for Improving Food Safety, A Report of the WHO Consultation*, © 1996, World Health Organization.

Appendix 3-A Reprinted with permission from Codex Alimentarious Commission, *Food Hygiene, Basic Texts*, © 1997, Joint FAO/WHO Food Standards Programme, Secretariat of the Joint FAO/WHO Food Standards Programme.

Figure 3-A Reprinted with permission from Codex Alimentarious Commission, *Food Hygiene, Basic Texts*, © 1997, Joint FAO/WHO Food Standards Programme, Secretariat of the Joint FAO/WHO Food Standards Programme.

CHAPTER 6

Figure 6-1a Reprinted from *International Journal of Food Microbiology*, Vol. 29, M.H. Silla-Santos, Biogenic Amines: Their Importance in Foods, pp. 213-231, Copyright 1996, with permission of Elsevier Science.

Figure 6-1b Reprinted from *International Journal of Food Microbiology*, Vol. 29, M.H. Silla-Santos, Biogenic Amines: Their Importance in Foods, pp. 213-231, Copyright 1996, with permission of Elsevier Science.

Figure 6-1c Reprinted from *International Journal of Food Microbiology*, Vol. 29, M.H. Silla-Santos, Biogenic Amines: Their Importance in Foods, pp. 213-231, Copyright 1996, with permission of Elsevier Science.

Figure 6-1d Reprinted from *International Journal of Food Microbiology*, Vol. 29, M.H. Silla-Santos, Biogenic Amines: Their Importance in Foods, pp. 213-231, Copyright 1996, with permission of Elsevier Science.

Table 6-1 Reprinted from *International Journal of Food Microbiology*, Vol. 29, M.H. Silla-Santos, Biogenic Amines: Their Importance in Foods, pp. 213-231, Copyright 1996, with permission of Elsevier Science.

Table 6-2 Reprinted from *International Journal of Food Microbiology*, Vol. 29, M.H. Silla-Santos, Biogenic Amines: Their Importance in Foods, pp. 213-231, Copyright 1996, with permission of Elsevier Science.

Table 6-3 Reprinted from *International Journal of Food Microbiology*, Vol. 29, M.H. Silla-Santos, Biogenic Amines: Their Importance in Foods, pp. 213-231, Copyright 1996, with permission of Elsevier Science.

Figure 6-3 Reprinted from *International Journal of Food Microbiology*, Vol. 29, M.H. Silla-Santos, Biogenic Amines: Their Importance in Foods, pp. 213-231, Copyright 1996, with permission of Elsevier Science.

Figure 6-4 Reprinted from *International Journal of Food Microbiology*, Vol. 29, M.H. Silla-Santos, Biogenic Amines: Their Importance in Foods, pp. 213-231, Copyright 1996, with permission of Elsevier Science.

Figure 6-5 Reprinted from *International Journal of Food Microbiology*, Vol. 29, M.H. Silla-Santos, Biogenic Amines: Their Importance in Foods, pp. 213-231, Copyright 1996, with permission of Elsevier Science.

CHAPTER 7

Table 7-1 Data from T.A. Roberts, A.C. Baird-Parker, and R.B. Tompkin, eds., *Microorganisms in Foods 5, Microbiological Specifications of Food Pathogens*, p. 513, © 1996, Blackie Academic Professional.

CHAPTER 9

Chapter 9 © Crown Copyright 1997. Published with the permission of the Controller of Her Britannic Majesty's Stationery Office. The views expressed are those of the author and do not necessarily reflect those of Her Britannic Majesty's Stationery Office or the VLA or any other government department.

Table 9-1 © Crown Copyright.

Figure 9-1 © Crown Copyright.

Table 9-2 © Crown Copyright.

Figure 9-2 © Crown Copyright.

Figure 9-3 © Crown Copyright.

Figure 9-4 © Crown Copyright.

Figure 9-5 © Crown Copyright.

Figure 9-6 © Crown Copyright.

Figure 9-7 © Crown Copyright.

Table 9-3 © Crown Copyright.

Table 9-4 © Crown Copyright.

Figure 9-8 © Crown Copyright.

Table 9-5 © Crown Copyright.

Table 9-6 © Crown Copyright.

CHAPTER 10

Figure 10-1 Reprinted from *Enzyme and Microbial Technology*, Vol. 26, T. Verrips, et al., pp. 812-818, Copyright 2000 with permission from Elsevier Science.

Table 10-1 Courtesy of RABO Bank.

Figure 10-2 Courtesy of RABO Bank.

Figure 10-3 Reprinted from *Ant. van Leeuwenhoek*, Vol. 70, 1996, pp. 299-316, Barriers to Application of Genetically Modified Lactic Acid Bacteria, C.T. Verrips and Van den Berg, with kind permission of Kluwer Academic Publishers.

Figure 10-6 Reprinted with permission from FAOSTAT, C.C. Mann, Crop Scientists Seek a New Revolution, *Science*, Vol. 283, pp. 310-314. Copyright 1999 American Association for the Advancement of Science.

Figure 10-7 Reprinted from *Ant. van Leeuwenhoek*, Vol. 70, 1996, pp. 299-316, Barriers to Application of Genetically Modified Lactic Acid Bacteria, C.T. Verrips and Van den Berg, with kind permission of Kluwer Academic Publishers.

Figure 10-10 Reprinted from *Ant. van Leeuwenhoek*, Vol. 70, 1996, pp. 299-316, Barriers to Application of Genetically Modified Lactic Acid Bacteria, C.T. Verrips and Van den Berg, with kind permission of Kluwer Academic Publishers.

CHAPTER 11

Exhibit 11-2 Data from S. Salminen and A. Von Wright, Current Probiotics-Safety Assured? *Microbial Ecology in Health and Disease*, Vol. 10, pp. 68-77, © 1998 and S. Salimen, et al., Demonstration of Safety of Probiotics: A Review, *International Journal of Food Microbiology*, Vol. 44, pp. 93-106, © 1998.

Table 11-1 Data from W.H. Holzapfel, et al., Taxonomy and Important Features of Probiotic Microorganisms in Food and Nutrition, *American Journal of Clinical Nutrition*, in press and T. Mitsuoka, Intestinal Flora and Ageing, *Nutrition Review*, Vol. 50, pp. 438-446, © 1992.

CHAPTER 12

Chapter 12 Portions of this chapter are adapted from *Fermentation: Assessment and Research. Report of a Joint FAO/WHO Workshop on Fermentation as a Household Technology to Improve Food Safety*, Pretoria, South Africa, December 11-15, 1995, © 1996, World Health Organization.

Figure 12-1 Courtesy of the World Health Organization, Geneva, Switzerland.

Figure 12-2 Courtesy of the World Health Organization, Geneva, Switzerland.

Table 12-2 Courtesy of the World Health Organization, Geneva, Switzerland.

Figure 12-3 Courtesy of the World Health Organization, Geneva, Switzerland.

Table 12-3 Data from ICMSF, *Microorganisms in Foods 4: Application of the Hazard Analysis Critical Control Point (HACCP) System to Ensure Microbiological Safety and Quality*, © 1988, Blackwell Scientific Publications.

Index

C

T

V

W

Y

Z

GPSR Compliance

The European Union's (EU) General Product Safety Regulation (GPSR) is a set of rules that requires consumer products to be safe and our obligations to ensure this.

If you have any concerns about our products, you can contact us on ProductSafety@springernature.com

In case Publisher is established outside the EU, the EU authorized representative is:

Springer Nature Customer Service Center GmbH
Europaplatz 3
69115 Heidelberg, Germany

Batch number: 10755805

Printed by Printforce, the Netherlands